1991 LECTURES IN COMPLEX SYSTEMS

1991 Lectures in Complex Systems

THE PROCEEDINGS OF THE
1991 COMPLEX SYSTEMS
SUMMER SCHOOL,
SANTA FE, NEW MEXICO,
JUNE, 1991

Editors

Lynn Nadel
Department of Psychology
University of Arizona

Daniel Stein
Department of Physics
University of Arizona

Lectures Volume IV

Santa Fe Institute
Studies in the Sciences of Complexity

Addison-Wesley Publishing Company, Inc.
The Advanced Book Program

Reading, Massachusetts   Menlo Park, California   New York
Don Mills, Ontario   Wokingham, England   Amsterdam   Bonn
Sydney   Singapore   Tokyo   Madrid   San Juan
Paris   Seoul   Milan   Mexico City   Taipei

Publisher: *David Miller*
Production Manager: *Michael Cirone*

Director of Publications, Santa Fe Institute: *Ronda K. Butler-Villa*
Publications Assistant, Santa Fe Institute: *Della L. Ulibarri*
Publications Secretary, Santa Fe Institute: *Joanne Carlsen*

This volume was typeset using T$_E$Xtures on a Macintosh II computer. Camera-ready output from a NEC Laser Printer.

Copyright © 1992 by Addison-Wesley Publishing Company, The Advanced Book Program, Jacob Way, Reading, MA 01867

All rights reserved. No part of this publication may be reproduced, stored in a retrieval system, or transmitted in any form or by any means, electronic, mechanical, photocopying, recording, or otherwise, without the prior written permission of the publisher. Printed in the United States of America. Published simultaneously in Canada.

1 2 3 4 5 6 7 8 9 10 - MA - 95 94 93 92
First printing, August 1992

About the Santa Fe Institute

The *Santa Fe Institute* (SFI) is a multidisciplinary graduate research and teaching institution formed to nurture research on complex systems and their simpler elements. A private, independent institution, SFI was founded in 1984. Its primary concern is to focus the tools of traditional scientific disciplines and emerging new computer resources on the problems and opportunities that are involved in the multidisciplinary study of complex systems—those fundamental processes that shape almost every aspect of human life. Understanding complex systems is critical to realizing the full potential of science, and may be expected to yield enormous intellectual and practical benefits.

All titles from the *Santa Fe Institute Studies in the Sciences of Complexity* series will carry this imprint which is based on a Mimbres pottery design (circa A.D. 950–1150), drawn by Betsy Jones.

Santa Fe Institute Editorial Board
June 1991

Dr. L. M. Simmons, Jr., *Chair*
Executive Vice President, Santa Fe Institute

Prof. Kenneth J. Arrow
Department of Economics, Stanford University

Prof. W. Brian Arthur
Dean & Virginia Morrison Prof. of Population Studies and Economics, Food Research Institute, Stanford University

Prof. Michele Boldrin
MEDS, Northwestern University

Dr. David K. Campbell
Director, Center for Nonlinear Studies, Los Alamos National Laboratory

Dr. George A. Cowan
Visiting Scientist, Santa Fe Institute and Senior Fellow Emeritus, Los Alamos National Laboratory

Prof. Marcus W. Feldman
Director, Institute for Population & Resource Studies, Stanford University

Prof. Murray Gell-Mann
Division of Physics & Astronomy, California Institute of Technology

Prof. John H. Holland
Division of Computer Science & Engineering, University of Michigan

Dr. Edward A. Knapp
President, Santa Fe Institute

Prof. Stuart Kauffman
School of Medicine, University of Pennsylvania

Prof. Harold Morowitz
University Prof., George Mason University

Dr. Alan S. Perelson
Theoretical Division, Los Alamos National Laboratory

Prof. David Pines
Department of Physics, University of Illinois

Prof. Harry L. Swinney
Department of Physics, University of Texas

Santa Fe Institute Studies in the Sciences of Complexity

PROCEEDINGS VOLUMES

Volume	Editor	Title
I	David Pines	Emerging Syntheses in Science, 1987
II	Alan S. Perelson	Theoretical Immunology, Part One, 1988
III	Alan S. Perelson	Theoretical Immunology, Part Two, 1988
IV	Gary D. Doolen et al.	Lattice Gas Methods for Partial Differential Equations, 1989
V	Philip W. Anderson, Kenneth Arrow, & David Pines	The Economy as an Evolving Complex System, 1988
VI	Christopher G. Langton	Artificial Life: Proceedings of an Interdisciplinary Workshop on the Synthesis and Simulation of Living Systems, 1988
VII	George I. Bell & Thomas G. Marr	Computers and DNA, 1989
VIII	Wojciech H. Zurek	Complexity, Entropy, and the Physics of Information, 1990
IX	Alan S. Perelson & Stuart A. Kauffman	Molecular Evolution on Rugged Landscapes: Proteins, RNA and the Immune System, 1990
X	Christopher Langton et al.	Artificial Life II, 1991
XI	John A. Hawkins & Murray Gell-Mann	Evolution of Human Languages, 1992
XII	Martin Casdagli & Stephen Eubank	Nonlinear Modeling and Forecasting, 1992

LECTURES VOLUMES

Volume	Editor	Title
I	Daniel L. Stein	Lectures in the Sciences of Complexity, 1989
II	Erica Jen	1989 Lectures in Complex Systems, 1990
III	Lynn Nadel & Daniel L. Stein	1990 Lectures in Complex Systems, 1991
IV	Lynn Nadel & Daniel L. Stein	1991 Lectures in Complex Systems, 1992

LECTURE NOTES VOLUMES

Volume	Author	Title
I	John Hertz, Anders Krogh, & Richard Palmer	Introduction to the Theory of Neural Computation, 1990
II	Gérard Weisbuch	Complex Systems Dynamics, 1990

REFERENCE VOLUMES

Volume	Author	Title
I	Andrew Wuensche & Mike Lesser	The Global Dynamics of Cellular Automata: An Atlas of Basin of Attraction Fields of One-Dimensional Cellular Automata, 1992

Contributors to This Volume

Jeremy J. Ahouse, Brandeis University
Paddy Andrews, Cambridge University
Maarten C. Boerlijst, University of Utrecht, The Netherlands
Erhard Bruderer, University of Michigan
Gail A. Carpenter, Boston University
Antonio C. R. Da Silva Filho, University of Sussex, England
Pedro P. B. de Oliveira, The University of Sussex, England
Angelica Gelovar-Santiago, Instituto de Fisica, Mexico
Barry N. Guinn, Indiana University
Norio Konno, Muroran Institute of Technology, Japan
Alfred Hübler, Beckmann Institute
Christopher G. Langton, Los Alamos National Laboratory and
    Santa Fe Institute
David Lazer, University of Michigan
Wentian Li, Santa Fe Institute and Rockefeller University
J. H. Lowenstein, New York University
Thomas Meyer, Beckmann Institute
George J. Mpitsos, Oregon State University and
    The Mark O. Hatfield Marine Science Center
Masatoshi Murase, Research Institute for Fundamental Physics, Japan
H. F. Nijhout, Duke University
George J. Papcun, Los Alamos National Laboratory
Carla J. Shatz, Stanford University
Michael F. Shlesinger, Office of Naval Research
Seppo Soinila, University of Helsinki, Helsinki, Finland
Sara Solla, AT&T Bell Laboratories
Timothy R. Thomas, Los Alamos National Laboratory
Jan D. van der Laan, University of Utrecht, The Netherlands
Stella Veretnik, University of Minnesota
Jin Wang, Beckmann Institute

Preface

The 1991 Complex Systems Summer School continued the traditions of its predecessors—a wide array of topics was discussed, students were by turns excited and exciting, and the editors of this volume of lectures were left with the task of finding something different to say about an event that has become almost stable by its fourth year. The alternative we have chosen is to be mercifully brief in this preface to the chapters based on the fourth summer school. We can start off by safely reporting that none of our participants cracked the problem of providing a completely satisfying definition of complexity, though not for want of trying.

As in the previous volumes, the contents of this book reflect the topics discussed in the 1991 Summer School. However, some of the lecturers given there do not appear within; some of those will appear in next year's proceedings. For completeness, we list here those lectures which are not present within this volume: Chaos (Predrag Cvitanovic), Statistical Mechanics of Neural Networks (Sara Solla), The Ecology of Computation (Bernardo Huberman), Neural Network Algorithms and Architectures (John Denker), Neural Basis of Vision in Insects (Nicholas Strausfeld), and Spin Glass Approaches to Protein Folding (Peter Wolynes).

Following an innovation begun last year, we are pleased to include a number of contributions from the participants themselves. These are the result of research by individuals or working groups set up during the school. The results are quite impressive.

ACKNOWLEDGMENTS

Many people contributed to the success of the summer school. The planning for the school, its day-by-day functioning, and the follow-up after the school finishes, are all a reflection of the efforts of a number of people at the Santa Fe Institute. Ed Knapp and Mike Simmons helped with the organization; Ginger Richardson and Andi Sutherland were indispensable from start to finish, as usual; Ronda Butler-Villa and Della Ulibarri played a major role in getting this volume together; Marcella Austin handled the rather complex financial side; and Robin Justice got the computational laboratory up and running, and kept it that way. Susan Coghlan helped organize the school's interaction with the laboratory at Los Alamos. Stuart Kauffman of the Santa Fe Institute committed much of his time to the students, and could often be seen in heated discussion with groups of them in the courtyard of the Institute. Most critical of all, Peter Hraber performed myriad tasks which kept the school running smoothly.

We thank also our advisory board, and several institutions that provided computers and associated peripherals. We also thank the University of Arizona, and its Center for the Study of Complex Systems, for permitting the two of us to spend time on this rewarding but time-consuming enterprise. Finally, we must thank those agencies that contributed the funds needed to make the school a reality—The Department of Energy, The National Science Foundation, the National Institute of Mental Health, and the Office of Naval Research.

      Lynn Nadel   Daniel Stein
      University of Arizona   University of Arizona
      Tucson, AZ 85721   Tucson, AZ 85721

May 29, 1992

Contents

Preface
   Lynn Nadel and Daniel Stein   xiii

Section 1. Lectures & Seminars   1

Neural Network Models for Pattern Recognition and
Associative Memory
   Gail A. Carpenter   3

Impulse Activity and the Patterning of Connections During
CNS Development
   Carla J. Shatz   43

In Search of a Unified Theory of Biological Organization:
What Does the Motor System of a Sea Slug Tell Us about
Human Motor Integration?
   George J. Mpitsos and Seppo Soinila   67

Mapping from Speech Acoustics to Tongue Dorsum
Movement: An Application of a Multilayer Perceptron
   Timothy R. Thomas, George J. Papcun, and Barry N. Guinn   139

Pattern Formation in Biological Systems
   H. F. Nijhout   159

Artificial Life
   Christopher G. Langton   189

Order Parameters, Broken Symmetry, and Topology
   James P. Sethna   243

Meissner Effects and Constraints
   James P. Sethna   267

Fractal Time Dynamics: From Glasses to Turbulence
Michael F. Shlesinger 289

Dynamics of Web Maps: Parameter Dependence of Stochastic Layers
J. H. Lowenstein 305

Non-Local Cellular Automata
Wential Li 317

Section 2. Student Contributions 329

Reality Kisses the Neck of Speculation: A Report From the NKC Workgroup
Jeremy John Ahouse, Erhard Bruderer, Angelica Gelovar-Santiago, Norio Konno, David Lazer, and Stella Veretnik 331

Complex Patterns, Simply Recognized
Paddy Andrews 353

Dynamical Behavior of a Pair of Spatially Homogeneous Neural Fields
Antonio C. Roque Da Silva Filho 371

A Cellular Automaton to Embed Genetic Search
Pedro Paulo Balbi de Oliveira 389

Some Mathematical Results on the NK Model
Norio Konno 409

Complex Dynamics of Flagella
Masatoshi Murase 415

Cellular Automata with Non-Uniform Rules: An Illustration of Kauffman's Boolean Network Theory
Jan D. van der Laan and Maarten C. Boerlijst 431

Random Boolean Networks: Comparison Between Randomly Connected and Lattice-Connected Networks
Stella Veretnik 441

The Production of Solitons By Optimal Driving Forces
Jin Wang, Thomas Meyer, and Alfred Hübler 457

Index 465

Lectures & Seminars

Gail A. Carpenter
Department of Cognitive and Neural Systems, Boston University, 111 Cummington Street, Boston, MA 02215

Neural Network Models for Pattern Recognition and Associative Memory

This review outlines some fundamental neural network modules for associative memory, pattern recognition, and category learning. Included are discussions of the McCulloch-Pitts neuron, perceptrons, adaline and madaline, back propagation, the learning matrix, linear associative memory, embedding fields, instars and outstars, the avalanche, shunting competitive networks, competitive learning, computational mapping by instar/outstar families, adaptive resonance theory, the cognitron and neocognitron, and simulated annealing. Adaptive filter formalism provides a unified notation. Activation laws include additive and shunting equations. Learning laws include back-coupled error correction, Hebbian learning, and gated instar and outstar equations. Also included are discussions of real-time and off-line modeling, stable and unstable coding, supervised and unsupervised learning, and self-organization.

1. INTRODUCTION

Neural network analysis exists on many different levels. At the highest level (Figure 1) we study theories, architectures, and hierarchies for big problems such as early vision, speech, arm movement, reinforcement, and cognition. Each architecture is typically constructed from pieces, or *modules*, designed to solve parts of a bigger problem. These pieces might be used, for example, to associate pairs of patterns with one another or to sort a class of patterns into various categories. In turn, for every such module there is a bewildering variety of examples, equations, simulations, theorems, and implementations, studied under various conditions such as fast or slow input presentation rates, supervised or unsupervised learning, and real-time or off-line dynamics. These variations and their applications are now the subject of hundreds of talks and papers each year. In this review I will focus on the middle level, on some of the fundamental neural network modules that carry out associative memory, pattern recognition, and category learning.

Even then this is a big subject. To help organize it further, I will trace the historical development of the main ideas, grouped by theme rather than by strict chronological order. But keep in mind that there is a much more complex history, and many more contributors, than you will read about here. I refer you to the Bibliography, in particular to the collection of articles in *Neurocomputing: Foundations of Research*, edited by James A. Anderson and Edward Rosenfeld (MIT Press, Cambridge, 1988).

2. THE McCULLOCH-PITTS NEURON

We would probably all agree to begin with the McCulloch-Pitts neuron (Figure 2(a)). The McCulloch-Pitts model describes a neuron whose activity x_j is the sum of inputs that arrive via weighted pathways. The input from a particular pathway is an incoming signal S_i multiplied by the weight w_{ij} of that pathway. These weighted inputs are summed independently. The outgoing signal $S_j = f(x_j)$ is typically a nonlinear function—binary, sigmoid, threshold-linear—of the activity x_j in that cell. The McCulloch-Pitts neuron can also have a bias term θ_j, which is formally equivalent to the negative of a threshold of the outgoing signal function.

3. ADAPTIVE FILTER FORMALISM

A very convenient notation for describing the McCulloch-Pitts neuron is the *adaptive filter*. It is this notation that I will here use to translate models into

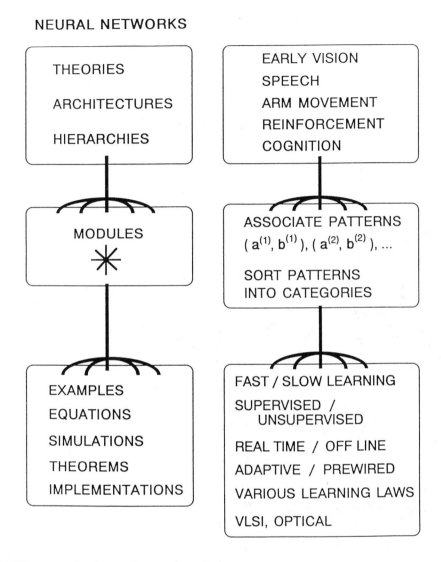

FIGURE 1 Levels of neural network analysis.

a common language so that we can compare and contrast them. The elementary adaptive filter depicted in Figure 2(b) has:

1. a level F_1 that registers an input pattern vector;
2. signals S_i that pass through weighted pathways; and

3. a second level F_2 whose activity pattern is here computed by the McCulloch-Pitts function:

$$x_j = \sum_i S_i w_{ij} + \theta_j. \tag{1}$$

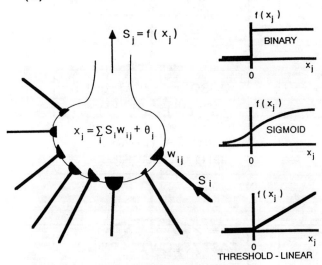

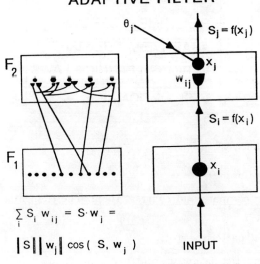

FIGURE 2 The McCulloch-Pitts model (a) as a neuron, with typical nonlinear signal functions; (b) as an adaptive filter.

The reason this formalism has proved so extraordinarily useful is that the F_2 level of the adaptive filter computes a pattern match, as in Eq. (2):

$$\sum_i S_i w_{ij} = \mathbf{S} \cdot \mathbf{w}_j = \|\mathbf{S}\|\|\mathbf{w}_j\| \cos(\mathbf{S}, \mathbf{w}_j). \tag{2}$$

The independent sum of the weighted pathways in Eq. (2) equals the dot product of the signal vector $\mathbf{S}$ times the weight vector $\mathbf{w}_j$. This term can be factored into the "energy," the product of the lengths of $\mathbf{S}$ and $\mathbf{w}_j$, times a dimensionless measure of "pattern match," the cosine of the angle between the two vectors. Suppose that the weight vectors $\mathbf{w}_j$ are normalized and the bias terms θ_j are all equal. Then the activity vector $\mathbf{x}$ across the second level describes the degree of match between the signal vector $\mathbf{S}$ and the various weighted pathway vectors $\mathbf{w}_j$: the F_2 node with the greatest activity indicates the weight vector that forms the best match.

4. LOGICAL CALCULUS AND INVARIANT PATTERNS

The paper that first describes the McCulloch-Pitts model is entitled "A Logical Calculus of the Ideas Immanent in Nervous Activity."[35] In that paper, McCulloch and Pitts analyze the adaptive filter without adaptation. In their models, the weights are constant. There is no learning. This 1943 paper shows that given the linear filter with an absolute inhibition term:

$$x_j = \sum_i S_i w_{ij} + \theta_j - [\text{inhibition}] \tag{3}$$

and binary output signals, these networks can be configured to perform arbitrary logical functions. And if you are looking for applications of neural network research, you need only read the memoirs of John von Neumann[47] to see how heavily the McCulloch-Pitts formalism influenced the development of present-day computer architectures.

In a sense, however, McCulloch and Pitts were looking backwards, to the early 20th century mathematics of *Principia Mathematica*.[43] A glance at the 1943 paper shows that it is written in notation with which few of us are now familiar. (This is a good example of revolutionary ideas being expressed in the language of a previous era. As the revolution comes about a new language evolves, making the seminal papers "hard to read.") McCulloch and Pitts also clearly looked forward toward present-day neural network research. For example, a later paper is entitled "How We Know Universals: The Perception of Auditory and Visual Forms."[38] There they examine ideas in pattern recognition and the computation of invariants. They thus took their research program into a domain distinctly different from the earlier analysis of formal network groupings and computation. Still, they considered only models without learning.

5. PERCEPTRONS AND BACK-COUPLED ERROR CORRECTION

The McCulloch-Pitts papers were extraordinarily influential, and it was not long before the next generation of researchers added learning and adaptation. One great figure of the next decade was Frank Rosenblatt, whose name is tied with the perceptron model.[39] Actually, "perceptron" refers to a large class of neural models. The models that Rosenblatt himself developed and studied are numerous and varied; see, for example, his book, *Principles of Neurodynamics*.[40]

The core idea of the perceptron is the incorporation of learning into the McCulloch-Pitts neuron model. Figure 3 illustrates the main elements of the perceptron, including, in Rosenblatt's terminology, the sensory unit (S); the association unit (A), where the learning takes place; and the response unit (R).

One of the many perceptrons that Rosenblatt studied, one that remains important to the present day, is the *back-coupled perceptron*.[40] Figure 4(a) illustrates a simple version of the back-coupled perceptron model, with a feedforward adaptive

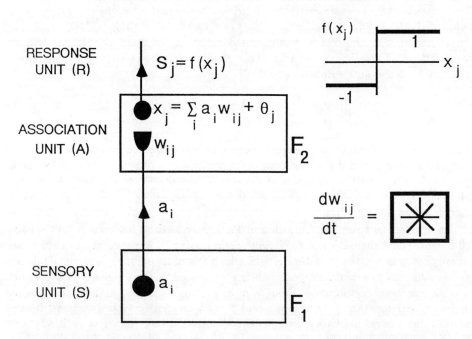

FIGURE 3 Principal elements of a Rosenblatt perceptron: sensory unit (S), association unit (A), and response unit (R).

filter and binary output signal. Weights w_{ij} are adapted according to whether the actual output S_j matches a target output b_j imposed on the system. The actual output vector is subtracted from the target output vector; their difference is defined as the error; and that difference is then fed back to adjust the weights, according to some probabilistic law. Rosenblatt called this process *back-coupled error correction*. It was well known at the time that these two-level perceptrons could sort linearly separable inputs, which can be separated by a hyperplane in vector space, into two classes. Figure 4(b) shows back-coupled error correction in more detail. In particular the error δ_j is fed back to every one of the weights converging on the jth node.

6. ADALINE AND MADALINE

Research in the 1960s did not stop with these two-level perceptrons, but continued on to multiple-level perceptrons, as indicated below. But first let us consider another development that took place shortly after Rosenblatt's perceptron formulations. This is the set of models used by Bernard Widrow and his colleagues, especially the *adaline* and *madaline* perceptrons. The adaline model has just one neuron in the F_2 level in Figure 5; the madaline, or many-adaline, model has any number of neurons in that level. Figure 5 highlights the principal difference between the adaline/madaline and Rosenblatt's two-level feedforward perceptron: an adaline/madaline model compares the *analog* output x_j with the target output b_j. This comparison provides a more subtle index of error than a law that compares the *binary* output with the target output. The error $b_j - x_j = \delta_j$ is fed back to adjust weights using a Rosenblatt back-coupled error correction rule:

$$\frac{dw_{ij}}{dt} = \alpha \delta_j \frac{a_i}{|a|^2}. \qquad (4)$$

This rule minimizes the mean squared error:

$$\sum_j \delta_j^2 \qquad (5)$$

averaged over all inputs.[50] It is therefore known as the *least mean squared* error correction rule, or LMS.

(a) BACK - COUPLED PERCEPTRON

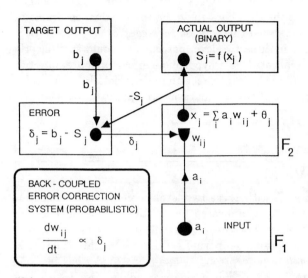

(b) BACK - COUPLED ERROR CORRECTION

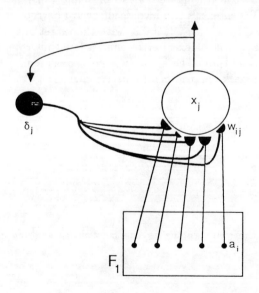

FIGURE 4 Back-coupled error correction. (a) The difference between the target output and the actual output is fed back to adjust weights when an error occurs. (b) All weights w_{ij} fanning in to the jth node are adjusted in proportion to the error d_j at that node.

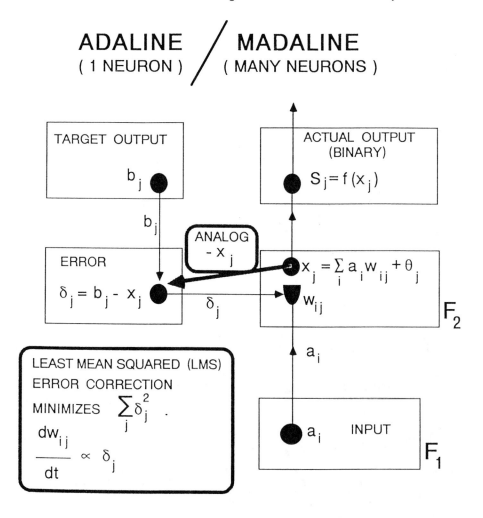

FIGURE 5 The adaline and madaline perceptrons use the analog output x_j, rather than the binary output S_j, in the back-coupled error correction procedure.

Once again, adaline and madaline provide many examples of the technological spin-offs already generated by neural network research. Some of these are summarized in an article by Widrow and Winter[51] in a *Computer* special issue on artificial neural systems. There the authors describe adaptive equalizers and adaptive echo cancellation in modems, antennae, and other engineering applications, all directly traceable to early neural network designs.

7. MULTI-LEVEL PERCEPTRONS: EARLY BACK PROPAGATION

We have so far been discussing only two-level perceptrons. Rosenblatt, not content with these, also studied multi-level perceptrons, as described in *Principles of Neurodynamics*. One particularly interesting section in that book is entitled "Back-Propagating Error Correction Procedures." The back-propagation model described in that section anticipates the currently used back-propagation model, which is also a multi-level perceptron. In Chapter 13, Rosenblatt defines a back-propagation algorithm that has, like most of his algorithms, a probabilistic learning law; he proves a theorem about this system; and he carries out simulations. His chapter, "Summary of Three-Layer Series-Coupled Systems: Capabilities and Deficiencies," is equally revealing. This chapter includes a hard look at what is lacking as well as what is good in Rosenblatt's back-propagation algorithm, and it puts the lie to the myth that all of these systems were looked at only through rose-colored glasses.

8. LATER BACK PROPAGATION

Let us now move on to what has become one of the most useful and well-studied neural network algorithms, the model we now call back propagation. This system was first developed by Paul Werbos,[48] as part of his Ph.D thesis "Beyond Regression: New Tools for Prediction and Analysis in the Behavioral Sciences"; and independently discovered by David Parker.[37] (See Werbos[49] for a review of the history of the development of back propagation.)

The most popular back-propagation examples carry out associative learning: during training, a vector pattern **a** is associated with a vector pattern **b**; and subsequently **b** is recalled upon presentation of **a**.[41] The back-propagation system is trained under conditions of *slow learning*, with each pattern pair (**a**,**b**) presented repeatedly during training. The basic elements of a typical back-propagation system are the McCulloch-Pitts linear filter with a sigmoid output signal function and Rosenblatt back-coupled error correction. Figure 6 shows a block diagram of a back-propagation system that is a three-level perceptron. The input signal vector converges on the "hidden unit" F_2 level after passing through the first set of weighted pathways w_{ij}. Signals S_j then fan out to the F_3 level, which generates the actual output of this feedforward system. A back-coupled error correction system then compares the actual output S_k with a target output b_k and feeds back their difference to all the weights w_{jk} converging on the kth node. In this process the difference $b_k - S_k$ is also multiplied by another term, $f'(x_k)$, computed in a "differentiator" step. One function of this step is to ensure that the weights remain in a bounded range: the shape of the sigmoid signal function implies that weights w_{jk} will stop growing if the magnitude of the activity x_k becomes too large, since then the derivative term $f'(x_k)$ goes to zero. Then there is a second way in

Neural Network Models for Pattern Recognition and Associative Memory

which the error correction is fed back to the lower level. This is where the term "back propagation" enters: the weights w_{jk} in the feedforward pathways from F_2 to F_3 are now used in a second place, to filter error information. This process is called *weight transport*. In particular, all the weights w_{jk} in pathways fanning *out* from the jth F_2 node are transported for multiplication by the corresponding error

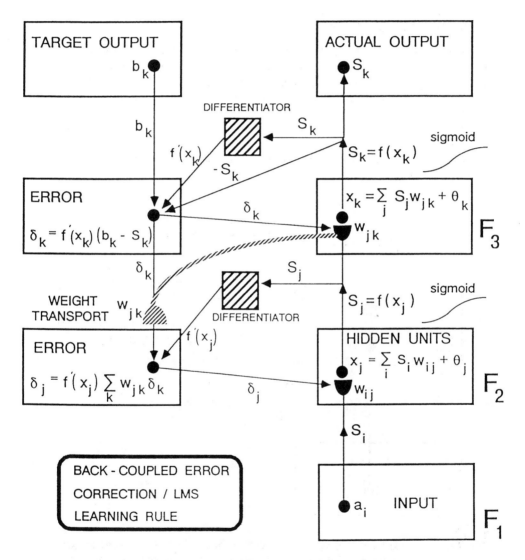

FIGURE 6 Block diagram of a back-propagation algorithm for associative memory. Weights in the three-level feedforward perceptron are adjusted according to back-coupled error correction rules. Weight transport propagates error information in F_2-to-F_3 pathways back to weights in F_1-to-F_2 pathways.

terms δ_k; and the sum of all these products, times the bounding derivative term $f'(x_k)$, is back-coupled to adjust all the weights w_{ij} in pathways fanning *in* to the jth F_2 node.

9. HEBBIAN LEARNING

This brings us close to the present in this particular line of perceptron research. I am now going to step back and trace another major neural network theme that goes under the name *Hebbian learning*. One sentence in a 1949 book, *The Organization of Behavior* by Donald Hebb, is responsible for the phrase Hebbian learning:

> "When an axon of cell A is near enough to excite a cell B and repeatedly or persistently takes place in firing it, some growth process or metabolic change takes place in one or both cells such that A's efficiency, as one of the cells firing B, is increased."[24]

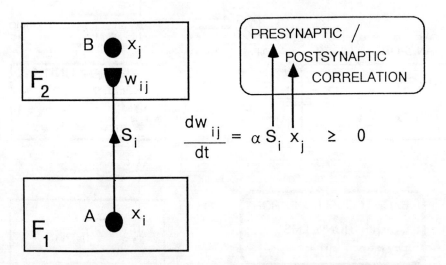

FIGURE 7 Donald Hebb[24] provided a qualitative description of increases in path strength that occur when cell A helps to fire cell B. In the adaptive filter formalism, this hypothesis is often interpreted as a weight change that occurs when a presynaptic signal S_i is correlated with a postsynaptic activity x_j.

Actually, "Hebbian learning" was not a new idea in 1949: it can be traced back to Pavlov and earlier. But in the decade of McCulloch and Pitts, the formulation of the idea in the above sentence crystallized the notion in such a way that it became widely influential in the emerging neural network field. Translated into a differential equation (Figure 7), the Hebbian rule computes a correlation between the presynaptic signal S_i and the postsynaptic activity x_j, with positive values of the correlation term $S_i x_j$ leading to increases in the weight w_{ij}.

The Hebbian learning theme has since evolved in a number of directions. One important development entailed simply adding a passive decay term to the Hebbian correlation term[13]:

$$\frac{dw_{ij}}{dt} = \alpha S_i x_j - w_{ij}. \tag{6}$$

Other developments are described below. In all these rules, changes in the weight w_{ij} depend upon a simple function of the presynaptic signal S_i, the postsynaptic activity x_j, and the weight itself, as in Eq. (6). In contrast, back-coupled error correction requires a term that must be computed away from the target node and then transmitted back to adjust the weight.

10. THE LEARNING MATRIX

Many of the models that followed the perceptron in the 1950s and 1960s can be phrased in Hebbian (plus McCulloch-Pitts) language. One of the earliest and most important is the learning matrix (Figure 8) developed by K. Steinbuch.[44] The function of the learning matrix is to sort, or partition, a set of vector patterns into categories. In the simple learning matrix illustrated in Figure 8(a), an input pattern **a** is represented in the vertical wires. During learning a category for **a** is represented in the horizontal wires of the crossbar: **a** is placed in category J when the Jth component of the output vector **b** is set equal to 1. During such an input presentation, the weight w_{iJ} is adjusted upward by a fixed amount if $a_i = 1$ and downward by the same amount if $a_i = 0$. Then during performance the weights w_{ij} are held constant; and an input **a** is deemed to be in category J if the weight vector $\mathbf{w}_J = (w_{1J}, \ldots w_{NJ})$ is closer than any other weight vector to **a**, according to some measure of distance.

Recasting the crossbar learning matrix in the adaptive filter format (Figure 8(b)) helps us to see that this simple model is the precursor of a fundamental module widely used in present-day neural network modeling, namely *competitive learning*. In particular, activity at the top level of the learning matrix corresponds to a category representation. Setting activity x_J equal to 1, while all other x_j's are set equal to 0, corresponds to the dynamics of a *choice*, or *winner-take-all*, neural network. Steinbuch's learning rule can also be translated into the Hebbian

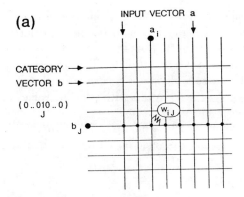

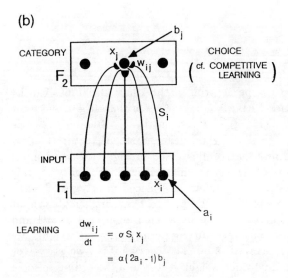

FIGURE 8 The learning matrix, for category learning. (a) Cross-bar architecture for electronic implementation. (b) The learning matrix in adaptive filter notation. The learning matrix was a precursor of the competitive learning paradigm.

formalism, with weight adjustment during learning a joint function of a presynaptic signal $S_i = (2a_i - 1)$ and a postsynaptic signal $x_j = b_j$. (This rule is not strictly Hebbian since weights can decrease as well as increase.) Then during performance, weight changes are prevented; a new signal function $S_i = a_i$ is chosen; and an F_2 choice rule is imposed based, for example, on the dot product measure illustrated in Figure 9(b).

A model comparative analysis of the learning matrix and the madaline models and their electronic implementations can be found in a paper by K. Steinbuch and B. Widrow.[45] This paper, entitled "A critical comparison of two kinds of adaptive classification networks," carries out a side-by-side analysis of the learning matrix and the madaline, tracing the two models' capabilities, similarities, and differences.

11. LINEAR ASSOCIATIVE MEMORY (LAM)

We will now move to a different line of research, namely the linear associative memory (LAM) models. Pioneering work on these models was done by J. Anderson,[5] T. Kohonen,[30] and K. Nakano.[36] Subsequently, many other linear associative memory models were developed and analyzed, for example by Kohonen and his collaborators, who studied LAM's with iteratively computed weights that converge to the Moore-Penrose pseudoinverse.[33] This latter system is optimal with respect to the LMS error (5), and so is known as the optimal linear associative memory (OLAM) model. Variations included networks with partial connectivity, probabilistic learning laws, and nonlinear perturbations.

At the heart of all these variations is a very simple idea, namely that a set of pattern pairs $(\mathbf{a}^{(p)}, \mathbf{b}^{(p)})$ can be stored as a correlation weight matrix:

$$w_{ij} = \sum_{p \text{ (all patterns)}} a_i^{(p)} b_j^{(p)}. \tag{7}$$

The LAM's have been an enduringly useful class of models because, in addition to their great simplicity, they embody a sort of perfection. Namely, perfect recall is achieved, provided the input vectors $\mathbf{a}^{(p)}$ are mutually orthogonal. In this case, during performance, presentation of the pattern $\mathbf{a}^{(p)}$ yields an output vector $\mathbf{x}$ proportional to $\mathbf{b}^{(p)}$, as follows:

$$\begin{aligned} x_j \equiv \mathbf{a}^{(p)} \cdot \mathbf{w}_j &= \sum_i a_i^{(p)} w_{ij} = \sum_i a_i^{(p)} \left(\sum_q a_i^{(q)} b_j^{(q)} \right) \\ &= \sum_q \left(\sum_i a_i^{(p)} a_i^{(q)} \right) b_j^{(q)} = \sum_q (\mathbf{a}^{(p)} \cdot \mathbf{a}^{(q)}) b_j^{(q)}. \end{aligned} \tag{8}$$

If, then, the vectors $\mathbf{a}^{(p)}$ are mutually orthogonal, the last sum in Eq. (8) reduces to a single term, with

$$x_j = \|\mathbf{a}^{(p)}\|^2 b_j^{(p)}. \tag{9}$$

Thus the output vector **x** is directly proportional to the desired output vector, $\mathbf{b}^{(p)}$. Finally, if we once again cast the LAM in the adaptive filter framework, we see that it is a Hebbian learning model (Figure 9).

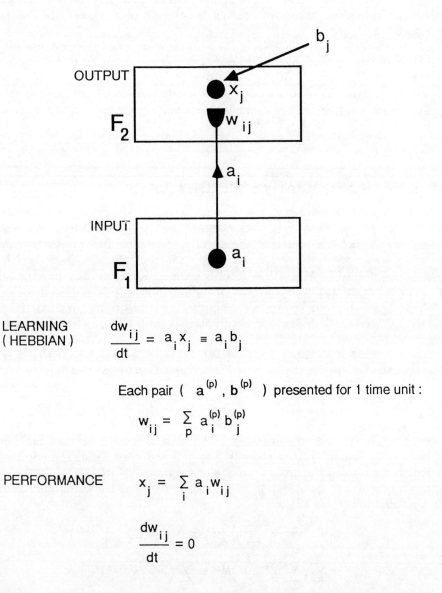

FIGURE 9 A linear associative memory network, in adaptive filter/Hebbian learning format.

12. REAL-TIME MODELS AND EMBEDDING FIELDS

Most of the models we have so far discussed require external control of system dynamics. In the back-propagation model shown in Figure 6, for example, the initial feedforward activation of the three-level perceptron is followed by error correction steps that require either weight transport or reversing the direction of flow of activation. In the linear associative memory model in Figure 9, dynamics are altered as the system moves from its learning mode to its performance mode. During learning, activity x_j at the output level F_2 is set equal to the desired output b_j, while the input $S_i a_i w_{ij}$ coming to that level from F_1 through the adaptive filter is suppressed. During performance, in contrast, the dynamics are reversed: weight changes are suppressed and the adaptive filter input determines x_j.

The phrase *real time* describes neural network models that require no external control of system dynamics. (*Real time* is alternatively used to describe any system that is able to process inputs as fast as they arrive.) Differential equations constitute the language of real-time models. A real-time model may or may not have an external teaching input, like the vector **b** of the LAM model; and learning may or may not be shut down after a finite time interval. A typical real-time model is illustrated in Figure 10. There, excitatory and inhibitory inputs could be either internal or external to the model, but, if present, the influence of a signal is not selectively ignored. Moreover, the learning rate $\epsilon(t)$ might, say, be constant or decay to 0 through time, but does not require algorithmic control. The dynamics of performance are described by the same set of equations as the dynamics of learning.

Real-time modeling has characterized the work of Stephen Grossberg over the past thirty years, work that in its early stages was called a theory of *embedding fields*.[12] These early real-time models, as well as the more recent systems developed by Grossberg and his colleagues at the Boston University Center for Adaptive Systems, portray the inextricable linking of fast nodal activation and slow weight adaptation. There is no externally imposed distinction between a learning mode and a performance mode.

13. INSTARS AND OUTSTARS

Two key components of embedding field systems are the *instar*[17,20,46] and the *outstar*.[13] Figure 11 illustrates the fan-in geometry of the instar and the fan-out geometry of the outstar.

Instars often appear in systems designed to carry out adaptive coding, or content-addressable memory (CAM).[31] For example, suppose that the incoming weight vector $(w_{1J}, \ldots, w_{NJ})$ approaches the incoming signal vector $(S_1, \ldots, S_N)$ while an input vector **a** is present at F_1; and that the weight and signal vectors are normalized. Then Eq. (2) implies that the filtered input $\sum_i S_i w_{iJ}$ to the Jth

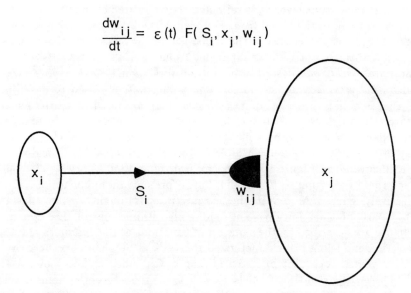

FIGURE 10 Elements of a typical real-time model, with additive activation equations.

F_2 node approaches its maximum value during learning. Subsequent presentation of the same F_1 input pattern **a** maximally activates the Jth F_2 node; that is, the "content addresses the memory," all other things being equal.

The outstar, which is dual to the instar, carries out spatial pattern learning. For example, suppose that the outgoing weight vector $(w_{J1}, \ldots, w_{JN})$ approaches the F_1 spatial activity pattern $(x_1, \ldots, x_N)$ while an input vector **a** is present. Then subsequent activation of the Jth F_2 node transmits to F_1 the signal pattern $(S_J w_{J1}, \ldots, S_J w_{JN}) = S_J(w_{J1}, \ldots, w_{JN})$, which is directly proportional to the prior F_1 spatial activity pattern $(x_1, \ldots, x_N)$, even though the input vector is now absent; that is, the "memory addresses the content."

The upper instar and outstar in Figure 11 are examples of *heteroassociative* memories, where the field F_1 of nodes indexed by i is disjoint from the field F_2 of nodes indexed by j. In general, these fields can overlap. The important special case in which the two fields coincide is called *autoassociative* memory, also shown in Figure 11. Powerful computational properties arise when neural network architectures are constructed from a combination of instars and outstars. We will later see some of these designs.

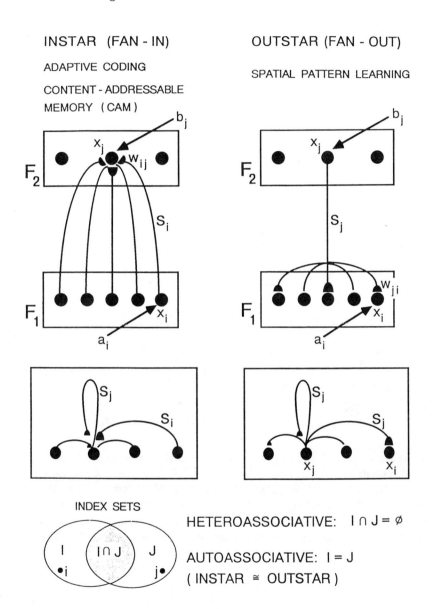

FIGURE 11 Heteroassociative and autoassociative instars and outstars, for adaptive coding and spatial pattern learning.

14. ADDITIVE AND SHUNTING ACTIVATION EQUATIONS

The outstar and the instar have been studied in great detail and with various combinations of activation, or short-term memory, equations and learning, or long-term memory, equations. One activation equation, the *additive model*, is illustrated in Figure 10. There, activity at a node is proportional to the difference between the net excitatory input and the net inhibitory input. Most of the models discussed so far employ a version of the additive activation model. For example, the McCulloch-Pitts activation equation (3) is the steady state of the additive equation (10):

$$\frac{dx_j}{dt} = -x_j + \left[\sum_i S_i w_{ij} + \theta_j\right] - [\text{inhibition}]. \tag{10}$$

Grossberg[23] reviews a number of neural models that are versions of the additive equation.

An important generalization of the additive model is the *shunting model*. In a shunting network, excitatory inputs drive activity toward a finite maximum, while inhibitory inputs drive activity toward a finite minimum, as in Eq. (11):

$$\frac{dx_i}{dt} = -x_i + (A-x_i)\sum\left[\text{excitatory inputs}\right] - (B+x_i)\sum\left[\text{inhibitory inputs}\right]. \tag{11}$$

In Eq. (11), activity x_i remains in the bounded range $(-B, A)$, and decays to the resting level 0 in the absence of all inputs. In addition, shunting equations display other crucial properties such as normalization and automatic gain control. Finally, shunting network equations mirror the underlying physiology of single nerve cell dynamics, as summarized by the Hodgkin-Huxley[27] equations:

$$\frac{dV}{dt} = -V + (V_{Na} - V)\overline{g}_{Na}m^3h - (V_K + V)\overline{g}_K n^4. \tag{12}$$

In this single nerve cell model, during depolarization, sodium ions entering across the membrane drive the potential V toward the sodium equilibrium potential V_{Na}; during repolarization, exiting potassium ions drive the potential toward the potassium equilibrium potential $-V_K$; and in the balance the cell is restored to its resting potential, which is here set equal to 0. In 1963 A. L. Hodgkin and A. F. Huxley won the Nobel Prize for their development of this classic neural model.

15. LEARNING EQUATIONS

A wide variety of learning laws for instars and outstars have also been studied. One example is the Hebbian correlation + passive decay equation (6). There, the weight w_{ij} computes a long-term weighted average of the product of presynaptic activity S_i and postsynaptic activity x_j.

A typical learning law for instar coding is given by Eq. (13):

$$\frac{dw_{ij}}{dt} = \epsilon(t)[S_i - w_{ij}]x_j. \qquad (13)$$

Suppose, for example, that the Jth F_2 node is to represent a given category. According to Eq. (13), the weight vector $(w_{1J}, \ldots, w_{NJ})$ converges to the signal vector $(S_1, \ldots, S_N)$ when the Jth node is active; but that weight vector remains unchanged when a different category representation is active. The term x_J thus buffers, or *gates*, the weights w_{iJ} against undesired changes, including memory loss due to passive decay. On the other hand, a typical learning law for outstar pattern learning is given by Eq. (14):

$$\frac{dw_{ji}}{dt} = \epsilon(t)[x_i - w_{ji}]S_j. \qquad (14)$$

In Eq. (14), when the Jth F_2 node is active the weight vector $(w_{J1}, \ldots, w_{JN})$ converges to the F_1 activity pattern vector $(x_1, \ldots, x_N)$. Again, a gating term buffers weights against inappropriate changes. Note that the pair of learning laws described by Eqs. (13) and (14) are non-Hebbian, and are also non-symmetric. That is, w_{ij} is generally not equal to w_{ji}, unless the F_1 and F_2 signal vectors S are identical to the corresponding activity vectors x.

A series of theorems encompassing neural network pattern learning by systems employing a large class of these and other activation and learning laws was proved by Grossberg in the late 1960s and early 1970s. One set of results falls under the heading *outstar learning theorems*. One of the most general of these theorems is contained in an article entitled "Pattern Learning by Functional-Differential Neural Networks with Arbitrary Path Weights."[16] This is reprinted in *Studies of Mind and Brain*,[22] which also contains articles that introduce and analyse additive and shunting equations (10) and (11); learning with passive and gated memory decay laws (6), (13), and (14); outstar and instar modules; and neural network architectures constructed from these elements.

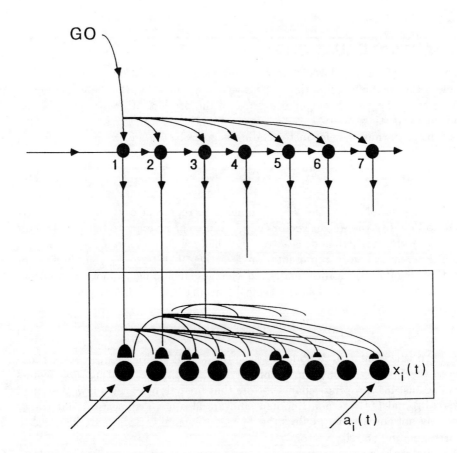

FIGURE 12 The avalanche: a neural network capable of learning and performing an arbitrary space-time pattern.

16. LEARNING SPACE-TIME PATTERNS: THE AVALANCHE

While most of the neural network models discussed in this article are designed to learn spatial patterns, problems such as speech recognition and motor learning require an understanding of space-time patterns as well. An early neural network model, called the *avalanche*, is capable of learning and performing an arbitrary space-time pattern.[14] In essence, an avalanche is a series of outstars (Figure 12). During learning, the outstar active at time t learns the spatial pattern $\mathbf{x}(t)$ generated by the input pattern vector $\mathbf{a}(t)$. It is useful to think of $\mathbf{x}(t)$ as the pattern determining finger positions for a piano piece: the same field of cells is used over and over, and the sequence ABC is not the same as CBA. Following learning, when no input patterns are present, activation of the sequence of outstars reads-out, or "performs," the space-time pattern it had previously learned. In its minimal form, this network can be realized as a single cell with many branches. Learning and performance can also be supervised by a nonspecific GO signal. The GO signal may terminate an action sequence at any time and otherwise modulate the performance energy and velocity. In general, the order of activation of the outstars, as well as the spatial patterns themselves, need to be learned. This can be accomplished using autoassociative networks, as in the theory of serial learning[15] or adaptive signal processing[25]

17. ADAPTIVE CODING AND CATEGORY FORMATION

Let us now return to the theme of adaptive coding and category formation, introduced earlier in our discussion of Steinbuch's learning matrix. As shown in Figure 8(b), the learning matrix can be recast in the adaptive filter formalism, with the dynamics of the F_2 level defined in such a way that only one node is active at a given time. The active node, or category representation, is selected by a "teacher" during learning. During performance the active node is selected according to which weight vector forms the best match with the input vector. Now compare the learning matrix in Figure 8(b) with the instar in Figure 11. The pictures, or network "anatomies," seem to indicate that the instar is identical to the learning matrix. The difference between the two models lies in the dynamics, or network "physiology." The fundamental characteristic of the instar that distinguishes it from the learning matrix and other early models is the constraint that instar dynamics occur in real time. In particular, the instar filtered input $\mathbf{S} \cdot \mathbf{w}_j$ influences x_j at all times, and is not artificially suppressed during learning. However, the desire to construct a category learning system that can operate in real time immediately leads to many questions. The most pressing one is: how can the categories be represented if the dynamics are not imposed by an external agent? For the choice case, for example, the *internal* system dynamics need to allow at most one F_2 node be active, even though other nodes may continue to receive large inputs, either internally, via the filter, or externally, via the vector **b**. Even when the category representation is a distributed pattern, this representation is generally a compressed, or contrast-enhanced, version of the highly distributed net pattern coming in to F_2 from all sources. This compression is, in fact, the step that carries out the process wherein some or many items are grouped into a new unit, or category.

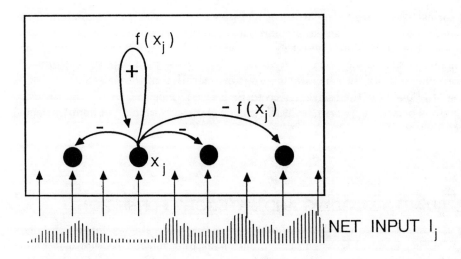

FIGURE 13 An on-center/off-surround shunting competitive network. Qualitative features of the signal function $f(x_j)$ determine the way in which the network transforms the input vector **I** into the state vector **x**.

18. SHUNTING COMPETITIVE NETWORKS

Fortunately, there is a well-defined class of neural networks ideally suited to play the role of the category representation field. This is the class of on-center/off-surround shunting competitive networks. Figure 13 illustrates one such system. There, the input vector **I** can be the sum of inputs from one or more sources and is, in general, highly distributed. *On-center* here refers to the feedback process whereby a cell sends net excitatory signals to itself and to its immediate neighbors; *off-surround* refers to the complementary process whereby the same cell sends net inhibitory signals to its more distant neighbors. In a 1973 article entitled "Contour Enhancement, Short-Term Memory, and Constancies in Reverberating Neural Networks," Grossberg carried out a mathematical characterization of the dynamics of various classes of shunting competitive networks. In particular he classified the systems according to the shape of the signal function $f(x_j)$. Depending upon whether this signal function is linear, faster-than-linear, slower-than-linear, or sigmoid, the networks are shown to quench or enhance low-amplitude noise, and to contrast-enhance or flatten the input pattern **I** in varying degrees. In particular, a faster-than-linear signal function implements the choice network needed for many models of category learning. A sigmoid signal function, on the other hand, suppresses noise and contrast-enhances the input pattern, without necessarily going to the extreme of concentrating all activity in one node. Thus an on-center/off-surround shunting competitive network with a sigmoid signal function is shown to be an ideal design for a category learning system with distributed code representations. This parametric analysis thus provided the foundation for constructing larger network architectures that use a competitive network as a component with well-defined functional properties.

19. COMPETITIVE LEARNING

A module of fundamental importance in recent neural network architectures is described by the phrase *competitive learning*. This module brings the properties of the into the real-time setting. The basic competitive learning architecture consists of an instar filter, from a field F_1 to a field F_2, and a competitive neural network at F_2 (Figure 14). The competitive learning module can operate with or without an external teaching signal **b**, and learned changes in the adaptive filter can proceed indefinitely or cease after a finite time interval. If there is no teaching signal at a given time, then the net input vector to F_2 is the sum of signals arriving via the adaptive filter. Then, if the category representation network is designed to make a choice, the node that automatically becomes active is the one whose weight vector best matches the signal vector, as in Eq. (2). If there is a teaching signal, the category representation decision still depends on past learning, but this is balanced against the external signal **b**, which may or may not overrule the past in the competition. In either case, an instar learning law such as Eq. (13) allows a chosen category to encode aspects of the new F_1 pattern in its learned representation.

20. COMPUTATIONAL MAPS

Investigators who have developed and analyzed the competitive learning paradigm over the years include K. Steinbuch[44]; S. Grossberg[17,19,20]; C. von der Malsburg[46]; S.-I. Amari[3]; S.-I. Amari and A. Takeuchi[4]; E. Bienenstock, L. Cooper, and P. Munro[6]; D. Rumelhart and D. Zipser[42]; and many others. Moreover, these and other investigators proceeded to embed the competitive learning module in higher-order neural network systems. In particular, systems were designed to learn computational maps, producing an output vector **b** in response to an input vector **a**. The core of many of these computational map models is an instar-outstar system.

INSTAR + CONTRAST ENHANCEMENT

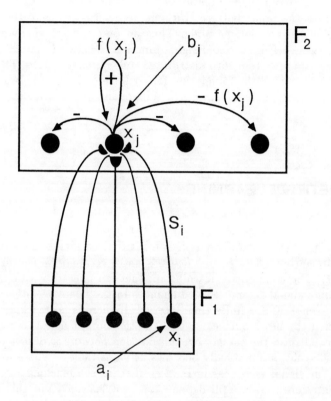

FIGURE 14 The basic competitive learning module combines the instar pattern coding system with a competitive network that contrast-enhances its filtered input.

Recognition of this common theme highlights the models' differences as well as their similarities. An early self-organizing three-level instar-outstar computational map model was described by Grossberg,[17] who later replaced the instar portion of this model with a competitive learning module.[20] The self-organizing feature map[32] and the counter-propagation network[26] are also examples of instar-outstar competitive learning models.

The basic instar-outstar computational map system is depicted in Figure 15. The first two levels, F_1 and F_2, form a competitive learning system. Included are the fan-in adaptive filter, contrast enhancement at the "hidden" level F_2, and a learning law for instar coding of the input patterns **a**. The top two levels then employ a fan-out adaptive filter for outstar pattern learning of the vector **b**. This three-level

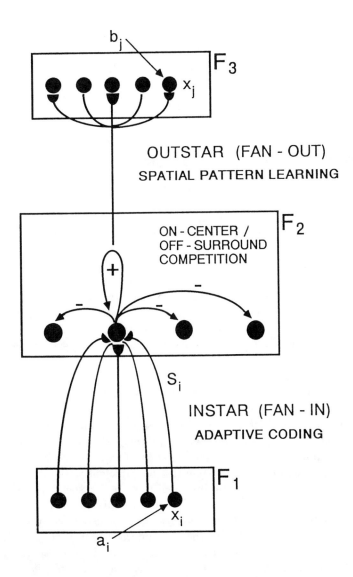

FIGURE 15 A three-level, feedforward instar-outstar module for computational mapping. The competitive learning module (F_1 and F_2) is joined with an outstar-type fan-out, for spatial pattern learning.

architecture allows, for example, two very different input patterns to map to the same output pattern: each input pattern can activate its own compressed representation at F_2, while each of these F_2 representations can learn a common output vector. In the extreme case where each input vector a activates its own F_2 node the system learns any desired output. The generality of this extreme case, which implements an arbitrary mapping from $\mathbf{R}^m$ to $\mathbf{R}^n$, is offset by its lack of generalization, or continuity, as well as by the fact that each learned pair $(\mathbf{a},\mathbf{b})$ requires its own F_2 node. Distributed F_2 representations provide greater generalization and efficiency, at a cost in complete *a priori* generality of the mapping.

21. INSTABILITY OF COMPUTATIONAL MAPS

The widespread use of instar-outstar families of computational maps attests to the power of this basic neural network architecture. This power is, however, diminished by the instability of feedforward systems: in general, recently learned patterns tend to erode past learning. This instability arises from two sources. First, even if a chosen category is the best match for a given input, that match may nevertheless be a poor one, chosen only because all the others are even worse. Established codes are thus vulnerable to recoding by "outliers." Second, learning laws such as Eq. (13) imply that a weight vector tends toward a new vector that encodes the presently active pattern, thereby weakening the trace of the past. Thus weight vectors can eventually drift far from their original patterns, even if learning is very slow and even if each individual input makes a good match with the past as recorded in the weights.

The many existing variations on the three-level instar-outstar theme illustrate some of the ways in which this family of models can be adapted to cope with the basic system's intrinsic instability. One stabilization technique causes learning to slow or cease after an initial finite interval, but then a subsequent unexpected pattern cannot be encoded, and instability could still creep in during the initial learning phase. Another approach is to restrict the class of input patterns to a stable set. This technique requires that the system can be sufficiently well analyzed to identify such a class, like the orthogonal inputs of the linear associative memory model (Figure 9), and that all inputs can be confined to this class. An often successful way to compensate for the instability of these systems is to slow the learning rate to such an extent that learned patterns are buffered against massive recoding by any single input. Of course, then, each pattern needs to be presented very many times for adequate learning to occur, a fact that was discussed, for example, by Rosenblatt in his critique of back propagation.

22. ADAPTIVE RESONANCE THEORY (ART)

It was analysis of the instability of feedforward instar-outstar systems that led to the introduction of adaptive resonance theory (ART)[21] and to the development of the neural network systems ART 1 and ART 2.[7,8] ART networks are designed, in particular, to resolve the *stability-plasticity dilemma*: they are stable enough to preserve significant past learning, but nevertheless remain adaptable enough to incorporate new information whenever it might appear.

The key idea of adaptive resonance theory is that the stability-plasticity dilemma can be resolved by a system in which the three-level network of Figure 15 is folded back on itself, identifying the top level (F_3) with the bottom level (F_1) of the instar-outstar mapping system. Thus the minimal ART module includes a bottom-up competitive learning system combined with a top-down outstar pattern learning system. When an input a is presented to an ART network, system dynamics initially follow the course of competitive learning (Figure 14), with bottom-up activation leading to a contrast-enhanced category representation at F_2. In the absence of other inputs to F_2, the active category is determined by past learning as encoded in the adaptive weights in the bottom-up filter. But now, in contrast to feedforward systems, signals are sent from F_2 back down to F_1 via a top-down adaptive filter. This feedback process allows the ART module to overcome both of the sources of instability described in Section 21, as follows.

First, as in the competitive learning module, the category active at F_2 may poorly match the pattern active at F_1. The ART system is designed to carry out a matching process that asks the question: should this input really be in this category? If the answer is no, the selected category is quickly rendered inactive, before past learning is disrupted by the outlier, and a search process ensues. This search process employs an auxiliary *orienting subsystem* that is controlled by the dynamics of the ART system itself. The orienting subsystem incorporates a dimensionless *vigilance parameter* that establishes the criterion for deciding whether the match is a good enough one for the input to be accepted as an exemplar of the chosen category.

Second, once an input is accepted and learning proceeds, the top-down filter continues to play a different kind of stabilizing role. Namely, top-down signals that represent the past learning meet the original input signals at F_1. Thus the F_1 activity pattern is a function of the past as well as the present, and it is this blend of the two, rather than the present input alone, that is learned by the weights in both adaptive filters. This dynamic matching during learning leads to stable coding, even with fast learning.

An example of the ART 1 class of minimal modules is illustrated in Figure 16. In addition to the two adaptive filters and the orienting subsystem, Figure 16 depicts gain control processes that actively regulate learning. Theorems have been proved to characterize the response of an ART 1 module to an arbitrary sequence of binary input patterns.[7] ART 2 systems were developed to self-organize recognition categories for analog as well as binary input sequences. One principal difference between the ART 1 and the ART 2 modules is shown in Figure 17. In examples so far developed, the stability criterion for analog inputs has required a three-layer feedback system within the F_1 level: a bottom layer where input patterns are read in; a top layer where filtered inputs from F_2 are read in; and a middle layer where the top and bottom patterns are brought together to form a matched pattern that is then fed back to the top and bottom F_1 layers.

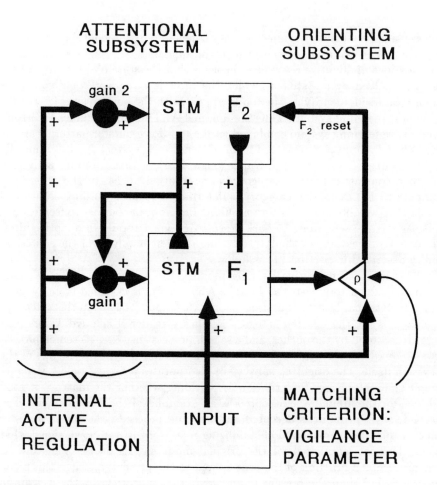

FIGURE 16 An ART 1 module for stable, self-organizing categorization of an arbitrary sequence of binary input patterns.

23. ART FOR ASSOCIATIVE MEMORY

A minimal ART module is a category learning system that self-organizes a sequence of input patterns into various recognition categories. It is not an associative memory system. However, like the competitive learning module in the 1970s, a minimal ART module can be embedded in a larger system for associative memory. A system such as an instar-outstar module (Figure 15) or a back-propagation algorithm (Figure 6) directly pairs sequences of individual *vectors* (**a**,**b**) during learning. If an ART system replaces levels F_1 and F_2 of the instar-outstar module, the associative learning system becomes self-stabilizing. ART systems can also be used to

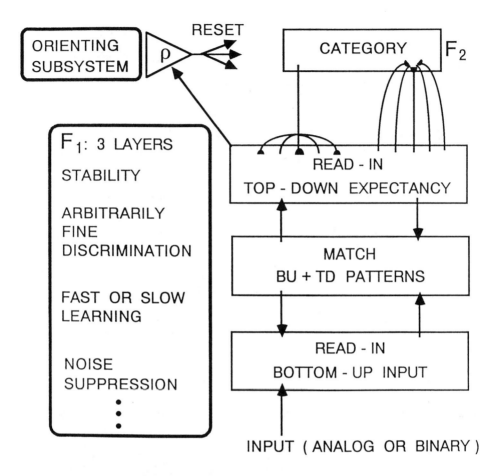

FIGURE 17 Principal elements of an ART 2 module for stable, self-organizing categorization of an arbitrary sequence of analog or binary input patterns. The F_1 level is a competitive network with three processing layers.

pair sequences of the *categories* self-organized by the input sequences (Figure 18). Moreover, the symmetry of the architecture implies that pattern recall can occur in either direction during performance. This scheme brings to the associative memory paradigm the code compression capabilities of the ART system, as well as its stability properties.

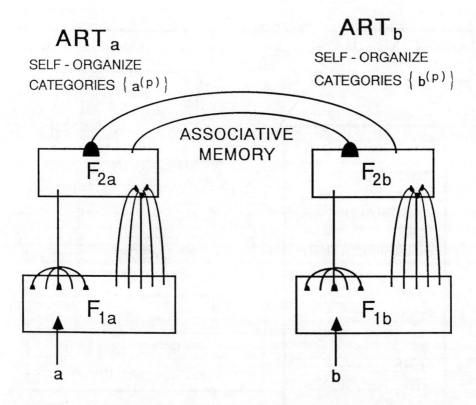

FIGURE 18   Two ART systems combined to form an associative memory architecture.

24. COGNITRON AND NEOCOGNITRON

In conclusion, we will consider two sets of models that are variations on the themes previously described. The first class, developed by Kunihiko Fukushima, consists of the cognitron[9] and the larger-scale neocognitron.[10,11] This class of neural models is distinguished by its capacity to carry out translation-invariant and size-invariant pattern recognition. This is accomplished by redundantly coding elementary features in various positions at one level; then cascading groups of features to the next level; then groups of these groups; and so on. Learning can proceed with or without a teacher. Locally the computations are a type of competitive learning that use combinations of additive and shunting dynamics.

25. SIMULATED ANNEALING

Finally, in addition to the probabilistic weight change laws which were a prominent feature of, for example, the modeling efforts of pioneers such as Rosenblatt and Amari, another class of probabilistic weight change laws appears in more recent work under the name *simulated annealing*, introduced by S. Kirkpatrick, C. D. Gellatt, and H. P. Vecchi.[29] The main idea of simulated annealing is the transposition of a method from statistical mechanics, namely the Metropolis algorithm,[34] into the general context of large complex systems. The Metropolis algorithm provides an approximate description of a many-body system, namely a material that anneals into a solid as temperature is slowly decreased. Kirkpatrick et al. drew an analogy between this system and problems of combinatorial optimization, such as the traveling salesman problem, where the goal is to minimize a cost function. The methods and ideas, as well as the large-scale nature of the problem, are so closely tied to those of neural networks that the two approaches are often linked. This link is perhaps closest in the Boltzmann machine,[1] which uses a simulated annealing algorithm to update weights in a binary network similar to the additive model studied by Hopfield.[28]

26. CONCLUSION

We have seen how the adaptive filter formalism is general enough to describe a wide variety of neural network modules for associative memory, category learning, and pattern recognition. Many systems developed and applied in recent years are variations on one or more of these modular themes. This approach can thus provide a core vocabulary and grammar for further analysis of the rich and varied literature of the neural network field.

ACKNOWLEDGMENTS

This chapter was originally published as a review article in *Neural Networks*, vol. 2, 243–257 (1989). It was based upon a tutorial lecture given on September 6, 1988 at the First Annual Meeting of the International Neural Network Society in Boston, Massachusetts. The author's research was supported in part by grants from the Air Force Office of Scientific Research (AFOSR F49620-86-C-0037 and AFOSR F49620-87-C-0018) and the National Science Foundation (NSF DMS-86-11959). I wish to thank these agencies for their long-term support of neural network research, and also to thank my colleagues at the Boston University Center for Adaptive Systems for their generous ongoing contributions of knowledge, skills, and friendship. In particular I thank Cynthia Bradford for her assistance in the preparation of the manuscript.

BIBLIOGRAPHY

Many of the articles cited above can be found in the collections listed below. In the references, the location of an article in one of these collections is indicated by the corresponding number in brackets.

A. COLLECTIONS OF ARTICLES

[1] Amari, S.-I., and M. A. Arbib, eds. *Competition and Cooperation in Neural Nets*. Lecture Notes in Biomathematics, vol. 45. Berlin: Springer-Verlag, 1982.

[2] Anderson, J. A., and E. Rosenfeld, eds. *Neurocomputing: Foundations of Research*. Cambridge, MA: MIT Press, 1988.

[3] Carpenter, G. A., and S. Grossberg. *Applied Optics*, Special Issue on Neural Networks, **26**(23) (December 1, 1987).

[4] Grossberg, S., ed. *Mathematical Psychology and Psychophysiology*. Providence, RI: American Mathematical Society, 1981.

[5] Grossberg, S. *Studies of Mind and Brain: Neural Principles of Learning, Perception, Development, Cognition, and Motor Control*. Boston: Reidel Press, 1982.

[6] Grossberg, S. *Neural Networks and Natural Intelligence*. Cambridge, MA: MIT Press, 1988.

[7] McCulloch, W. S., ed. *Embodiments of Mind*. Cambridge, MA: MIT Press, 1965.

[8] Rumelhart, D., J. McClelland, and the PDP Research Group. *Parallel Distributed Processing*. Cambridge, MA: MIT Press, 1986.

[9] Sanders, A. C., and Y. Y. Zeevi, eds. *IEEE Transactions on Systems, Man, and Cybernetics*, Special Issue on Neural and Sensory Information Processing, **SMC-13(5)** September/October (1983).

[10] Shriver, B., ed. *Computer*, Special Issue on Artificial Neural Systems, **21**(3), March (1988).

[11] Szu, H. H., ed. *Optical and Hybrid Computing*. SPIE–634. Bellingham, WA: The Society of Photo-Optical Instrumentation Engineers, 1987.

B. JOURNALS

[12] *Kybernetik* (1961–1974); *Biological Cybernetics* (1975–).
[13] *Neural Networks* (1988–).

C. REVIEWS

[14] Grossberg, S. "Nonlinear Neural Networks: Principles, Mechanisms, and Architectures." *Neural Networks* **1** (1988): 17–61.
[15] Kohonen, T. (1987). "Adaptive, Associative, and Self-Organizing Functions in Neural Computing." In *Applied Optics*, edited by G. A. Carpenter and S. Grossberg. Special Issue on Neural Networks, **26**(23) (December 1, 1987).
[16] . *Applied Optics* **26**, 4910–4918.
[17] Levine, D. "Neural Population Modeling and Psychology: A Review." *Mathematical Biosciences* **66** (1983): 1–86.
[18] Simpson, P. K. *Artificial Neural Systems: Foundations, Paradigms, Applications, and Implementations.* Elmsford, NY: Pergamon Press, 1990.

REFERENCES

1. Ackley, D. H., G. E. Hinton, and T. J. Sejnowski. "A Learning Algorithm for Boltzmann Machines." *Cognitive Science* **9** (1985): 147–169. Reprinted in *Neurocomputing: Foundations of Research*, edited by J. A. Anderson and E. Rosenfeld. Cambridge, MA: MIT Press, 1988.
2. Amari, S.-I. "Learning Patterns and Pattern Sequences by Self-Organizing Nets of Threshold Elements." *IEEE Transactions on Computers* **C-21** (1972): 1197–1206.
3. Amari, S.-I. "Neural Theory of Association and Concept-Formation." *Biological Cybernetics* **26** (1977): 175–185.
4. Amari, S.-I., and A. Takeuchi. "Mathematical Theory on Formation of Category Detecting Nerve Cells." *Biological Cybernetics* **29** (1978): 127–136.
5. Anderson, J. A. "A Simple Neural Network Generating an Interactive Memory." *Mathematical Biosciences* **14** (1972): 197–220. Reprinted in *Neurocomputing: Foundations of Research*, edited by J. A. Anderson and E. Rosenfeld. Cambridge, MA: MIT Press, 1988.
6. Bienenstock, E., L. N. Cooper, and P. W. Munro. "A Theory for the Development of Neuron Selectivity: Orientation Specificity and Binocular Interaction in the Visual Cortex." *J. Neurosci.* **2** (1982): 32–48. Reprinted in *Neurocomputing: Foundations of Research*, edited by J. A. Anderson and E. Rosenfeld. Cambridge, MA: MIT Press, 1988.

7. Carpenter, G. A., and S. Grossberg. "A Massively Parallel Architecture for a Self-Organizing Neural Pattern Recognition Machine." *Computer Vision, Graphics, and Image Processing* **37** (1987): 54–115. Reprinted in *Neural Networks and Natural Intelligence*, edited by S. Grossberg. Cambridge, MA: MIT Press, 1988.
8. Carpenter, G. A., and S. Grossberg. "ART 2: Self-Organization of Stable Category Recognition Codes for Analog Input Patterns." *Applied Optics* **26** (1987): 4919–4930. In *Applied Optics*, edited by G. A. Carpenter and S. Grossberg. Special Issue on Neural Networks, **26**(23) (December 1, 1987).
9. Fukushima, K. "Cognitron: A Self-Organizing Multilayered Neural Network." *Biological Cybernetics* **20** (1975): 121–136.
10. Fukushima, K. "Neocognitron: A Self-Organizing Neural Network Model for a Mechanism of Pattern Recognition Unaffected by Shift in Position." *Biological Cybernetics* **36** (1980): 193–202.
11. Fukushima, K. "Neocognitron: A Hierarchical Neural Network Capable of Visual Pattern Recognition." *Neural Networks* **1** (1988): 119–130.
12. Grossberg, S. *The Theory of Embedding Fields with Applications to Psychology and Neurophysiology*. New York: Rockefeller Institute for Medical Research, 1964.
13. Grossberg, S. "Some Nonlinear Networks Capable of Learning a Spatial Pattern of Arbitrary Complexity." *Proceedings of the National Academy of Sciences USA* **59** (1968): 368–372.
14. Grossberg, S. "Some Networks that Can Learn, Remember, and Reproduce any Number of Complicated Space-Time Patterns I." *J. Math. & Mech.* **19** (1969): 53–91.
15. Grossberg, S., and J. Pepe. "Schizophrenia: Possible Dependence of Associational Span, Bowing, and Primacy vs. Recency on Spiking Threshold." *Behavioral Science* **15** (1970): 359–362.
16. Grossberg, S. "Pattern Learning by Functional-Differential Neural Networks with Arbitrary Path Weights." In *Delay and Functional Differential Equations and Their Applications*, edited by K. Schmitt, 121–160. New York: Academic Press, 1972. Reprinted in *Studies of Mind and Brain: Neural Principles of Learning, Perception, Development, Cognition, and Motor Control*, edited by S. Grossberg. Boston: Reidel Press, 1982.
17. Grossberg, S. "Neural Expectation: Cerebellar and Retinal Analogs of Cells Fired by Learnable or Unlearned Pattern Classes." *Kybernetik* **10** (1972): 49–57. Reprinted in *Studies of Mind and Brain: Neural Principles of Learning, Perception, Development, Cognition, and Motor Control*, edited by S. Grossberg. Boston: Reidel Press, 1982.
18. Grossberg, S. "Contour Enhancement, Short-Term Memory, and Constancies in Reverberating Neural Networks." *Studies in Applied Mathematics* **52** (1973): 217–257. Reprinted in *Studies of Mind and Brain: Neural Principles of Learning, Perception, Development, Cognition, and Motor Control*, edited by S. Grossberg. Boston: Reidel Press, 1982.

19. Grossberg, S. "On the Development of Feature Detectors in the Visual Cortex with Applications to Learning and Reaction-Diffusion Systems." *Biological Cybernetics* **21** (1976): 145–159.
20. Grossberg, S. "Adaptive Pattern Classification and Universal Recoding, I: Parallel Development and Coding of Neural Feature Detectors." *Biological Cybernetics* **23** (1976): 121–134. Reprinted in *Neurocomputing: Foundations of Research*, edited by J. A. Anderson and E. Rosenfeld. Cambridge, MA: MIT Press, 1988; and in *Studies of Mind and Brain: Neural Principles of Learning, Perception, Development, Cognition, and Motor Control*, edited by S. Grossberg. Boston: Reidel Press, 1982.
21. Grossberg, S. "Adaptive Pattern Classification and Universal Recoding, II: Feedback, Expectation, Olfaction, and Illusions." *Biological Cybernetics* **23** (1976): 187–202.
22. Grossberg, S. *Studies of Mind and Brain: Neural Principles of Learning, Perception, Development, Cognition, and Motor Control.* Boston: Reidel Press, 1982.
23. Grossberg, S. *Neural Networks and Natural Intelligence.* Cambridge, MA: MIT Press, 1988.
24. Hebb, D. O. *The Organization of Behavior.* New York: Wiley, 1949. Reprinted, in part, in *Neurocomputing: Foundations of Research*, edited by J. A. Anderson and E. Rosenfeld. Cambridge, MA: MIT Press, 1988.
25. Hecht-Nielsen, R. "Neural Analog Information Processing." *Proceedings of the Society of Photo-Optical Instrumentation Engineers* **298** (1981): 138–141.
26. Hecht-Nielsen, R. "Counterpropagation Networks." *Applied Optics* **26** (1987): 4979–4984. In *Applied Optics*, edited by G. A. Carpenter and S. Grossberg. Special Issue on Neural Networks, 26(23) (December 1, 1987).
27. Hodgkin, A. L., and A. F. Huxley. "A Quantitative Description of Membrane Current and Its Application to Conduction and Excitation in Nerve." *J. Physiology* **117** (1952): 500–544.
28. Hopfield, J. J. "Neural Networks and Physical Systems with Emergent Collective Computational Abilities." *Proceedings of the National Academy of Sciences USA* **79** (1982): 2554–2558.
29. Kirkpatrick, S., C. D. Gelatt, Jr., and M. P. Vecchi. "Optimization by Simulated Annealing." *Science* **220** (1983): 671–680. Reprinted in *Neurocomputing: Foundations of Research*, edited by J. A. Anderson and E. Rosenfeld. Cambridge, MA: MIT Press, 1988.
30. Kohonen, T. "Correlation Matrix Memories." *IEEE Transactions on Computers* **C-21** (1972): 353–359. Reprinted in *Neurocomputing: Foundations of Research*, edited by J. A. Anderson and E. Rosenfeld. Cambridge, MA: MIT Press, 1988.
31. Kohonen, T. *Content-Addressable Memories.* Berlin: Springer-Verlag, 1980.
32. Kohonen, T. *Self-Organization and Associative Memory.* Berlin: Springer-Verlag, 1984.
33. Kohonen, T., and M. Ruohonen. "Representation of Associated Data by Matrix Operators." *IEEE Transactions on Computers* **C-22** (1973): 701–702.

34. Metropolis, N., A. W. Rosenbluth, M. N. Rosenbluth, A. H. Teller, and E. Teller. "Equations of State Calculations by Fast Computing Machines." *J. Chem. Phys.* **21** (1953): 1087–1091.
35. McCulloch, W. S., and W. Pitts. "A Logical Calculus of the Ideas Immanent in Nervous Activity." *Bulletin of Mathematical Biophysics* **9** (1943): 127–147. Reprinted in *Neurocomputing: Foundations of Research*, edited by J. A. Anderson and E. Rosenfeld. Cambridge, MA: MIT Press, 1988; and Reprinted in *Embodiments of Mind*, edited by W. S. McCulloch. Cambridge, MA: MIT Press, 1965.
36. Nakano, N. "Associatron: A Model of Associative Memory." *IEEE Transactions on Systems, Man, and Cybernetics* **SMC-2** (1972): 381–388.
37. Parker, D. "Learning Logic." Invention Report, S81-64, File 1, Office of Technology Licensing, Stanford University, 1982.
38. Pitts, W., and W. S. McCulloch. "How We Know Universals: The Perception of Auditory and Visual Forms." *Bulletin of Mathematical Biophysics* **9** (1947): 127–147. Reprinted in *Neurocomputing: Foundations of Research*, edited by J. A. Anderson and E. Rosenfeld. Cambridge, MA: MIT Press, 1988; and Reprinted in *Embodiments of Mind*, edited by W. S. McCulloch. Cambridge, MA: MIT Press, 1965.
39. Rosenblatt, F. "The Perceptron: A Probabilistic Model for Information Storage and Organization in the Brain." *Psychological Review* **65** (1958): 386–408. Reprinted in *Neurocomputing: Foundations of Research*, edited by J. A. Anderson and E. Rosenfeld. Cambridge, MA: MIT Press, 1988.
40. Rosenblatt, R. *Principles of Neurodynamics*. Washington, DC: Spartan Books, 1962.
41. Rumelhart, D. E., G. E. Hinton, and R. J. Williams. "Learning Internal Representations by Error Propagation." In *Parallel Distributed Processing: Explorations in the Microstructures of Cognitions, I*, edited by D. E. Rumelhart and J. L. McClelland, 318–362. Cambridge, MA: MIT Press, 1986. In *Parallel Distributed Processing*, edited by D. Rumelhart, J. McClelland, and the PDP Research Group. Cambridge, MA: MIT Press, 1986.
42. . Reprinted in *Neurocomputing: Foundations of Research*, edited by J. A. Anderson and E. Rosenfeld. Cambridge, MA: MIT Press, 1988.
43. Rumelhart, D. E., and D. Zipser. "Feature Discovery by Competitive Learning." *Cognitive Science* **9** (1985): 75–112.
44. Russell, B., and A. N. Whitehead. *Principia Mathematica*, vols. I–III. Cambridge: Cambridge University Press, 1910/1912/1913.
45. Steinbuch, K. "Die Lernmatrix." *Kybernetik* **1** (1961): 36–45.
46. Steinbuch, K., and B. Widrow. "A Critical Comparison of Two Kinds of Adaptive Classification Networks." *IEEE Transactions on Electronic Computers* **EC-14** (1965): 737–740.
47. von der Malsburg, C. "Self-Organization of Orientation Sensitive Cells in the Striate Cortex." *Kybernetik* **14** (1973): 85–100. Reprinted in *Neurocomputing: Foundations of Research*, edited by J. A. Anderson and E. Rosenfeld. Cambridge, MA: MIT Press, 1988.

48. von Neumann, J. *The Computer and the Brain.* New Haven: Yale University Press, 1958. Reprinted, in part, in *Neurocomputing: Foundations of Research*, edited by J. A. Anderson and E. Rosenfeld. Cambridge, MA: MIT Press, 1988.
49. Werbos, P. J. "Beyond Regression: New Tools for Prediction and Analysis in the Behavioral Sciences." Ph.D. Thesis, Harvard University, 1974.
50. Werbos, P. J. "Generalization of Backpropagation with Application to a Recurrent Gas Market Model." *Neural Networks* **1** (1988): 339–356.
51. Widrow, B., and M. E. Hoff. "Adaptive Switching Circuits." 1960 IRE WESCON Convention Record, part 4, 1960, 96–104. Reprinted in *Neurocomputing: Foundations of Research*, edited by J. A. Anderson and E. Rosenfeld. Cambridge, MA: MIT Press, 1988.
52. Widrow, B., and R. Winter. "Neural Nets for Adaptive Filtering and Adaptive Pattern Recognition." *Computer* **21** (1988): 25–39. Reprinted in *Computer*, Special Issue on Artificial Neural Systems, edited by B. Shriver, **21(3)**, March (1988).

Carla J. Shatz
Department of Neurobiology, Stanford University School of Medicine, Stanford, CA 94305

Impulse Activity and the Patterning of Connections During CNS Development

Reprinted from *Neuron* **5** (1990): 745–756. Permission granted by Cell Press.

How are the highly ordered sets of axonal connections so characteristic of organization in the adult vertebrate central nervous system formed during development? Many problems must be solved to achieve such precise wiring: axons must grow along the correct pathways and must select their appropriate target(s). Even once the process of target selection is complete, however, the many axons that comprise a particular projection must still arrange themselves in an orderly and highly stereotyped pattern, typically one in which nearest-neighbor relations are preserved so that the terminal arbors of neighboring projection neurons are also neighbors within the target. Here, I would like to consider the process by which this final patterning of neuronal connections comes about during development. Studies of the vertebrate visual system, reviewed here, have provided extensive evidence in favor of the hypothesis that an activity-dependent competition between axonal inputs for common postsynaptic neurons is responsible in good part for the establishment of orderly sets of connections.

COMPETITION IN THE FORMATION OF OCULAR DOMINANCE COLUMNS IN THE MAMMALIAN PRIMARY VISUAL CORTEX.

Many insights into developmental mechanisms underlying the formation of orderly connections have come from studies of the mammalian visual system, in which the clear-cut patterning of connections is exemplified in the highly topographic ordering of projections and strict segregation of inputs from the two eyes at successive levels of visual information processing (for reviews, see Rodieck[54] and Sherman and Spear[68]). Ganglion cell axons from each eye project to the lateral geniculate nucleus (LGN) on both sides of the brain. However, within the LGN, axons from the two eyes terminate in a set of separate, alternating eye-specific layers that are strictly monocular[22] (see Figure 1). Neurons in the LGN project, in turn, to layer 4 of the primary visual cortex where, again, axons are segregated according to eye of origin into alternating monocularly innervated patches that represent the system of ocular dominance columns within cortical layer 4.[23,26,59,60]

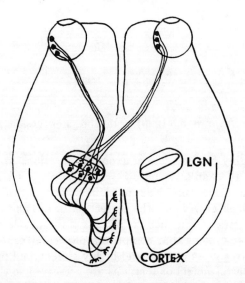

FIGURE 1 A simplified diagram of the mammalian visual pathways. Only connections from each eye to the left side of the brain are shown. Retinal ganglion cell axons from the two eyes travel to the lateral geniculate nucleus (LGN) of the thamalus, where their terminals are segregated in separate eye-specific layers. The axons of neighboring retinal ganglion cells within each eye terminate in neighboring regions within the appropriate layers, establishing a topographically ordered map. LGN neurons, in turn, project to layer 4 of the primary visual cortex where again axonal terminal arbors of LGN neurons representing the two eyes are segregated into alternating ocular dominance patches.

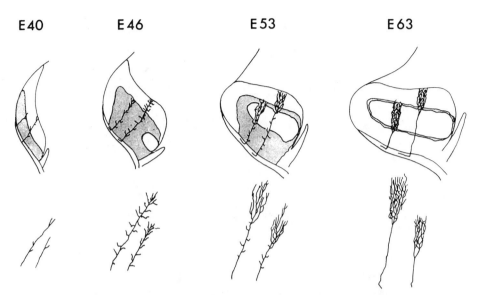

FIGURE 2 Summary of the prenatal development of the eye-specific layers in the cat's LGN. Shaded areas indicate regions within the LGN simultaneously occupied by ganglion cell axons from the two eyes at different times in development, as derived by the anterograde transport of intraocularly injected tracers. Stick figures show the appearance of representative ganglion cell axons from the ipsilateral (shorter axons at each age) and contralateral (longer) eyes, based on studies of the morphology of individual axons filled with horseradish peroxidase *in vitro* (see Shatz,[67] for more details; reproduced with permission from Shatz[65]). The eye-specific layers emerge as retinal ganglion cell axons, withdraw delicate sidebranches from inappropriate regions, and elaborate complex terminal arbors within appropriate regions of the LGN. E=embryonic age; gestation in the cat is 65 days.

Remarkably, neither the layers within the LGN nor the columns within the cortex are present initially during development (for reviews, see Sretavan and Shatz,[69] Shatz,[67] and Miller and Stryker[42]). When retinal ganglion cell axons from the two eyes first grow into the LGN, they are intermixed with each other throughout a good portion of the nucleus; the eye-specific layers emerge as axons from the two eyes gradually remodel by withdrawing modest branches from inappropriate territory and growing extensive terminal arbors within appropriate territory[69] (Figure 2). Physiological studies *in vitro*[62] and electron microscopic examination of identified retinal ganglion cell axons[8,67] suggest that this remodeling is accompanied by the reorganization of synapses from the two eyes such that initial binocular convergence is replaced by monocular inputs.

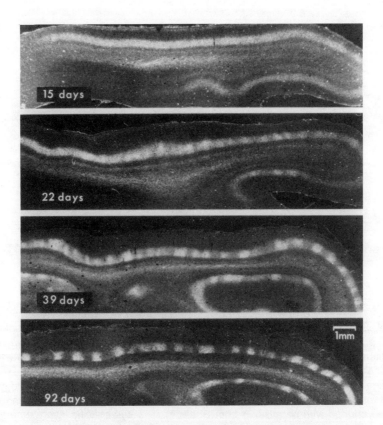

FIGURE 3 The postnatal development of the ocular dominance patches within layer 4 of the primary visual cortex of the cat is summarized. The location of LGN axons is monitored by means of the transneuronal transport through the LGN of radioactively labeled material (which appears white in these darkfield photographs) injected into one eye. The adult pattern of layer 4 labeling—patches separated by gaps of roughly equal size—can be seen by 92 days postnatal. However, at 2 weeks postnatal, the pattern of labeling within layer 4 is continuous, indicating that LGN axons representing the two eyes are intermixed with each other. (Based on experiments presented in LeVay et al.[34])

Ocular dominance columns in layer 4 form from extensively intermixed LGN inputs representing the two eyes (Figure 3), presumably also by a process of axonal remodeling and synapse elimination. At present, little is known about the exact morphological details because few individual axons have been successfully labeled for study, but microelectrode recordings have shown that initially the majority of neurons in cortical layer 4 receive functional inputs from LGN afferents representing both eyes.[34,35] Thus, here too, ocular segregation emerges from an initial condition

Impulse Activity and the Patterning of Connections During CNS Development

(a)

(b)

(c)

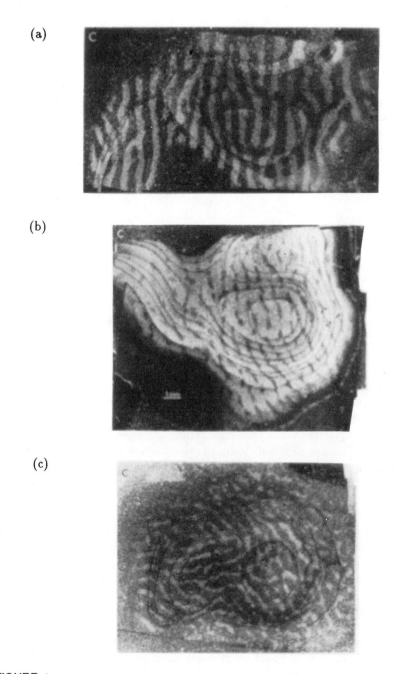

FIGURE 4

FIGURE 4 (cont'd.) The effects of monocular eye closure at birth on the subsequent (adult) organization of the ocular dominance columns in layer 4 of the monkey visual cortex, as revealed by the transneuronal transport method (see Figure 3). (a) The normal tangential organization of LGN afferents within layer 4 into alternating stripes of equal width representing the injected and uninjected eye. (b) The representation of the open eye within layer 4 following monocular deprivation—LGN axons occupy most of layer 4, with only small unlabeled regions remaining for the LGN axons representing the closed eye. (c) The pattern of transneuronal labeling resulting from injection of the closed eye is complementary to that shown in (b), indicating a shrinkage of territory devoted to the representation of the closed eye within layer 4. Reprinted with permission from Hubel et al.[26]

of functional synaptic convergence of inputs representing the two eyes onto common (layer 4 cortical) neurons (See Figure 5: compare neonate and adult). In higher mammals, the formation of the LGN layers occurs largely, if not entirely, prenatally and precedes the onset of ocular dominance column formation within the cortex, which occurs largely (monkey) or entirely (cat) postnatally.[34,35,47,61]

How do inputs representing the two eyes segregate from each other to form layers or columns? The first clues came from the pioneering studies of Hubel and Wiesel on the effects of visual deprivation on the functional organization of the primary visual cortex. In the normal adult visual cortex, the majority of neurons are binocular: that is, they respond to visual stimulation of either eye. Even binocular neurons tend to be dominated by one eye or the other, and as mentioned above, layer 4 neurons tend to be exclusively driven by stimulation of one eye only so that the cortex is evenly divided into ocular dominance columns for both eyes.[23,26,60] However, if one eye is deprived of vision by closing the eyelids at birth for several days to weeks, the ocular dominance distribution of neurons in visual cortex is drastically shifted: as shown in Figure 5 (MD), now, the majority (90%) of neurons are monocularly driven only by stimulation of the open eye.[25,26] (Neurons in the retina and LGN remain responsive to their normal inputs.[78]) The physiological shift in ocular dominance within the cortex is paralleled by a profound change in the anatomical organization of LGN axons within layer 4: LGN axons representing the open eye now occupy most of layer 4, while those representing the closed eye are relegated to very small patches[26] (see Figure 4).

The observation that the wiring of LGN axons and the eye preference of cortical neurons can be influenced by early visual experience sets the stage for the idea that *use* of the visual system is required for its normal development and for the maintenance of its connections, at least during an early period of susceptibility called the "Critical Period."[25,35] But how might abnormal use, such as the occlusion of one eye, result in such profound changes in connectivity at the level of the visual cortex? The most reasonable explanation is that a use-dependent synaptic competition between LGN axons serving the two eyes for layer 4 neurons normally drives the formation of the ocular dominance columns during the critical period.

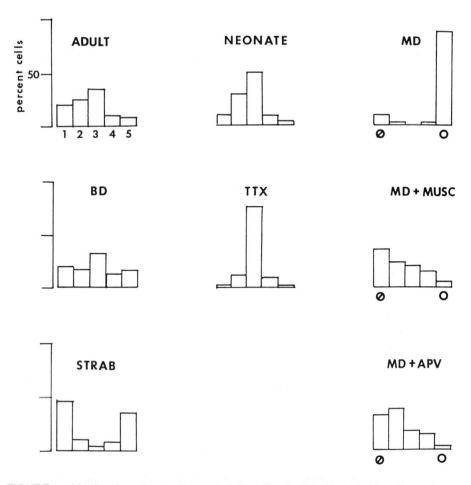

FIGURE 5 Idealized summary diagram of the effects of various manipulations that alter the pattern or levels of visually driven activity on the ocular dominance of visual cortical neurons as assessed physiologically. In the normal adult cortex (ADULT), the majority of neurons are binocularly driven, with a roughly even distribution of neurons representing each eye (group 1 = neurons responding exclusively from the right eye; group 2 = neurons responding predominantly to the right eye but some also from the left eye; group 3 = neurons responding equally to the two eyes; group 4 = neurons responding predominantly to the left eye; and group 5 = neurons responding exclusively to the left eye).[60,73] The majority of group 1 and 5 neurons are found within layer 4. In neonates, there are very few monocularly driven neurons, presumably since inputs from the two eyes are extensively intermixed even within layer 4.[34] If the right eye is closed at birth (Ø), then an ocular dominance shift in favor of the left eye (O) occurs with long-term monocular deprivation (MD).[25,60] However, if MD is combined with cortical infusion of muscimol (MD+MUSC)[52] during the critical period, then the shift in favor of the open eye is prevented close to the infusion site and instead a shift in favor of the closed

FIGURE 5 (cont'd.) eye results. Binocular deprivation (BD)[79,73] during the critical period, however, does not have an obvious effect on cortical ocular dominance, whereas intraocular injections of TTX during the same period retains, or possibly exaggerates, the highly binocular distribution present in neonates[73] (cf. TTX and Neonates). In contrast, alternating monocular deprivation or strabismus (STRAB) causes a complete loss of binocular neurons within the cortex.[77] See text for further details.

Consequently, unequal use caused by monocular deprivation could bias the outcome in favor of the open eye. Many lines of evidence support the suggestion that competitive interactions are involved. Binocular deprivation leaves the ocular dominance of cortical neurons unaltered (Figure 5: BD),[79] although neurons eventually do not respond briskly to visual stimulation. In a clever experiment, Guillery[19] demonstrated that competitive interactions are not only present, but must occur locally within the cortex, between LGN axons subserving corresponding regions of the visual field. He sutured one eye closed and then made just a small lesion in the open eye, destroying a localized group of ganglion cells there. As a consequence, the effects of monocular deprivation were manifested everywhere *except* within the small region receiving LGN axons representing the lesioned area of the open eye and the corresponding region of the closed eye. Thus equal use of the two eyes during the critical period subserves competitive interactions whose outcome is manifested in the even distribution of ocular dominance columns. Similar competitive interactions are thought to operate even earlier in development to drive the formation of LGN layers, as discussed more fully at the conclusion of this article.

THE ROLE OF PATTERNED NEURAL ACTIVITY IN COMPETITIVE INTERACTIONS

Signalling by neurons is, of course, via action potentials and synaptic transmission; hence, the effects of visual experience on cortical organization must be a consequence of alterations in either the level or patterning (or both) of neural activity within the visual pathways. The most graphic demonstration that this must be the case comes from experiments in which the inputs from both eyes are completely silenced by injecting Tetrodotoxin (TTX), a blocker of the sodium channel, for several weeks postnatally during the critical period and then examining the consequences on the formation of ocular dominance columns in layer 4.[73] Intraocular application of TTX conveniently silences the entire pathway from retina to cortex since there is very little spontaneously generated activity in central visual pathways in the absence of the eyes.[73] Segregation of LGN axons into patches within cortical layer 4 was prevented completely, and neurons in layer 4, normally monocularly driven, were instead binocularly driven (Figure 5: TTX), reminiscent of the initial period of normal postnatal development.[34,35] Indeed, at present it is not known whether the effect of the TTX treatment is to simply arrest development or to permit continued but undirected growth of LGN axon terminals within layer 4. Examination of axonal morphology should eventually clarify this issue. Analogous results are obtained when cortical activity alone is blocked (both pre- and postsynaptically), by infusing TTX locally via an osmotic minipump[51]: such treatment, when performed during the critical period, prevents the shift in cortical ocular dominance produced by monocular eye closure.

These experiments indicate that neural activity is necessary for ocular dominance columns to form during development (and for them to be perturbed with monocular deprivation), but they do not reveal how an activity-dependent signal might permit the selection of appropriate inputs from each eye to generate the segregated pattern characteristic of the adult geniculocortical projection. Experiments in which the use of the two eyes remains equal, but is never synchronous, provide some clues. During the critical period, if artificial strabismus is produced by cutting the extraocular muscles of one eye, thereby disrupting normal eye alignment, or if the eyes are closed alternately so that the total amount of vision received by each eye is the same, but vision is never binocular, then essentially every neuron in the primary visual cortex becomes exclusively monocularly innervated, with cells of like ocular dominance grouped into entirely "monocular" columns (see Figure 5: AMD) (recall that in normal animals, only layer 4 is monocular).[24,77] These results suggest that information concerning the relative *timing* of activity in the two eyes is somehow used to distinguish inputs at the cortical level: asynchrony leads to ocular segregation; synchrony maintains binocularity.

The conclusion that the formation of ocular dominance columns is influenced by the timing and patterning of neuronal activity within the retinae is underscored by the results of an experiment by Stryker and Strickland[72] in which retinal activity was first blocked by intraocular injections of TTX, but then experimentally controlled by electrically stimulating the optic nerves either synchronously or asynchronously. Synchronous stimulation of the two nerves prevented the formation of ocular dominance columns, whereas asynchronous stimulation permitted them to form. The only difference between the two experiments was the timing of stimulation, thereby demonstrating directly that the patterning of neural activity provides sufficient information for ocular segregation to occur, at least at the level of the primary visual cortex.

The above considerations can also explain why ocular dominance columns can develop even when animals are binocularly deprived or reared in the dark during the critical period.[73] In the absence of visual stimulation, ganglion cells in the mammalian retina of adults[36,53] and even in fetal animals[18,39] fire action potentials spontaneously. Such spontaneous firing could supply activity-dependent cues, provided that ganglion cell firing in the two eyes is asynchronous.[43,82]

Before examining further how the timing and patterning of impulse activity might lead to segregation of geniculocortical afferents, it is worth considering briefly why an alternative hypothesis for the formation of segregated inputs, one that invokes the existence of eye-specific molecular labels within the cortex, is at odds with most experimental observations. First, geniculocortical axons segregate to form ocular dominance columns whose precise locations within the visual cortex are unpredictable, although the global arrangement of the columns is similar from one animal to the next.[26] Second, blockade of neural activity within the eyes prevents segregation of geniculocortical axons (themselves not directly affected by TTX). Thus, if eye-specific labels were present within the cortex, they should still have been recognized by LGN axons. Moreover, such markers should operate to form columns regardless of the *pattern* of electrical stimulation of the optic nerves (synchronous vs. asynchronous). Third, there is no obvious tendency for axons representing the right or left eyes to be grouped together prior to segregation[34,63]; indeed, activity-dependent models of ocular dominance column formation can easily produce segregated inputs from an initially randomly intermixed condition (see Miller et al.,[43] for more details). It should be noted that the absence of such labels with respect to eye of origin in no way argues against the existence of specific molecular cues that could initially guide axons to their appropriate targets (LGN, visual cortex) during development, or that could help to establish coarse retinotopic projections within these targets. By analogy with studies in lower vertebrates (for review see Udin and Fawcett[76]), such cues are highly likely to be present in the mammalian CNS as well. However, once axons reach their correct target and establish a coarse topographic projection, activity-dependent interactions could provide the major cues necessary for segregation.

Finally, a set of creative experiments performed in the amphibian visual system also argues against the presence of intrinsic eye-specific labels within the postsynaptic targets of retinal ganglion cells. In amphibians during larval development, the projections from retinal ganglion cells to their principal target, the optic tectum, are entirely crossed. Consequently, each tectum receives a map from the whole contralateral retina. The map is topographically orderly, such that the axons of neighboring retinal ganglion cells terminate in neighboring regions of the optic tectum. In frogs, it is possible to perform experimental manipulations in the embryo to transplant an extra eye onto one side of the head. Axons from both the normal and transplanted eyes are then capable of growing into the optic tectum, artificially creating a competitive situation. Constantine-Paton and her colleagues have shown that axons from both eyes segregate into eye-specific stripes reminiscent of the stripe-like pattern of the mammalian geniculocortical projection[32] (see Figure 6(a)). Thus, segregation of eye-specific inputs can occur in an experimentally manipulated system that normally never forms a segregated projection and therefore is highly unlikely to contain intrinsic eye-specific labels within the postsynaptic target. Moreover, blockade of action potential activity with TTX causes ganglion cell axons from the two eyes to desegregate[6,40,50] (see Figure 6(b)). In amphibia, connections between retina and tectum continue to grow throughout larval and early postmetamorphic life, in a process involving the continual reshaping of synaptic connections.[14,49] These experiments suggest an analogous conclusion, that the maintenance of segregated inputs in these three-eyed frogs is a dynamic ongoing process that requires neural activity (presumably asynchronous) in the two eyes.

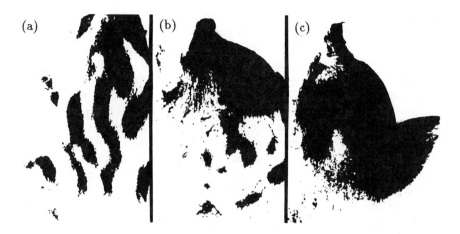

FIGURE 6 The organization of ganglion cell axon projections to a dually innervated optic tectum in three-eyed frogs, as revealed in tectal wholemounts by injecting one of the two eyes with horseradish peroxidase. (a) Axons from the two eyes segregate into alternating stripes reminiscent of the system of mammalian ocular dominance columns in cortical layer 4. Desegregation occurs when either TTX (not shown) or APV ((b) after 2.5 weeks; (c) after 4 weeks treatment) is infused into the tectum. Modified, with permission, from Cline et al.[9]

CELLULAR CORRELATES OF ACTIVITY-DEPENDENT COMPETITION

The finding that the synchronous activation of afferents prevents them from segregating, while asynchronous activation promotes segregation, indicates that the timing of presynaptic activity is crucial to the process. Studies also suggest that involvement of the postsynaptic cell is necessary. For example, in the mammalian visual cortex, when visual stimulation through one eye is paired simultaneously with postsynaptic depolarization produced by extracellular stimulation, the strength of inputs from the stimulated eye can be enhanced in some cells from minutes to hours.[17] The effect is quite variable and is more frequently produced in young animals during the critical period than in adults; nevertheless, this experiment serves to illustrate the point that coincidence of pre- with postsynaptic activity can, at least under certain circumstances, enhance visually driven inputs.

Manipulations that block postsynaptic activity exclusively can also alter the outcome of competition in the visual system. Reiter and Stryker[52] have shown that when cortical neurons are silenced during the critical period by the intracortical infusion of muscimol, a GABA-A receptor agonist, monocular eye closure has surprising consequences for the inputs from the two eyes: within the silenced region of cortex, inputs from the *closed* eye come to dominate over inputs from the open eye (see Figure 5: MD and muscimol), whereas, of course, the reverse is true outside the silenced zone. This observation shows that the activity of postsynaptic cortical cells is highly likely to be involved in the synaptic reorganization occurring during the critical period, since the same patterning of presynaptic activity produces different outcomes depending on the state of activation of the postsynaptic cell.

The requirement for the participation of both pre- and postsynaptic partners in activity-dependent rearrangements, and the fact that coincident activation can strengthen coactivated inputs, is consistent with the idea that a Hebb rule may govern the process of synapse rearrangements during ocular dominance column development in mammals (and in three-eyed frogs) (for review see Brown et al.[7]; see Kossel et al.[30] for an alternate view). Hebb[21] suggested that when pre- and postsynaptic neurons are coactivated, their synaptic connections are strengthened, whereas connections are weakened with the lack of coincident activation. In this context, the muscimol experiment described above[52] is also consistent with a Hebb rule in the sense that the levels of presynaptic activity in geniculocortical axons representing the closed eye are better matched to the silenced postsynaptic neurons than those inputs representing the open eye. Thus, the correlated firing of nearby ganglion cells within one eye, and the lack of synchronous firing of ganglion cells in the other eye could provide appropriate signals to produce the regional strengthening and weakening of synaptic inputs needed for segregation to take place.

These activity-dependent properties of visual cortical synapses during the critical period are very reminiscent of some of the well-known characteristics of synapses in the adult mammalian hippocampus that are capable of undergoing long-term potentiation (LTP): that is, a long lasting increase in synaptic strength produced with the appropriate matching of pre- and postsynaptic activation.[7,45] In the CA1 region of the hippocampus, many lines of experimentation indicate that activation of the NMDA receptor (N-methhyl-D-aspartate) on postsynaptic neurons by means of the presynaptic release of glutamate is required for LTP.[80] The consequent strengthening of synaptic transmission appears to be due at least in part to a presynaptic change: an increase in transmitter release from the presynaptic terminals.[5,37,81] The wealth of information on LTP, and its similarities with activity-dependent development, has prompted many recent experiments in the visual system designed to learn whether the two forms of synaptic change share similar cellular mechanisms. Of course, it should be noted that in at least one respect, the two must differ ultimately in that in development major structural changes occur not only in individual synapses but also in the overall morphology of presynaptic terminals, since some terminals are actually eliminated while others are newly formed. Moreover, the physiological properties of developing synapses are very different from those of adult, suggesting that the parameters for patterned activity to produce synaptic change may also differ.

A major question is whether NMDA-receptor activation is necessary for developmental plasticity. This is a reasonable question to pose since glutamate is thought to be the excitatory neurotransmitter released by retinal ganglion cells in all vertebrates[27,31] and also by LGN neurons in mammals.[20] The most compelling evidence in favor of the specific involvement of NMDA-receptors in activity-dependent development comes from recent studies of the retinotectal system in fish and frogs. For instance, in three-eyed frogs, the ocular dominance stripes desegregate in the presence of the NMDA receptor antagonist APV (2-amino-5-phosphonovaleric acid),[9] suggesting that activation of this receptor is necessary for the maintenance of segregated inputs (see Figures 6(b) and (c)).

NMDA receptor activation is also apparently necessary for the maintenance of two other activity-dependent processes known to occur in the retinotectal system. The first is in the refinement of topographic projections from retina to tectum that occurs during regeneration of the optic nerve in goldfish. Ganglion cell axons can establish coarse topographic projections even when activity is blocked with TTX, presumably because activity-independent molecular cues are unaltered.[12,16] However, the fine-tuning of axon terminal arbors necessary for the re-establishment of highly refined connections is prevented.[41,56] In this case, topographic fine-tuning would be expected to occur if the activity of neighboring retinal ganglion cells was highly correlated, while that of distant ganglion cells was not—a situation naturally produced with visual stimulation. Consistent with this suggestion, rearing animals in stroboscopic light, which causes all ganglion cells to fire in near synchrony, prevents the fine-tuning of topography during regeneration.[57] Recent experiments have demonstrated that infusion of APV also blocks the fine-tuning of the retinotectal map.[58] Moreover, Schmidt has demonstrated that during the period of map refinement following optic nerve regeneration, low frequency electrical stimulation of the optic nerve causes long-term potentiation of the postsynaptic tectal response which is also blocked by APV.

Another example demonstrating the involvement of NMDA receptors involves the process by which binocular neurons are normally created and maintained in the frog optic tectum. Although the optic tectum only receives direct input from the retinal ganglion cells in the opposite eye, an indirect pathway from one tectum to the other via a relay nucleus, the isthmo-tectal nucleus, does convey input from the other eye to create binocular neurons. Here too, the maintenance of the binocular map is activity dependent, as demonstrated by the fact that rotation of one eye in its orbit leads to a systematic and anatomically demonstrable re-wiring of isthmo-tectal connections so as to preserve ocular correspondence.[75] The rewiring induced by eye rotation is prevented by infusion of APV into the optic tectum.[55] An essential finding in these studies is that the levels of APV necessary to prevent the activity-dependent rearrangements apparently do not block appreciably retinotectal synaptic transmission or the excitability of the postsynaptic neuron.[9,58] Thus, the APV treatment does not act like TTX to block neural activity generally, but more likely acts specifically to prevent whatever cascade of events is triggered by NMDA receptor activation.

The specific involvement of the NMDA receptor in the synaptic alterations occurring during the critical period in the mammalian visual cortex is more controversial, but there is no doubt that NMDA receptors are present throughout (but also after) the relevant times in the cat visual system. Physiological studies of cortical neurons demonstrate that both their spontaneous firing and their responses to visual stimulation can be decreased by APV, and that lower doses of APV are needed in younger animals.[15,44,74] Moreover, Fox et al.[15] found that there is a systematic change in the laminar distribution of responsiveness to iontophoretic application of APV with age: in neonates, neurons in all cortical layers are sensitive to APV, whereas by the end of the critical period, the visually evoked responses of neurons in the deeper cortical layers (layers 4, 5, and 6) are not affected by APV iontophoresis. Thus, the changing susceptibility of cortical neurons in layer 4 to APV application is generally correlated with the period in which segregation of the geniculocortical afferents occurs. However, the fact that the superficial cortical layers remain highly sensitive to NMDA receptor blockade after the critical period draws to a close is difficult to reconcile with a simple view for the participation of NMDA receptors in the events of activity-dependent segregation and visual cortical plasticity.

If NMDA receptors are to contribute to the mechanism underlying synaptic rearrangements during the critical period, then pharmacological blockade of the receptor might be expected to prevent the segregation of LGN axons into ocular dominance patches within layer 4 in a fashion analogous to that found for the desegregation of stripes in three-eyed frogs. At present, this possibility has not been investigated in the mammalian visual system. A correlate, that receptor blockade might prevent the shift in ocular dominance toward the open eye caused by monocular eye closure, has been studied by using minipumps to infuse APV into the cat visual cortex during the critical period.[4,28] Within the infusion zone, a shift towards the open eye was prevented and, in fact, there was an unanticipated shift in favor of the *closed* eye—reminiscent of the results obtained in a similar experiment decribed above in which muscimol[52] was infused in order to silence selectively the postsynaptic cortical neurons without also blocking presynaptic afferent inputs. At first glance, then, these results would seem to conform nicely to the hypothesis that NMDA receptors play a specific role in activity-dependent cortical development and plasticity. Unfortunately, the alternative interpretation exists, namely that APV acts in a nonselective fashion to block postsynaptic activity, much as muscimol does; that is, current flowing through an NMDA-gated channel is not exclusively a "plasticity" signal. This alternate interpretation seems quite likely in view of the results of the iontophoresis experiments described above demonstrating that activation of NMDA receptors is necessary for cortical neurons to respond normally to visual stimulation. Thus, at present, it is not possible to draw strict parallels between the requirement for NMDA receptor activation in hippocampal LTP and an analagous role in visual cortical plasticity during development.

Even if a specific role for the NMDA receptor during development of the visual cortex is eventually clearly established, the synaptic basis for its mode of action remains to be elucidated. Clues come from experiments performed on rat visual cortical slices *in vitro* which suggest that synaptic connections can undergo LTP following appropriate tetanic stimulation of the white matter (which contains the incoming LGN axons), both during neonatal life and in adulthood.[1,29,30,46] LTP is much more difficult to induce in cortical neurons than in the hippocampus (only about 30% of all recorded neurons demonstrate the phenomenon in cortex), and, in fact, frequently requires a concommitant reduction in local inhibitory influences by application of a GABA antagonist; nevertheless, as in the hippocampus, LTP can be blocked consistently by iontophoresis of APV.[1] However, unlike the hippocampus, the circuitry of the cortex makes it difficult to stimulate an isolated excitatory pathway in order to separate monosynaptic from polysynaptic inputs. Thus, while the LTP studied in hippocampus clearly involves a change in the efficacy of a single type of excitatory synapse, what is called LTP in cortical slices may involve a mixture of several effects, both excitatory and inhibitory. Nevertheless, these observations raise the possibility that a cascade of physiological and biochemical events similar to those known to occur during hippocampal LTP might also take place during activity-dependent strengthening of visual cortical connections during development.

In the formation of ocular dominance columns, both normally during development and when perturbed by abnormal visual experience, some connections are strengthened, but others must be weakened and likely even eliminated in order for neurons in layer 4 to become monocularly driven. While a mechanism such as LTP could help to explain synaptic strengthening in the visual cortex, what about the reverse? A recent experiment by Artola et al.[2] suggests that it may be possible to produce a weakening, or long-term depression (LTD), of synaptic transmission in neurons in slices of rat visual cortex. These authors suggest that a level of membrane depolarization above resting level but below the greater level required for the induction of LTP can produce LTD in active synapses. (A similar phenomenon has been described in the hippocampus by Stanton and Sejnowski[71]). Moreover, Artola et al. report that LTD can be produced even in the presence of APV, consistent with the idea that activation of an NMDA-gated channel is not involved. This result may help to explain why monocular deprivation combined with cortically infused muscimol[52] or APV[4] causes an ocular dominance shift in favor of the closed eye within the infusion zone. Perhaps in the presence of these agents, activation of inputs from the open eye brings cortical neurons only to a level of membrane potential critical for LTD, consequently weakening those inputs. While these experiments provide a convenient conceptual framework for thinking about how activity-dependent synaptic change may occur during visual cortical development, it will be essential first to understand these effects *in vitro* at the level of single identified synapses and next to demonstrate that similar alterations in synaptic efficacy indeed take place *in vivo* during the critical period as a consequence of natural visual stimulation.

GENERALITY OF ACTIVITY-DEPENDENT DEVELOPMENT IN THE CENTRAL NERVOUS SYSTEM

The experiments discussed thus far provide compelling evidence in favor of the idea that activity-dependent competitive interactions in the visual system can account for the establishment of highly segregated and topographically ordered sets of connections during ocular dominance column development in mammals, and in the regeneration and maintenance of retinotectal connections in lower vertebrates. Moreover, a variety of new experiments has begun to draw exciting parallels between the cellular bases for these events and those thought to underlie LTP in the hippocampus. A common thread in all these examples is that synaptic change can be produced by the appropriate patterning of presynaptic activity and its conjunction with postsynaptic activity. In the hippocampus, when these requirements are met, evidence suggests that the resulting alterations may subserve memory and learning.[7] In the postnatal visual system, they subserve synaptic rearrangements that are generally dependent upon visual stimulation in order to provide the presynaptic correlations in neural activity necessary to preserve topographic relations, and the asynchrony required for ocular segregation.

Studies of the development of connections between retinal ganglion cells and their target neurons in the LGN suggest that structured activity may even play a role long before vision is possible. As mentioned early in this article, in the adult, ganglion cell axons from the two eyes project to each LGN, where they terminate in strictly segregated eye-specific layers. These layers are not present initially in development but rather emerge as retinal ganglion cell axons from the two eyes remodel their terminals[67] (see Figure 2). In the cat and monkey visual system, the period during which the layers form is entirely prenatal. It begins before all photoreceptor cells become postmitotic and is complete before photoreceptor outer segments are present.[13] Nevertheless, many lines of evidence suggest that here too segregation comes about by a process of activity-dependent synaptic competition. The idea that competitive interactions of some form might govern layer formation originates with observations that removal of one eye during development permits axons from the other eye to occupy the entire LGN.[48,69] Hints that the competition might be mediated by synaptic interactions comes from physiological observations that individual LGN neurons initially receive binocular inputs when the optic nerves are electrically stimulated *in vitro*[62] and that retinal ganglion cell axons from one eye can make synaptic contacts in regions later exclusively innervated by axons from the other eye.[8,67] These observations provide evidence to suggest that synaptic remodelling accompanies the formation of the eye-specific layers in the LGN.

What might be the source of activity-dependent signals during these early times in development when vision is not possible? The most likely source is the spontaneously generated activity of retinal ganglion cells. In a technically remarkable experiment, Galli and Maffei[18] succeeded in making microelectrode recordings from fetal rat retinal ganglion cells *in vivo* and found that they fired spontaneously, sometimes correlated with each other when several cells were recorded together on the same electrode.[36] Recently it has been possible to examine the spatial and temporal pattern of firing of up to 100 retinal ganglion cells simultaneously by removing fetal and neonatal retinae and recording *in vitro* using a multielectrode array; results show that even in the absence of photoreceptor function, ganglion cells fire in a very stereotyped bursting pattern, with neighboring cells firing in near synchrony.[39] These two experiments together provide evidence that the spontaneous activity of retinal ganglion cells may have the appropriate spatiotemporal patterning to provide necessary activity-dependent cues for the formation of topographically ordered and segregated inputs to the LGN and other central visual targets of ganglion cell axons.

If spontaneous activity does play a role in the segregation of retinal ganglion cell axons into the eye-specific layers within the LGN, then blockade of such activity should prevent the formation of the layers. Minipump infusions of TTX into the thalamus of fetal cats, indeed, block layer formation[66] and correspondingly perturb the branching pattern of individual retinal ganglion cell axons so that branches are no longer restricted to appropriate zones within the LGN.[70] Indeed, the effects of TTX on the shapes of retinal ganglion cell axons in the cat are remarkably similar to its effects on ganglion cell axons in the optic tectum of three-eyed frogs,[50] as shown in Figure 7. However, a criticism of the results is that TTX may have acted in a non-specific fashion to cause unregulated growth of the axons.[10] Definitive proof that this is not the case requires an experiment analogous to that performed by Stryker and Strickland,[72] in which the patterning of neural activity is specifically perturbed. This should be possible in future, when the mechanisms for the generation of synchronous bursting among retinal ganglion cells are better understood. Meanwhile, it should be noted that ganglion cell axon growth is not entirely unregulated in the presence of TTX: the axons are still capable of detecting and stopping their growth at the LGN boundaries.

The results of the experiments described above permit an important generalization concerning the universality of activity-dependent synaptic interactions. During normal development, such interactions may be driven not only by the normal pattern of use (e.g., visually evoked activity), but even earlier before vision begins by patterned spontaneously generated activity. This suggestion raises the possiblility that spontaneously generated activity elsewhere in the CNS during development may play a similar role in establishing orderly sets of connections. If so, then the synaptic changes produced by activity-dependent interactions early in development may be at one end of a continuum of synaptic change, the other end of which are the use-dependent alterations in synaptic strength associated with learning and memory. Although the changes occurring during development require major anatomical restructuring of axons, whereas those occurring during learning and memory are more likely to be confined to individual synapses,[3] evidence presented here suggests that the two types of change may not be all that different in terms of cellular mechanisms. Future experiments will reveal the extent to which the two areas of investigation converge, and whether there are similarities at the molecular level as well. The existence of similar mechanisms could represent an extremely elegant solution to the complex problem of establishing and maintaining specific synaptic connections throughout life.

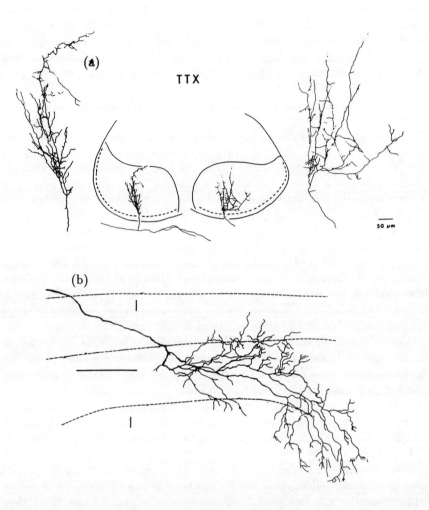

FIGURE 7 A comparison of the morphology of retinal ganglion cell axons in the fetal cat at E57 (a) and the three-eyed frog (b) following TTX treatment. In both cases, the terminal arbors of the axons are not as restricted as usual: in fetal cats, retinal ganglion cell axons normally have terminal arbors that branch only in the inner or outer half of the LGN rather than throughout (compare with Figure 2). In three-eyed frogs, the arbors are usually restricted to one stripe and do not cross stripe boundaries (indicated by dashed lines). Adapted from Sretavan et al.,[70] and Reh and Constantine-Paton.[50]

REFERENCES

1. Artola, A., and W. Singer. "Long-Term Potentiation and NMDA Receptors in Rat Visual Cortex." *Nature* **330** (1987): 649–652.
2. Artola, A., S. Brocher, and W. Singer. "Different Voltage-Dependent Thresholds for Inducing Long-Term Depression and Long-Term Potentiation in Slices of Rat Visual Cortex." *Nature* **347** (1990): 69–72.
3. Bailey, C. H., and M. Chen. "Structural Plasticity at Identified Synapses During Long-Term Memory in *Aplysia*." *J. Neurobiol.* **20** (1989): 356–372.
4. Bear, M. F., A. Kleinschmidt, Q. Gu, and W. Singer. "Disruption of Experience-Dependent Synaptic Modifications in Striate Cortex by Infusion of an NMDA Receptor Antagonist." *J. Neurosci.* **10** (1990): 909–925.
5. Bekkers, J. M., and C. F. Stevens. "Presynaptic Mechanism for Long-Term Potentiation in the Hippocampus." *Nature* **346** (1990): 724–729.
6. Boss, V. C., and J. T. Schmidt. "Activity and the Formation of Ocular Dominance Patches in Dually Innervated Tectum of Goldfish." *J. Neurosci.* **4** (1984): 2891–2905.
7. Brown, T. H., E. W. Kairiss, and C. Keenan. "Hebbian Synapses: Mechanisms and Algorithms." *Ann. Rev. Neurosci.* **13** (1990): 475–511.
8. Campbell, G., and C. J. Shatz. "Synapses Formed by Identified Retinogeniculate Axons During The Segregation of Eye Input." *J. Neurosci.* (1992): in press.
9. Cline, H. T., E. A. Debski, and M. Constantine-Paton. "NMDA Receptor Antagonist Desrgregates Eye-Specific Stripes." *PNAS* **84** (1987): 4342–4345.
10. Cohan, C. S., and S. B. Kater. "Suppression of Neurite Elongation and Growth Cone Motility by Electrical Activity." *Science* **232** (1986): 1638–1640.
11. Constantine-Paton, M., H. T. Cline, and E. Debski. "Patterned Activity, Synaptic Convergence, and the NMDA Receptor in Developing Visual Pathways." *Ann. Rev. Neurosci.* **13** (1990): 129–154.
12. Cox, E. C., B. Muller, and F. Bonhoeffer. "Axonal Guidance in the Chick Visual System: Posterior Tectal Membranes Induce Collapse of Growth Cones from the Temporal Retina." *Neuron* **2** (1990): 31–37.
13. Donovan, A. "Postnatal Development of the Cat Retina." *Exp. Eye Res.* **5** (1966): 249–254.
14. Easter, S. S. Jr., and C. A. G. Stuermer. "An Evaluation of the Hypothesis of Shifting Terminals in the Goldfish Optic Tectum." *J. Neurosci.* **4** (1984): 1052–1063.
15. Fox, K., H. Sato, and N. W. Daw. "The Location and Function of NMDA Receptors in Cat and Kitten Visual Cortex." *J. Neurosci.* **9** (1989): 2443–2454.
16. Fraser, S. E., and D. H. Perkel. "Competitive and Positional Cues in the Patterning of Nerve Connections." *J. Neurobiol.* **21** (1990): 51–72.

17. Fregnac, Y., D. Schulz, S. Thorpe, and E. Bienenstock. "A Cellular Analogue of Visual Cortical Plasticity." *Nature* **333** (1988): 367–370.
18. Galli, L., and L. Maffei. "Spontaneous Inpulse Activity of Rat Retinal Ganglion Cells in Prenatal Life." *Science* **242** (1988): 90–91.
19. Guillery, R. W. "Binocular Competition in the Control of Geniculate Cell Growth." *J. Como. Neurol.* **144** (1972): 117–130.
20. Hagihara, K., T. Tsumoto, H. Sato, and Y. Hata. "Actions of Excitatory Amino Acid Antagonists on Geniculo-Cortical Transmission in the Cat's Visual Cortex." *Exp. Brain Res.* **69** (1988): 407–416.
21. Hebb, D. O. *The Organization of Behavior.* New York: John Wiley & Sons, 1949.
22. Hickey, T. L., and R. W. Guillery. "An Autoradiographic Study of Retinogeniculate Pathways in the Cat and Fox." *J. Comp. Neurol.* **156** (1974): 239–254.
23. Hubel, D. H., and T. N. Wiesel. "Shape and Arrangement of Columns in Cat's Striate Cortex." *J. Physiol.* **165** (1963): 559–568.
24. Hubel, D. H., and T. N. Wiesel. "Binocular Interaction in Striate Cortex of Kittens Reared with Artificial Squint." *J. Neurophysiol.* **28** (1965): 1041–1059.
25. Hubel, D. H., and T. N. Wiesel. "The Period of Susceptibility to the Physiological Effects of Unilateral Eye Closure in Kittens." *J. Physiol.* **206** (1970): 419–436.
26. Hubel, D. H., T. N. Wiesel, and S. LeVay. "Plasticity of Ocular Dominance Columns in the Monkey Striate Cortex." *Phil. Trans. R. Soc. Lond. B* **278** (1977): 377–409.
27. Kemp, J. A., and A. M. Sillito. "The Nature of the Excitatory Transmitter Mediating X and Y Cell Inputs to the Cat Dorsal Lateral Geniculate Nucleus." *J. Physiol.* **323** (1982): 377–391.
28. Kleinschmidt, A., M. Bear, and W. Singer. "Blockade of NMDA Receptors Disrupts Experience-Dependent Modifications in Kitten Striate Cortex." *Science* **238** (1987): 355–358.
29. Komatsu, Y., K. Fujii, J. Maeda, H. Sakaguchi, and K. Toyama. "Long Term Potentiation of Synaptic Transmission in Kitten Visual Cortex." *J. Neurophysiol.* **59** (1988): 124–141.
30. Kossel, A., T. Bonhoeffer, and J. Bolz. "Non-Hebbian Synapses in Rat Visual Cortex." *Neuroreport* **1** (1990): 115–118.
31. Langdon, R. B., and J. A. Freeman. "Pharmacology of Retinotectal Transmission in Goldfish: Effects of Nicotinic Ligands, Strychnine and Kynurenic Acid." *J. Neurosci.* **7** (1987): 760–773.
32. Law, M. I., and M. Constantine-Paton. "Anatomy and Physiology of Experimentally Produced Striped Tecta." *J. Neurosci.* **1** (1981): 741–759.
33. LeVay, S., D. H. Hubel, and T. N. Wiesel. "The Pattern of Ocular Dominance Columns in Macaque Visual Cortex Revealed by a Reduced Silver Stain." *J. Comp. Neurol.* **159** (1975): 559–576.

34. LeVay, S., M. P. Stryker, and C. J. Shatz. "Ocular Dominance Columns and Their Development in Layer IV of the Cat's Visual Cortex." *J. Comp. Neurol.* **179** (1978): 223–244.
35. LeVay, S., T. N. Wiesel, and D. H. Hubel. "The Development of Ocular Dominance Columns in Normal and Visually Deprived Monkeys." *J. Comp. Neurol.* **191** (1980): 1–51.
36. Maffei, L., and L. Galli-Resta. "Correlation in the Discharges of Neighboring Rat Retinal Ganglion Cells During Prenatal Life." *PNAS* **87** (1990): 2861–2864.
37. Malinow, R., and R. W. Tsien. "Presynaptic Enhancement Shown by Whole-Cell Recordings of Long Term Potentiation in Hippocampal Slices." *Nature* **346** (1990): 177–180.
38. Mastronarde, D. N. "Correlated Firing of Cat Retinal Ganglion Cells. I. Spontaneously Active Inputs to X and Y Cells." *J. Neurophysiol.* **49** (1983): 303–324.
39. Meister, M., R. O. L. Wong, D. A. Baylor, and C. J. Shatz. "Synchronous Bursts of Action Potentials in Ganglion Cells of the Developing Mammalian Retina." *Science* **252** (1991): 939–943.
40. Meyer, R. L. "Tetrodotoxin Blocks the Formation of Ocular Dominance Columns in the Goldfish." *Science* **218** (1982): 589–591.
41. Meyer, R. L. "Tetrodotoxin Inhibits the Formation of Refined Retinotopographyin Goldfish." *Dev. Brain Res.* **6** (1983): 293–298.
42. Miller, K. D., and M. P. Stryker. "Development of Ocular Dominance Columns: Mechanisms and Models." In *Connectionist Modeling and Brain Function: The Developing Interface*, edited by S. J. Hanson and C. R. Olson, 255–305. Cambridge, MA: MIT Press, 1990.
43. Miller, K. D., J. B. Keller, and M. P. Stryker. "Ocular Dominance Column Development: Analysis and Simulation." *Science* **245** (1989): 605–615.
44. Miller, K. D., B. Chapman, and M. P. Stryker. "Visual Responses in Adult Cat Visual Cortex Depend on N-Methyl-D-Aspartate Receptors." *PNAS* **856** (1989): 5183–5187.
45. Nicoll, R. A., J. A. Kauer, and R. C. Malenka. "The Current Excitement in Long-Term Potentiation." *Neuron* **1**: 97–103.
46. Perkins, A. T., and T. J. Teyler. "A Critical Period for Long Term Potentiation in the Developing Rat Visual Cortex." *Brain Res.* **439** (1988): 222–229.
47. Rakic, P. "Prenatal Development of the Visual System in the Rhesus Monkey." *Phil Trans. R. Soc. Lond. B* **278** (1977): 245–260.
48. Rakic, P. "Mechanism of Ocular Dominance Segregation in the Lateral Geniculate Nucleus: Competitive Elimination Hypothesis." *TINS* **9** (1986): 11–15.
49. Reh, T. A., and M. Constantine-Paton. "Retinal Ganglion Cells Change Their Projection Sites During Larval Development of *Rana Pipiens*." *J. Neurosci.* **4** (1983): 442–457.
50. Reh, T. A., and M. Constantine-Paton. "Eye-Specific Segregation Requires Neural Activity in Three-Eyed *Rana Pipiens*." *J. Neurosci.* **5** (1985): 1132–1143.

51. Reiter, H. O., D. M. Waitzman, and M. P. Stryker. "Cortical Activity Blockade Prevents Ocular Dominance Plasticity in the Kitten Visual Cortex." *Exp. Brain Res.* **65** (1986): 182–188.
52. Reiter, H. O., and M. P. Stryker. "Neural Plasticity Without Postsynaptic Action Potentials: Less-Active Inputs Become Dominant When Kitten Visual Cortical Cells are Pharmacologically Inhibited." *PNAS* **85** (1988): 36230–3627.
53. Rodieck, R. W., and P. S. Smith. "Slow Dark Discharge Rhythms of Cat Retinal Ganglion Cells." *J. Neurophysiol.* **29** (1966): 942–953.
54. Rodieck, R. W. "Visual Pathways." *Ann. Rev. Neurosci.* **2** (1979): 193–255.
55. Scherer, W. S., and S. B. Udin. "N-Methyl-D-Aspartate Antagonists Prevent Interaction of Binocular Maps in *Xenopus* Tectum." *J. Neurosci.* **9** (1989): 3837–3843.
56. Schmidt, J. T., and D. L. Edwards. "Activity Sharpens the Map During the Regeneration of the Retinotectal Projection in Goldfish." *Brain Res.* **209** (1983): 29–39.
57. Schmidt, J. T., and L. E. Eisele. "Stroboscopic Illulmination and Dark Rearing Block the Sharpening of the Regenerated Retinotectal Map in Goldfish." *Neruosci.* **14** (1985): 535–546.
58. Schmidt, J. T. "Long Term Potentiation and Activity-Dependent Retinotopic Sharpening in the Regenerating Retinotectal Projection of Goldfish: Common Sensitive Period and Sensitivity to NMDA Blockers." *J. Neurosci.* **10** (1990): 233–246.
59. Shatz, C. J., S. Lindstrom, and T. N. Wiesel. "The Distribution of Afferents Representing the Right and Left Eyes in the Cat's Visual Cortex." *Brain Res.* **131**: 103–116.
60. Shatz, C. J., and M. P. Stryker. "Ocular Dominance in Layer IV of the Cat's Visual Cortex and the Effects of Monocular Deprivation." *J. Physiol.* **281** (1978): 267–283.
61. Shatz, C. J. "The Prenatal Development of the Cat's Retinogeniculate Pathway." *J. Neurosci.* **3** (1983): 482–499.
62. Shatz, C. J., and P. A. Kirkwood. "Prenatal Development of Functional Connections in the Cat's Retinogeniculate Pathway." *J. Neurosci.* **4** (1984): 1378–1397.
63. Shatz, C. J., and M. B. Luskin. "The Relationship Between Geniculocortical Afferents and Their Cortical Target Cells During Development of the Cat's Primary Visual Cortex." *J. Neurosci.* **6** (1986): 3655–3668.
64. Shatz, C. J., and D. W. Sretavan. "Interactions Between Ganglion Cells During the Development of the Mammalian Visual System." *Ann. Rev. Neurosci.* **9** (1986): 171–207.
65. Shatz, C. J. "The Role of Function in the Prenatal Development of Retinogeniculate Connections." In *Cellular Thalamic Mechanisms*, edited by M. Bentivoglio and R. Spreafico, 435–446. New York: Elsevier, 1988.
66. Shatz, C. J., and M. P. Stryker. "Prenatal Tetrodotoxin Infusion Blocks Segregation of Retinogeniculate Afferents." *Science* **242** (1988): 87–89.

67. Shatz, C. J. "Competitive Interactions Between Retinal Ganglion Cells During Prenatal Development." *J. Neurobiol.* **21** (1990): 197–211.
68. Sherman, S. M., and P. D. Spear. "Organization of the Visual Pathways in Normal and Visually Deprived Cats." *Physiol. Rev.* **62** (1982): 738–855.
69. Sretavan, D. W., and C. J. Shatz. "Prenatal Development of Retinal Ganglion Cell Axons: Segregation into Eye-Specific Layers." *J. Neurosci.* **6** (1986): 234–251.
70. Sretavan, D. W., C. J. Shatz, and M. P. Stryker. "Modification of Retinal Ganglion Cell Axon Morphology by Prenatal Infusion of Tetrodotoxin." *Nature* **336** (1988): 468–471.
71. Stanton, P. K., and T. J. Sejnowski. "Associative Long Term Depression in the Hippocampus: Induction of Synaptic Plasticity by Hebbian Covariance." *Nature* **339** (1989): 215–218.
72. Stryker, M. P., and S. L. Strickland. "Physiological Segregation of Ocular Dominance Columns Depends on the Pattern of Afferent Electrical Activity." *Invest. Ophthalmol. and Vis. Sci. (suppl.)* **25** (1984): 278.
73. Stryker, M. P., and W. Harris. "Binocular Impulse Blockade Prevents the Formation of Ocular Dominance Columns in Cat Visual Cortex." *J. Neurosci.* **6** (1986): 2117–2133.
74. Tsumoto, T., H. Hagihara, H. Sato, and Y. Hata. "NMDA Receptors in the Visual Cortex of Young Kittens are More Effective Than Those of Adult Cats." *Nature* **327** (1987): 513–514.
75. Udin, S. B. "Abnormal Visual Input Leads to Development of Abnormal Axon Trajectories in Frogs." *Nature* **301** (1983): 336–338.
76. Udin, S. B., and J. W. Fawcett. "Formation of Topographic Maps." *Ann. Rev. Neurosci.* **11** (1990): 289–237.
77. Van Sluyters, R. C., and F. B. Levitt. "Experimental Strabismus in the Kitten." *J. Neurophysiol.* **43** (1980): 686–699.
78. Wiesel, T. N., and D. H. Hubel. "Effects of Visual Deprivation on Morphology and Physiology of Cells in the Cat's Lateral Geniculate Body." *J. Neurophysiol.* **26** (1963): 978–993.
79. Wiesel, T. N., and D. H. Hubel. "Comparison of the Effects of Unilateral and Bilateral Eye Closure on Cortical Unit Responses in Kittens." *J. Neurophysiol.* **28** (1965): 1029–1040.
80. Wigstrom, H., and B. Gustafsson. "On Long-Lasting Potentiation in the Hippocampus: A Proposed Mechanism for its Dependence on Coincident Pre- and Post-Synaptic Activity." *Acta Physiol. Scand.* **123** (1985): 519–522.
81. Williams, J. H., M. L. Errington, M. A. Lynch, and T. V. P. Bliss. "Arachidonic Acid Induces a Long-Term Activity-Dependent Enhancement of Synaptic Transmission in the Hippocampus." *Nature* **341** (1989): 739–742.
82. Willshaw, D. J., and C. Von der Marlsberg. "How Patterned Neural Connections can be Set Up by Self-Organization." *Proc. R. Soc. Lond. (Biol.)* **194** (1976): 431–445.

George J. Mpitsos† and Seppo Soinila‡
†Department of Pharmacology, Oregon State University and The Mark O. Hatfield Marine Science Center, Newport, OR 97365 USA and ‡Department of Anatomy, Neurobiological Research Unit, University of Helsinki, Helsinki, Finland

In Search of a Unified Theory of Biological Organization: What Does the Motor System of a Sea Slug Tell Us About Human Motor Integration?

Printed with permission from *Variability and Motor Control*, edited by K. M. Newell and D. Corcos. Champaign: Human Kinetics, 1992.

We summarize the behavioral, electrophysiological, and immunohistochemical findings in the sea slug, *Pleurobranchaea*, and compare these finding to those obtained in other invertebrate animals, in higher animals, and in humans. The findings show that there is "massive" distribution and sharing of information occurring, respectively, through diverging and converging network connections.

We examine the findings of reductionist approaches and find them inadequate to answer the problems arising from such widely distributed, multifunctional, and highly converging networks whose activity may be variable. Such findings indicate that "cooperative" actions among groups of neurons may arise dynamically and nonlinearly in shifting contexts or "consensuses" of response in which individual neurons may have different functions, even during times when the behaviors are similar. Control of these systems is emergent, "fuzzy," and error-prone rather than being reflexive or following explicit causes and effects that can be read from the "switchboard" circuit of the connections between neurons.

A unified theoretical perspective is needed that accounts for both the emergent and switch-board systems. Two problems apply in both cases: First, animals may have evolved highly specialized behaviors whose underlying neural networks may not necessarily reflect generally applicable principles. Second, owing to their complexity, it may not be possible to characterize biological networks in sufficient detail to permit an understanding of the system through simulation of the system itself. Thus, we use biological information only as indications or points of departure to identify first principles that are not initially intended to account for a particular behavior, but to provide insights into generally applicable self-organizing processes at the local-neuron level that can then be used to understand how large-group action emerges.

We discuss a number of these avenues to examine computationally and biologically, e.g., (1) error and variation may not only be products of but may be causally related to the generation system dynamics. (2) The possibility that attractors provide avenues for energy or error minimization yields mechanisms from which emerge many important building blocks, e.g., the ability of groups of synapses to encode different categories of information simultaneously; threshold effects that enhance system function; and input signal dynamics which not only carry encoded information but also provide a variety of search strategies for locating attractor basins. (3) Minimal network architectures may be identified that permit bifurcation into different dynamical states. (4) Computer graphical analysis of spatio-temporal activity may show how different attractors are established and move and merge in space and time. (5) Competition between synapses may continuously sculpt and readjust network connections to changing conditions.

CONTENTS

1 Introduction: Grand Unification Theories	71
2 Findings in a Sea Slug	75
2.1 Behavior	75
2.2 Neurophysiology	76
2.2.1 Key Features of all Mount-Related Behaviors Can Be Examined Through a Small Population of Neurons, the BCNs	76
2.2.2 Connectivity of the BCNs	76
3 Distributed Function, Multifunctionality, and Variation	79
3.1 Rationale for Change in Conceptual Framework: Single Cells to Contextual Groups	79
3.2 Context of Neuronal Group Action: Inferences from Behavioral Choice	80
3.3 Context of Action in the Buccal-Oral System	81
3.4 Definitions: Modes of Cooperativity	83
3.4.1 Attractors as Dissipative Structures	83
3.4.2 Low Dimensionality in High-Dimensional Systems	84
3.4.3 Turbulence, "Attracting States, and Self-Organizing Criticality"	84
3.5 Chaos and Other Forms of Variation	85
3.5.1 Bifurcation Parameters and Chaos	85
3.5.2 Bifurcation-Induced Variations	87
3.5.3 Random Noise	89
3.5.4 Variation-Dependent Optimization in Muntifunctional Systems	90
3.6 Error as an Integrative Principle	90
3.7 Definitions: Dynamics, Behavior, and Multifunctional Systems	91
3.8 Definition of Contexts in Group Action: Linear and Nonlinear Organization	92
4 Behavioral & Neurophysiological Findings in Other Animals	92
4.1 Invertebrates	92
4.1.1 Overview of Multifunctionality and Variability	92
4.1.2 Bifurcation and Response Modality in the Lobster Stomatogastric System	94
4.2 Mammals	95
4.3 Divisions of the Mammalian Motor System: Relationship to Divergence and Convergence	96

5 Neuromodulation — 98
  5.1 Convergence and Divergence of Neurotransmitter Systems — 98
    5.1.1 Invertebrates — 98
    5.1.2 Vertebrates — 101
6 Reduction & Emergence in Control Mechanisms — 104
  6.1 Transmitters Control Network Function and Architecture — 104
  6.2 Control of Whole-Animal Behavior: Reductionist Explanation of Learning in *Aplysia* — 106
    6.2.1 Synapse-Specific Control of Behavior — 106
    6.2.2 Complications — 107
  6.3 Emergent Control of *Aplysia* Behavior: Parallel Distributed Processing — 108
    6.3.1 Don't Worry, Be Happy: New Synthesis — 108
    6.3.2 The Locus of Learning May Not Be at A Unique Cellular Site — 109
    6.3.3 The Neurocircuit May Not Be Definable — 110
    6.3.4 Different Levels of Learning within Definable Sets of Synapses — 111
  6.4 "Fuzzy" Control — 112
  6.5 Is Our View Holistic? — 113
7 Does A Theory Exist? — 114
8 Computer Simulations: Minimal Multifunctional Networks — 116
  8.1 Nonlinearities and Bifurcations in Simple Network Architectures — 116
    8.1.1 Nonlinearity and Bifurcation in Model Systems — 116
    8.1.2 Relationship between Bifurcation Dynamics and Network Architectures — 118
    8.1.3 Continuous vs. Discrete Processes — 119
  8.2 Response Optimization, Energy Gradients, and Attractors in Biological Networks — 120
    8.2.1 Attractors, From Sea Slugs to Bees — 120
    8.2.2 Comparison through Analogy in Principles, Not in Identity of Mechanisms — 121
  8.3 Local Error-Minima in Biological Adaptation — 123
  8.4 Visualization of Spatio-Temporal Dynamics — 124
9 Conclusion — 124
References — 126

1. INTRODUCTION: GRAND UNIFICATION THEORIES

Much of our discussion here will address the functional meaning of divergence and convergence of connections among neurons. At the simplest level, both are anatomically definable: divergence occurs when a single neuron sends synaptic[1] projections to many other neurons, and convergence occurs when many neurons send projections onto a common follower neuron. A more functional definition is to say that divergence distributes information, whereas convergence produces sharing of information. The consequence of divergence is to increase the size of the co-functional group of neurons, but this alone would only produce a set of independent processors. In parallel programming, the programmer breaks down a problem into different components and then assigns each component to a different processor; the programmer distributes the components, but the processors act independently. Similarly, there may be multiple sites of learning, perhaps arising from divergence of input-stimulus pathways onto many different cells, and each site may involve different cellular mechanisms, but unless there is some interaction or convergence, each site processes information independently. Because of its potential for sharing information, convergence forces many neural sites to work interdependently. Thus, convergence lies at the heart of our definition of parallel processing in biological systems,[139,143] as it does in simple connectionist neural networks[185] that have little resemblance to biological ones.

In attempting to understand the functional implications of divergence and convergence even in small networks, Pribram's[157] analogy to holography for distributed memory storage seemed a possibility,[143] particularly, as Mpitsos and Cohan[139] later reported, since some networks are able to reorganize similar motor output patterns of activity after neurons are removed that appear to control the pattern of activity going to motor neurons. In these studies, the neuron was removed from taking part in the motor pattern by hyperpolarizing it below its firing threshold. This produced two types of errors: cessation of firing in the motor neurons that it controlled, and cessation of all motor activity. Eventually, the original pattern recovered even though the hyperpolarized neuron, and the motor neuron(s) it drove, did not take part in the reformed motor pattern. Since the overall firing pattern in the reformed activity in the motor roots appeared similar to the original pattern, it seems reasonable that the error was somehow distributed throughout the generator network. By analogy to holography, the "picture" of activity emerging from the memory distributed among the pattern-generating neurons exhibited graininess when bits of information were lost rather than exhibiting holes or gaps in some regions while retaining high resolution in others as would occur in some neural networks.[136] We use "graininess" here because fewer neurons became involved in the reformed pattern than in the original one, yet the overall structure of the pattern seemed the same.

---

[1] We use the terms "synaptic projections" and "connections" to refer both to well-defined pre- and postsynaptic structures involving localized transmitter release and to morphologically indistinct structure involving diffuse transmitter release.

There are problems even with the notion of holography, and in carrying the analogy of graininess too far, but for the present purposes, the real question that these studies point to is one of memory storage and control in high-dimensional systems. The high dimensionality that we refer to is not just in the number of interacting components. It also includes, as we shall discuss, the storage of different forms of information within the same set of synapses and nonlinear ways of addressing it.

While it is easy to see high dimensionality, and the consequences of it, in the human cortex, it has not been so easy to admit that it exists in animals that neuroscience persists in calling "simple." A world view that polarizes animals into simple and complex (into generalizations relating to invertebrate and vertebrate phyla) emerged; e.g., see comment in Edelman.[51] A wide variety of factors, including the technology of intracellular microelectrode recordings,[110] the ability to use these recording methods on cells that can be identified in different experimental preparations, findings showing that activity is encoded within the central nervous system itself for generating patterned motor activity,[190] the importation of the ethologist's[114] fixed-action pattern (FAP), identification of functional types of cells such as command neurons that control central pattern generators and stereotyped behaviors or FAPs,[46,104] and the related findings showing that much of this activity is genetically encoded,[19] worked together to entrench reductionism. Though each finding remains useful in its own right, concepts developed from reductionist single-neuron methods have proved inadequate to understand distributed, multifunctional, and variable systems.

It is an interesting discovery that many biological systems, being potentially high dimensional, may generate complex behavior that is governed by relatively low-dimensional dynamics.[2] Choatic systems fall into this category, and, because of their complex response dynamics, have been a subject of considerable attention over the past ten years.[144,154,158,172] We shall summarize some of these efforts. But rather than dealing with the verifiability of chaos itself or of any dynamic process, which has already been addressed sufficiently elsewhere,[133] what we wish to do here is to address common features of all nervous systems which give rise to or exclude the ability of the systems to produce particular response dynamics. This is to say that the important features are not so much whether repetitive activity, as one example, is generated by limit-cycle or chaotic dynamics, as it is of the system characteristics that permit different activities to arise.

It may be useful to forewarn the reader that our own perspective of brain function, or of the function of systems composed of aggregates of nonlinearly interacting components, has two parts, one experimental and the other philosophical. It is essential, of course, that the philosophy or theory one holds about the actions of a system must have a foundation on hard biological fact. However, problems arise when doing only that. Take just one example: All visual systems use on-responses to respond to

---

[2] There is often no need to go beyond its definition of dynamics simply as "time-dependent variations of activity," though there are different forms of dynamics. Rather than presenting a formal definition, we shall introduce various ideas that modify our standard working definition as they arise in the course of the discussion.

the onset of light, and off-responses to respond to the off-set of light. But knowing the cellular and physiological mechanisms that generate off-responses in some molluscs would lead one completely astray about the mechanisms that produce them in vertebrate animals.[132,188] Evolutionary selection mechanisms tend to optimize the adaptive[3] mechanisms in each organism. Thus, owing to diversification and optimization, it is often difficult to determine what features permit generalization across organisms, or for that matter, across integrative systems within an organism because the various systems may have developed under different evolutionary constraints. It is possible to argue in favor of comments one might find in print, which go something like this: Owing to the observation that evolution conserves mechanisms, what we understand of mechanisms of learning in a simple animal such as a sea slug will allow us to understand the mechanism of learning in humans. But to take that argument is to forget the equally important fact that diversification is a crucially important driving force in biological evolution, not only through variations arising from random factors, but also through deterministic low-dimensional factors whose dynamics gives them a life of their own.

As neurobiologists, we are interested in the integrative mechanisms of sea slugs, crayfish, insects, leeches, lampreys, or humans. But from a broader perspective, we wish to ask whether there are scale-independent principles, namely, ones that apply to different levels of organization, from chemical processes to cellular, organismal, and social ones. The question is: Can we identify unifying principles, as one might say of the attempts to establish grand unification theories (GUTs) in physics? Unfortunately biological systems are too complex and uncontrollable to permit such a synthesis presently, as we shall try to show in the present paper. One possibility is to conduct computer simulations of models that reduce a particular biological system within the bounds of definable characteristics. While this may give insight into mechanisms pertaining to that system, it may not provide much insight into general principles.

An alternative simulation approach is to use biological information as "points of departure" to conduct computer simulations that do not necessarily attempt to replicate the structure or function of any particular biological system. We go further to suggest that it might be useful to use simulation systems that are actually extreme caricatures of biology, but which nonetheless might generally give insight into biology. Eventually, what we hope to do is to obtain some idea about how network architecture incorporates various linear and nonlinear interactions between neurons to allow the network, as a whole, to generate different types of response dynamics. We want also to understand how these fundamental network principles

---

[3] The term "adaptive" implies some conformation of a system (biological or computational) that allows it to survive in its environment. The process of conforming, as we shall discuss in detail in Section 7, may represent a gradient descent in the error of the response with respect to the response required for survival, or in the energy required to generate the response. That there may be local minima in such conformations indicates that there may be non-optimal ways of responding, and, conversely, it indicates that there may also be an absolute minimum representing some optimal way that the system might respond for a given environmental demand, though local minima may be sufficient for survival.

become sculpted selectively to produce the neural responses observed in individual animals. The neural architecture in individual organisms may retain more or less of these primal features, as required or permitted by the tasks presented for adaptive fitness. Thus, by seeking to identify common principles from which different mechanisms may emerge, we are joining a call to reconsider the importance of comparative biology,[22] a subject which has suffered as research has become entrenched in animal-specific encampments. But, as we hope will become apparent, our efforts will not be to determine, for example, whether command processes are the same in different animals or to define the command process more exactly. As important as such issues are, we shall nonetheless aim to address comparisons at a broader or more abstract level. Much of our discussion here will center on making analogies through commonality in dynamical principles rather than in mechanisms.

There are, of course, many people who, in one way or another, have addressed the question of how cooperative action arises among groups of intercommunicating individuals. The works of Grossberg, for example, on neural networks and the mathematical foundation of many of psychological phenomena are too numerous even to summarize adequately.[71,72] It is a theme of modern neural network connectionism,[152] in studies of chemical dynamics,[6,53,167] and in mammalian nervous system.[172] In many biological aspects, it can be traced back to Darwin,[40] and to Aristotle.[182] Such works notwithstanding, we shall attempt to show in the present discussion that a unifying theory of how neurons (or individuals of any type) act cooperatively within a group is presently lacking. Along the way we shall also attempt to identify ways for continuing the search for unifying principles.

In the course of this paper we shall first describe the behavioral, physiological, and immunohistochemical studies in our experimental system the sea slug *Pleurobranchaea*, and then compare these results to those obtained in other invertebrate animals and in vertebrates. Another gastropod mollusc, the sea slug *Aplysia*, has been the focus of reductionist researches in many laboratories that have attempted to explain animal behavior and associative learning in terms of definable reflexes. Section 6 deals with reductionism; we examine these findings, show the difficulties that have arisen, and then reassess them from the point of view of parallel-distributed processing. Given growing interest in nonlinear dynamics in model mathematical and physical models, we examine the viability of applying tools arising from these studies to biological systems. In Section 8, we suggest computer methods which might give some insight into how the integrated activity of large numbers of neurons might arise from interactions occurring locally between individual neurons. Thanks to the work of René Thom,[182] we use a call from Aristotle[7] to summarize the intent of our own work begun two decades ago: "''Αλλην ἀρξὴν ἀρξόμενοι," namely, "Now let us make a fresh start," at least to point out what it is that traditional thinking in neurobiology does not address sufficiently, and what the problems are in progressing further.

2. FINDINGS IN A SEA SLUG
2.1 BEHAVIOR

Pleurobranchaea is a large sea slug, a member of the opisthobranch gastropod molluscs, ranging in size from a few millimeters to tens of centimeters, depending on its age. Its general body features resemble a snail, though like land slugs, it has no shell (see photographs in Mpitsos[142,143,145,147]). The animal exhibits a relatively large repertoire of behaviors,[4] including, righting when turned upside down, defensive withdrawal, mating, egg-laying, feeding and a variety of other mouth-related behaviors involving the mouth, lips, jaws, and radula (a structure analogous to a tongue). Feeding behavior usually has dominance over the other behaviors. For example, animals normally withdraw from tactile stimuli applied to their head regions, but in the presence of food, withdrawal responses are suppressed in feeding-motivated animals.[45,145] The most obvious feature of the feeding behavior is the rapid bite-strike response in which the entire jaw structures comprising the proboscis are rapidly thrust out to bite at a food object and then rapidly withdrawn. Feeding also consists of bite-ingestion movements in which food is grasped and then sequentially drawn into the mouth cavity largely through cyclical inward and outward movements of the radula and coordinated movements of the anterior regions of the jaws and mouth. A third stage of feeding consists of swallowing movements in which food is passed from the buccal cavity through the esophagus and then into the stomach. The bite-ingestion and swallow components of feeding[44] are excellent for neurophysiological work because of their oscillatory characteristics, much as might happen in humans during opening and closing of the jaws and related movements of the tongue. Because of the sequence of oscillations, the behavior persists and is amenable to analysis, whereas single-shot behaviors such as withdrawal are more difficult to analyze. However, as in humans, the number of cycles that the animal may exhibit during a single bout of bite-ingestion and swallow is often short and possibly nonstationary in its temporal characteristic, which, as discussed below, pose difficult problems in studies aimed at understanding the dynamics of the behavior.

The jaws, radula, mouth, and lips of the animal generate many different and variable behaviors.[138] These include several components of feeding, regurgitation, defensive biting, among others.[34,124,125,126,127,138] The animal also exhibits self- and inter-animal gill grooming,[148] but we presently have no way to evoke gill-grooming behavior reliably. However, of all its behaviors, inter-animal gill-grooming is particularly interesting because *Pleurobranchaea* is cannibalistic, raising questions into

---

[4] The ensuing discussion also relies on the term "behavior," and identifies a number of behaviors within the repertoire of what the animal can do. For the moment, we use "behavior" to refer specifically to a definable response of the animal, or generically to some unspecified but potentially identifiable response. We shall see by subsection 3.7, however, that the definition of behavior, of behavioral repertoire, and of behaviorally multibehavioral or multifunctional systems (ones that can produce different behaviors using the same sets of neurons) needs to be revised to take into account the consequences of variation in "contexts" of neuronal group action.

the mechanisms that turn carnivorous feeding mouth, radula, and jaw movements into cleaning movements.

2.2 NEUROPHYSIOLOGY

2.2.1 KEY FEATURES OF ALL MOUTH-RELATED BEHAVIORS CAN BE EXAMINED THROUGH A SMALL POPULATION OF NEURONS, THE BCNS

The cerebropleural ganglion ("brain") of *Pleurobranchaea* innervates the mouth and anterior head regions, whereas the buccal ganglion innervates the muscles that move the jaws and radula. Thus, coordination of buccal-oral behaviors, namely ones that involve both the buccal structures and the mouth and lips, must happen through these ganglia.

The only way this can happen is through the buccal-cerebral neurons (BCNs), of which there are approximately 15–20 in each half of the two buccal hemiganglia. The BCNs are unique because they are: (1) the only cells in the buccal ganglion that project to the brain, except for two bilaterally paired giant neurons whose function is presently unknown, and (2) that are either directly involved in generating the central pattern generator for the buccal behaviors or intimately involved in controlling it.[33,34,139] There may be other oscillators located in the brain, but by comparison to the effect of the BCN oscillator, other oscillators have weak effects. The BCNs and the two giant cells are the only sources of information to the brain about processes in the buccal ganglion. All of the behaviors involving movements of the mouth and lips in coordination with the tongue and jaws must act through BCNs, and since the BCNs are part of the central pattern generator, they do more than perform coordination of the different motor centers.

Although the various mouth-related behaviors may involve thousands of neurons, key features of the information required to generate these behaviors may be obtained from much smaller subsets of neurons consisting primarily of the BCNs and some of the neurons with which they interconnect. Thus, the BCNs acting individually and as a group are *multifunctional* because they must generate activity pertaining to multiple behaviors.

2.2.2 CONNECTIVITY OF THE BCNS

Figure 1 summarizes the BCN connections. The evidence for these connections has been described in several publications.[32,33,34,139] The present evidence indicates that they connect with one another primarily polysynaptically, as indicated by the interneurons in Figure 1; however, many of these polysynaptic connections may be through other BCNs. In a few cases there may be mutual inhibitory connections between the BCNs, but the exact connectivity, if it can be defined, remains for further study. As indicated schematically in Figure 1, many BCNs converge onto the same target motor neurons, and individual BCNs diverge onto different motor neurons. In turn, the motor neurons neurons feed back to the BCNs that drive them. An identified group of neurons in the brain, the paracerebral neurons (PCNs), converge onto the BCNs, and the BCNs feed back to the PCNs.[65,66,139]

In Search of a Unified Theory of Biological Organization

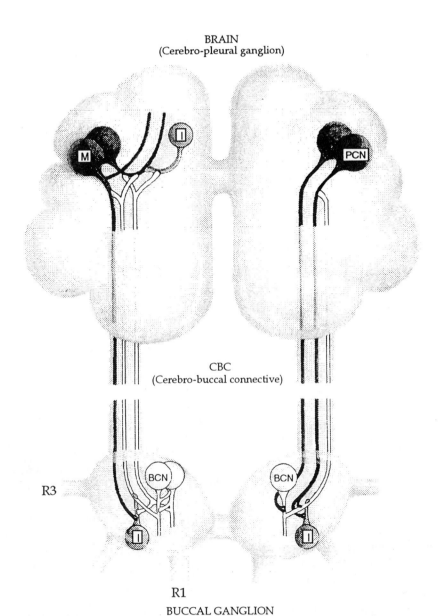

FIGURE 1 Cartoon showing central features of converging and diverging connections in *Pleurobranchaea* nervous system. BCN: buccal-cerebral neurons. I: interneuron. M: Motor neuron. PCN: Paracerebral neuron. Size of each of these pools of neurons is about 10 to 20 units each. There are many more motor neuron pools, one for (cont'd.)

FIGURE 1 (cont'd.) for each motor root; some cells send axons out multiple roots. R1: motor root that innervates muscles for opening jaws. R3: motor root for closing jaws. Motor roots of brain are not shown. For clarity of presentation, the BCN-motor neuron connections are shown on the left, and BCN-PCN connections are shown on the right. Reprinted with permission from *Brain Res. Bull.* **21** (1988): 529–538.

The actual biological network is much larger and more interconnected than shown in Figure 1. For example, there are different pools of neurons that send axons out of the brain through the various motor roots, of which there are approximately a dozen on each side of the brain, though some motor neurons send axons out different roots. Additionally, it is necessary to consider that there are numerous pools of interneurons. Thus, the number of converging and diverging connections in the brain and buccal ganglion is quite large. Moreover, just as there are interactions between the brain and buccal ganglion, there are interconnections between the brain and other ganglia. Therefore, the extended network consisting of neurons affecting the BCNs, and ones that the BCNs affect, involves hundreds of neurons.

What we hope to achieve in our present line of work is to add neuron pools to the core model shown in Figure 1. We want especially to obtain the temporal relationships in the firing of as many of the neurons as possible, partly to use the data to reassess the conclusions we have already reached, and partly to use it to obtain some insight into how such large numbers of neurons interact with one another. The time of firing of all BCNs and PCNs is being extracted from multiple recordings conducted simultaneously at different extracellular sites along the nerves that connect the brain and buccal ganglia (the cerebro-buccal connectives, CBCs). Since activity occurs in both directions in the CBC, the multiple recording sites allows us to determine the direction of propagation of firing in different nerve fibers, and thereby to distinguish between the BCNs and other neurons. It is only a matter of extended labor to include the time of firing of motor neurons in the different motor roots.

The point of all of this work, however, is not to obtain a complete network, but to use the data to assure that our computer simulations of different model assumptions will provide activity that reflects the activity in the biological system. A particularly important aspect of this work will be to obtain an indication of the types of variations and motor pattern blending that the system generates.

Owing to similarities in their gross neuroanatomical features, which distribute different functions to the buccal ganglion and to the brain, the principles obtained in *Pleurobranchaea* may hold in many other snails and slugs. Moreover, it is likely, though not demonstrated sufficiently, that neurons analogous to the BCNs in *Pleurobranchaea* may have similar functions in all snails and slugs. But it is not clear presently whether other snails and slugs generate as many mouth-related behaviors as *Pleurobranchaea*, and whether the behaviors in these other animals are as variable.

3. DISTRIBUTED FUNCTION, MULTIFUNCTIONALITY, AND VARIATION
3.1 RATIONALE FOR CHANGE IN CONCEPTUAL FRAMEWORK: SINGLE CELLS TO CONTEXTUAL GROUPS

Our initial aim for studying this "simple" sea slug was to understand the cellular basis of learning. The many control experiments in the studies of Mpitsos and Collins[142] and Mpitsos, Collins, and McClellan[143] were the first to demonstrate that sea slugs are capable of Pavlovian and avoidance associative learning, and even earlier work, though not as extensively controlled, promised that associative learning could be examined in isolated nervous systems.[145] However, work begun in the mid 1970s closely examined the motor patterns and behaviors, and showed that networks are multifunctional in being capable not only of generating different behaviors and that similar motor patterns can yield different behaviors.[32,124,125,126,127,138,139] More importantly the motor patterns of different behaviors often blend with one another and the underlying motor patterns of neural and muscular activity are quite variable[32,138,139] As discussed below, rather than a definable reflex system, it seemed possible that networks of neurons work by flexible contexts of action. The variations in the contexts might involve linear regroupings or might arise from nonlinearities that cause rapid shifts or bifurcations in the patterns of activity generated by the network. It became apparent that attempts to attribute specific function to a given neuron, or to locate the engram of a learned behavior to a particular synapse could fail.

Consequently, we had to backtrack, to reassess how it is that even innate or "unlearned" motor patterns arise in such systems before we could address the problem of how newly learned information is incorporated into the network. Although we continued to conduct learning studies after the observations made in the mid to late 1970s, our rationale for doing them has not been to find the locus of learning at specific synapses, but to determine whether learning could actually be identified in the responses of reduced preparations.[138,140,141] Additionally, given the indication that information may be distributed over many neurons it was necessary to develop the technology for identifying populations of neurons that are involved in specific aspects of learning among which we could examine how learning affected cooperative actions among neurons in the population.[147,150,151,174]

The idea of cooperativity, which Freeman and coworkers[172] have used to advantage in their studies of rabbit olfactory bulb, resembles what we refer to as *"contexts"* in neuronal group function. Much of the discussion in this paper will attempt to present our understanding of functional contexts. Early in the development of the idea of command neurons (cells that evoke stereotypic behaviors), Davis and Kennedy[42,43] showed that each command neuron of the lobster swimmeret system produces characteristically different effects and selectively controls different motor neurons, indicating that the command process arises from group action in which each command neuron performs specific subtasks of the command process and activates a particular set of motor neurons. Later work, such as the finding in *Pleurobranchaea* that command neurons receive feedback connections from the motor network that they drive,[65] blurred functional distinctions that may be attributed to single neurons because function seemed to be shared. Davis[41] used the term "consensus" to refer to the emergent actions that might arise among groups of interacting neurons. In studies on locust walking, Kien[90,92] used "consensus" to refer to variable activity in ensembles of neurons. Our thinking on the ability of groups of neurons to act contextually includes variation in the effects produced by individual neurons, by the group as a whole, and in the neurons that constitute the group. For the present discussion we use the idea of "contexts" interchangeably with "consensus," partly because we, too, are inclined to believe that its meaning of "all or most" is descriptive of what may often take place in the number of neurons that become active during normal behavior.

Although there are similarities between our use of the "contexts/consensus" and Davis' and Kien's use of "consensus," there are also some important differences which we shall address. Our definition relies on many factors other than the number of neurons that become active. Therefore, we hold off a definition, which is given in subsection 3.8, until have first presented behavioral examples, and provided discussions of principles relating to variation, dynamics, and nonlinear function.

3.2 CONTEXT OF NEURONAL GROUP ACTION: INFERENCES FROM BEHAVIORAL CHOICE

The following example may help to explain our use of the term "consensus" (or "context"): One of the original purposes for studying *Pleurobranchaea* was to examine how animals "choose" to perform a particular behavior when confronted simultaneously by many stimuli that often require conflicting responses, as might occur in the natural environment.[44] For example, turning an animal upside down evokes righting behavior having a definable duration. Presenting food to the animal produces several components of feeding behavior at definable thresholds. When turning the animal upside down and presenting food simultaneously, righting times significantly increase, but feeding thresholds remain constant. By such simultaneous presentations of different stimuli to evoke pairs of behaviors, it is possible to define a behavioral hierarchy,[45] and to view the process of establishing the hierarchy as a reflex system where one behavior inhibits another.[100]

It is necessary, however, to go one step further. Early studies on behavioral "choice"[148] indicated that some behaviors seem to blend into one another, as Kirsti Bellman[18] was to show later in lizards. In *Pleurobranchaea*, for example, the anterior portion of the foot may start to twist in order to right, but, at the same time, it may begin to cup around the descending solution of the food stimulus. The anterior foot appears to be attempting to perform two contradictory behaviors at the same time. Even when righting behavior starts, it is slowed because the foot's motor-system is still receiving conflicting activities, one for righting and one for feeding. We do not deny that reflexes involving inhibition can be found, but doing that alone places one's concepts on the side of the razor's edge in which behavior, and the underlying neurointegrations, are viewed as set and repeatedly definable structures. The important issue to us is the process of forming the behavioral "choice" during the time that the animal is presented multiple stimuli rather than a stereotyped behavioral hierarchy. The two approaches speak about the same behaviors but give different explanations. The contextual approach views behavior as arising fluidity among many different and blendable behaviors. The reflex approach views the animal as a generator of a set of fixed-action patterns (FAPs; e.g., Gillette[64]), each relating to definable and repeatedly identifiable responses in the animal. The definition of behavioral hierarchy forces one to think of behaving animals as concatenations of reflexes or FAPs that are repeatedly definable. In the extreme situation in which an inverted animal lies motionless, neither feeding nor righting, the definition of behavioral hierarchy would lead one to develop experiments showing inhibition between feeding and righting sensory-motor systems, as shown for the interaction between feeding and withdrawal.[100] It would also lead one to identify a particular locus in the nervous system at which such inhibition takes place. The variability of activity in *Pleurobranchaea*, and the high degree of converging and diverging connections in its nervous system lead us to believe that such localization of mechanism may be misleading. By contrast, when taking these factors into account, one's focus is directed to dynamically shifting contexts of activity in which the identity and location of the underlying mechanism for a behavior is not fixed, just as the behavior may not be fixed and always distinguishable from others. One is more apt to think of variably emerging networks rather than "switchboard" reflexes.

Thus, although the definition of behavioral hierarchy is useful for categorization, and although it is defined using the behavioral choice paradigm, it dangerously excludes the dynamics within choice-making processes. To be sure, reflex actions are indications of a process, but the reflex approach leads one to examine the structure of the network itself whereas an approach that deals with the dynamics of interactions leads one to examine principles of interaction from which networks emerge not only variably but also nonlinearly, as we shall try to illustrate in Section 6, when dealing with reductionism, and in Section 8 when dealing with computer simulations. Inhibitory interactions between motor systems may be used by both explanations, but the dynamical approach uses inhibition either as a potential explanation that may or may not actually take place, or as a participating variable in a system that expresses the dynamics. In either of these non-reflex explanations, the role of inhibition may not be discernible from the structure of the network itself, though dynamical explanations must also account for conditions that actually express reflexes.

3.3 CONTEXT OF ACTION IN THE BUCCAL-ORAL SYSTEM

The buccal-oral system of *Pleurobranchaea*, consisting of the lips, mouth, radula, and jaws, seems to magnify variation and behavioral blending because, as noted above, it is capable of generating many different behaviors and variants within individual behaviors. Moreover, blending happens among the various mouth-related behaviors themselves, as well as with behaviors produced by other motor systems. A number of studies have provided criteria for identifying motor patterns relating to particular buccal-oral behaviors. McClellan[124,126,127,128] and Croll and Davis[38,39] have established specific motor-pattern differences in electrical recordings made from muscles and nerves to distinguish between feeding, regurgitation, and rejection behaviors, but even McClellan's studies demonstrated that different behaviors can be generated by similar motor patterns.

Having observed considerable motor-pattern variations, Mpitsos and Cohan[138,140,141] devised a series of associative learning experiments to determine whether a learned response persisted in even minimally dissected animals. The results clearly showed that the behaviors of the undissected and dissected, behaving animals were identical, as determined by direct observation of what the animal did in response to the applied experimental and control stimuli that were used in training. However, when examining the electromyographic data alone, obtained simultaneously while observing the behaviors, it was not possible to identify consistent differences in the firing patterns of muscles during feeding, regurgitation, and rejection. The information had to reside within these patterns, but the information itself could not be read simply by examining the temporal orchestration of activity in the recorded motor patterns. An alternative explanation is that the information resides in the dynamics of the neuromuscular system as a whole, i.e., in the combination of interactions between the motor output, in the nonlinear loading presented by the muscles and mouth and jaw structures, and in the effect of sensory feedback to the central nervous systems. Such systems may have qualities similar to damped-driven oscillators whose dynamics are sensitive to changes in parameter-constants that control the effects of different variables (e.g., see the description of the Duffing oscillator in Thompson[183]). Not inconsistent with this is that the animal can perform a given behavioral effect successfully using combination of patterns. In neural activity, it may be sufficient to have reached an approximating and variable "consensus" or "context" of action rather than requiring an explicit stereotyped pattern.

The neural sources of some of this variation were identified in studies of isolated nervous systems that were used in order to remove the influence of sensory perturbations. For example, neural patterns reemerge even when BCNs that were initially responsible for generating patterned activity are reversibly removed from the coactive networks (Figure 5 in Mpitsos[139]), showing that different combinations of neurons generate similar responses. Similarly, the firing of some BCNs shift variably between completely opposite phases of the cycle of opening and closing of the jaws (Figure 16 in Mpitsos[139]). Graded intermediates may occur as the nervous system generates patterns of rhythmic activity and spontaneously shifts into another pattern.

Our view is that the intermediate and variable forms of activity give crucial information about integrative mechanisms. Variations that occur within group action must arise from variations at the level of individual neurons. To present these ideas, the next two subsections discuss "attractors" and "attracting states," and the role that different forms of variation and error have in the response properties of biological systems.

3.4 DEFINITIONS: MODES OF COOPERATIVITY

3.4.1 ATTRACTORS AS DISSIPATIVE STRUCTURES
An intuitive definition of attractor may be given by examining the property of attraction. Suppose for the moment that we are dealing with a process governed by three variables. The state of the system at any given time is represented by the values of the these variables. The progression of these values over time define the parameter state space of the activity of the system. Plots of these variables, one variable in each coordinate of three-dimensional space, defines the phase space. The flow or trajectory from one point to another provides a view of the phase portrait of the dynamics of the activity. For continuous periodic activity, the trajectory is a closed loop. A brief external perturbation, applied to one or any combination of the variables, will move the state of the system away from the closed loop. If the trajectory then collapses asymptotically back toward the closed loop, the system may be considered to be governed by an attractor. The set of all possible perturbations, and subsequent dissipative responses shown by the asymptotic recovery, define the inset to the attractor or its *basin of attraction*. In the case of periodic activity the attractor is a *limit cycle*. The activity could also be generated by *chaotic attractors* whose trajectories are not represented by a limit set either before or after perturbations, but by an attracting set. An indication of this set may be viewed through the geometry of the topological manifold in which the trajectories mix. Examples of the mixing geometry of attractors in *Pleurobranchaea* responses and model systems in our own work may be found in Mpitsos[137,144] and Andrade et al.,[6] respectively. Though we have used phase portraits to obtain an intuitive view of attractors, a single dynamical system may have phase portraits containing multiple, competing attractors.[183]

The above-cited work from our laboratory also discusses a variety of geometrical and computational tools that may be used to determine whether the activity is generated by limit-cycle or chaotic attractors. In either case, the most useful for determining whether the system is generated by an attractor is to conduct the perturbation experiments described above, which a major focus of our present efforts in both biological and model systems. Much experimental work needs to be done in this way, but it is quite likely that attractors underlie much biological function, as shown, for example, by perturbation experiments designed to test for resetting of the phase of oscillatory activity (an example of an externally applied current pulses to one of the BCNs in *Pleurobranchaea* is shown in Figure 3 in Mpitsos[139]).

3.4.2 LOW DIMENSIONALITY IN HIGH-DIMENSIONAL SYSTEMS As the system evolves to dissipate perturbations, one would observe that the ensemble of points in state space decreases over time, i.e., that there is volume contraction. Volume contraction simplifies the topology of the structure defined by the trajectories, and as pointed out by Thompson and Stewart,[183] *"This can often mean that a complex dynamical system with even infinite-dimensional phase space...can settle to final behavior in a subspace of only a few dimensions"* (p. 1).

This phenomenon is particularly important in biological systems because they are inherently high dimensional. A single cell in the visual cortex of the mouse, for example, receives inputs from approximately 5000 other cells,[21] each of which may be a controlling variable. Numerical analyses of spontaneous cortical neuron activity,[158] of EEGs in olfactory bulb,[172] cortex,[4,9,173] and of motor patterns in *Pleurobranchaea*,[137,144] all indicate that the activity is generated by relatively few variables. One of the tasks facing work in animals such as *Pleurobranchaea*, and of correlative computer simulations, is to identify the variables, out of the many available, that become active in low-dimensional activity, and to identify the conditions among these variables that permit low dimensionality to arise. Part of the goal of our computer simulation is to define minimal structures that permit the generation of different types of attractors, and to determine how different attractors might arise at different times within the same high-dimensional space. An interesting possibility is that what determines which sub-space is occupied may simply be a matter of what attractor becomes established first. In a sense, there may be a type of competition such that the same behavior at some different times may be generated by a somewhat different attractors arising from variable subsets of the available high-dimensional possibilities.

3.4.3 TURBULENCE, "ATTRACTING STATES, AND SELF-ORGANIZING CRITICALITY"
Given weak connections, which are common in the *Pleurobranchaea* nervous system,[139] it is not inconceivable that different limit-cycle and chaotic attractors may emerge simultaneously within the same network, moving and blending in space and time, and giving rise to the blending seen in whole-animal behavior[138] and in some motor patterns.[139] These conditions may provide the opportunity for analogs of *turbulence* to occur.[133] As discussed in the computer studies described in Section 8, we believe that large groups of neurons need not all act in a coordinated fashion, particularly when a large number of relatively weak synapses are distributed throughout the network. The statistical properties of the network and the effect of weak coupling may permit conditions under which different subsets of the extended network are able to begin acting cooperatively within themselves. Yet owing to extensive convergence and divergence of the underlying connectivity, one subset of neurons may influence the coordinated firing of other subsets. In this way, small foci of coordinated firing may move spatially, blend, or separate in to different foci, much as one might envision of vortices in hydrodynamic turbulence. Instructive examples of such phenomena in physical models have been presented in laboratory simulation[177] and computer simulations of the formation of the large red spot of Jupiter.[117] Videotapes showing the evolution of vortices in the hydrodynamic model

and in the computer simulations were seminal in solidifying our own intuition about what may happen in neural systems.[180] In considering the possibility of turbulence in neural systems, our own feeling is that the definition of "attractor" in such cases may not be as suitable as in more definable spatio-temporal structures. We prefer to use the term *"attracting states."*

Attracting states may have some resemblance to mechanisms of *self-organizing criticality* (SOC) proposed by Bak and coworkers.[10,11,12,15,31,187] The ideas have been applied to models of turbulence in forest fires[14] and the production of unpredictable avalanches that occur when attempting to build mounds of sand by piling one grain of sand over another.[13] Local effects are deterministic and easily observed, but the global effects are not predictable from such local information, and partly for these reasons, systems governed by SOC seem to be acting near the "boarder of chaos."[10] To our knowledge, SOC has not been applied to nervous systems. We envision that conditions that would allow SOC to take place would retain the deterministic character of monosynaptic actions between neurons, but given weak interactions, would also permit statistical or random spatio-temporal long-range effects through polysynaptic action.

3.5 CHAOS AND OTHER FORMS OF VARIATION

3.5.1 BIFURCATION PARAMETERS AND CHAOS

We shall examine bifurcation parameters in more detail in a Section 8. It is sufficient to state briefly that they are parameter constants that control how a system (or its defining set of equations) expresses its nonlinear characteristics. When the system is far from critical points, changes in bifurcation constants have relatively little effect on the dynamics of the system. At or near critical points, small changes in bifurcation parameters produce rapid changes (bifurcations) in the response of the system. Within certain ranges in the values of these parameters, the system may exhibit rapid shifts between different types of periodic activity and chaos as the parameter is successively changed.[6,183]

The simplest definition of chaos is that it is completely deterministic at each step of its temporal evolution, yet over the long term, its response is not predictable. An example we shall discuss later is the logistic equation, given by $X_{n+1} = R(1 - X_n)X_n$ where R is the bifurcation constant. This equation has no random factor in it, yet, for certain values of R, it is not possible to predict the evolution of the time series several iterations into the future given some initial starting value. Despite its long-term equivalence to random noise, the organized geometry in plots of X_n versus X_{n+1} clearly show the deterministic, non-random character of chaos.[123,136,183]

It is difficult to prove that biological systems generate chaotic attractors, owing primarily to their short-lived and apparently nonstationary behavior.[133] However, computer simulations have clearly shown that Hodgkin-Huxley membranes[29,30] and the parabolic burster neuron, R_{15}, in the abdominal ganglion of *Aplysia*[25] may be capable of bifurcating into a broad spectrum of simple periodic and chaotic activity. Our previous studies on the implications of attractors and variation, and of their implication in the generation of contexts of interrelated firing in groups of neurons, have been discussed in behavioral and neurophysiological studies.[138,139,146] And there is some evidence for chaos in the responses of individual BCNs and motor neurons in *Pleurobranchaea*.[137,144] Other activity of single neurons is more consistent with noisy limit cycles.[135,148]

The lessons to be gained from chaos are: (1) as illustrated by the logistic equation, variations arising from chaos are not "noise" superimposed on the information-carrying signal; they themselves represent the information. (2) The information in chaotic systems is always increasing with respect to information available at a given initial time. This is to say that if chaos is to represent behavior, it is necessary to use the long-term phase-space geometry of the attractor driving the system to gain a view of what the behavior is like. Given equal noise-free conditions, the behavior represented by periodic activity can be defined in a single orbit. (3) Periodic or limit-cycle activity dissipates perturbations differently than chaotic systems. As pointed out by Conrad,[37] limit cycles in biological motor systems dissipate perturbations in ways equivalent to heat loss through the body structures innervated by the neural system in question, whereas chaotic attractors dissipate the perturbations by generating new variations. Limit-cycle attractors always return to doing behaviors in the same stereotyped ways. Chaotic attractors generate new variations naturally in response to perturbations because their sensitivity to initial conditions always forces them to generate the behaviors in different ways, which is to say that behaviors are always different in chaotic systems. (4) Mpitsos and Burton[136] have shown that chaotic discrete processes, much as might occur in spike trains communicating between networks, allow simple networks to perform complicated tasks that would require considerably more complex networks to perform if the signals were generated by nonchaotic discrete processes or by continuous periodic or continuous chaotic processes. (5) It was also shown that the inherent variations of chaotic discrete processes permits networks that receive such signals to optimize their responses either in transmitting the signal one-for-one or in performing computations on them. That is, the deterministic character of chaotic discrete processes allows them to convey information, yet their long-term randomness provides sufficient variation to allow the responding network to learn rapidly. As we shall discuss below, random noise may be used advantageously to perform such optimizations. But random noise has the disadvantage of being high dimensional, and high-dimensional processes are difficult to generate because they must represent many degrees of freedom. Chaotic processes are long-term equivalent to random noise, yet the expression of chaos can be easily controlled using low-dimensional systems and simple adjustments to a single control parameter, as in the logistic equation. In multibehavioral systems such as *Pleurobranchaea*, the combined informational content and variation of chaos may be useful in accessing the different response possibilities.[136]

3.5.2 BIFURCATION-INDUCED VARIATIONS Another form of low-dimensional variation arises when systems approach bifurcation points. An intuitive understanding for this may be given by recalling the above discussion on the demonstration of attractors lying in three-dimensional space, and using this example to understand what happens to Lyapunov exponents as the system approaches bifurcation points. In a system governed by three variables, there are three exponents. (A useful discussion of Lyapunov exponents and numerical methods for estimating them are presented in Wolf[192]). A negative Lyapunov exponent indicates that there is contraction in a given direction in phase space. If all three exponents were negative, the flow of points in phase space would collapse in all directions into a single point. For continuous, bounded systems not at a fixed point, at which the system remains at equilibrium at some non-changing parameter state (see definition in Thompson and Stewart,[183] p. 194), Haken[73] has shown that one of the exponents must be zero. In a simple limit cycle governed by three variables, the remaining exponents must be negative. The negativity in the sum of the exponents assures that there is an overall contraction in the flow of points in phase space to keep the system bounded. The summed negativity also assures that the system will dissipate perturbations if they are not so large as to push the state beyond the attractor's basin of attraction. Bifurcations into chaos introduce a positive exponent, but retain the criteria of one zero-valued exponent and that the sum of the exponents be negative. The positive exponent shows that the state of the system in the corresponding dimension of phase space is always expanding. Having a zero-valued Lyapunov exponent indicates that the growth in phase space is neither contracting nor expanding over time. Thus, the rate of growth of a three-variable[5] system in phase space is given by $2^{(\lambda_1+\lambda_2+\lambda_3)t}$, where λ_1, λ_2, and λ_3 are the corresponding Lyapunov exponents for growth in each direction of phase space, and t is time. Since the exponential change is given as base 2, the exponents express the rate of change of growth in phase space as information in bits per second. Thus limit cycles lose information as they evolve with respect to some initial state, whereas chaotic systems gain information.

[5] The need for three variables in continuous systems that can generate chaos may be viewed intuitively by examining the flow of trajectories in phase space and their ability to mix as they course through the attractor surface; a typical trajectory will visit every vacinity. Evidence for mixing can be obtained by cutting a Poincaré section through the phase portrait and noting the interrelated positions of the crossings of the trajectory through the section.[183] If one places a string on a flat surface defined by two variables, it is possible to conform the shape of the string to flow to a fixed point, to form a variety of self-similar spirals,[168] or to connect the two ends of the string to form limit cycles (also see a discussion of the Jordan curve theorem and the theorem of Poincaré-Bendixon in Hofbauer and Sigmund[79]). However, it is not possible to have nearby lengths of the string diverge from one another and eventually mix in their interrelated positions without causing the string to cross on itself somewhere unless the trajectories flow into a third dimension and then fold back onto a thickened plane; i.e., however imperceptible, there must be a thickness to the surface of the attractor composed of countless layers arising from continuous stretching and folding which brings distant trajectories close together. Discrete processes, on the other hand, can generate chaos in a single dimension, as shown by the logistic equation.

As a system approaches bifurcation points, some of the Lyapunov exponents approach zero values, as we show herein for the catalytic network model of Andrade et al.[6,148] Setting the bifurcation parameter, μ, to a value of .02, generates a one-period limit cycle far from a bifurcation point, and λ_1, λ_2, and λ_2, have values, respectively, of 0, -2.8, and -43. Adjusting μ to .0125, well past the bifurcation into a two-period limit cycle, the exponents have values of 0, -3.6, and -43. However, setting μ to .0149, which is near the bifurcation point, the exponents are 0, -.05, and -46; λ_2 vanishes. Thus, as the system approaches bifurcation points, a greater number of Lyapunov exponents approach zero than when the system is farther away from these points. Perturbations in directions of phase space governed by exponents having small negative values would be dissipated slowly. Even in model systems having no extraneous injected noise, transient variations are often difficult to remove when attempting to locate bifurcation points.

Kelso, Schlultz, and Schöner[89] have given the term "critical fluctuations" to the variations observed in human finger movements during phase transitions, or, in our terminology, at critical bifurcation conditions. We have observed similar fluctuations in our own studies using sinusoidal current to drive individual neurons in *Pleurobranchaea* and *Aplysia*.[67] Moreover, since the *Pleurobranchaea* buccal-oral system appears to sit metastably near transitions into different patterns of activity (as shown, for example, by frequent spontaneous transitions of activity in isolated nervous systems; e.g., see Mpitsos[144]), we should expect to see variations in activity simply because of the tendency of the system to pass through bifurcation conditions. In model networks, it is possible to generate activity in the system long enough to get rid of transients. But biological systems, which generally do not have such long-term luxury, should exhibit considerable variation simply because of bifurcation effects, unless they lie far from critical points.

A rather interesting problem of bifurcation-induced variations occurs in regions of the controlling parameter that cause chaos. Such regions are filled with sub-regions that lead to periodic activity, as can easily be demonstrated by examining the bifurcation parameter of the logistic equation at expanded scales.[183] Therefore, small changes in a control parameter may actually lead to rapid shifts between chaos and periodicity, with each state being accompanied by transient variations. Clearly, there is a need to understand how biological systems cope with the sensitivity in the adjustment of bifurcation parameters and with the different forms of variations that arise from such adjustments. One possibility may be that the large number of converging and diverging connections among neurons may buffer unwanted bifurcation conditions by lifting the controlling effect from residing in single neuron or a few of them and distributing it over a large number of neurons. In this way, the bifurcation conditions emerge from group action, though individual neurons may exhibit near critical behavior. This may also be a reason for the observation of the wide distribution and convergence of neurotransmitters and modulators.

3.5.3 RANDOM NOISE Other variability in *Pleurobranchaea* seems to be high-dimensional, or even random, as shown by the response of a single neuron in Figure 1 of Mpitsos[135] and by the analysis of electromyograms in Mpitsos.[138] It has long been known that a little random noise may help systems to avoid local minima which may be defined for the present purposes as non-optimal responses (see Figure 8 in Burton and Mpitsos[23] for a diagrammatic demonstration of local minima). The physicochemical properties of DNA provide an example of one use of noise in biological studies.[49] Heating solutions of DNA (injecting noise into the system) breaks the two complementary strands apart. If the solution is cooled too rapidly, the original complementary bonds between base pairs is not completely restored; i.e., the system has fallen into a local minimum. If the solution is cooled slowly, the strands recombine optimally, forming the absolute minimum. Thus, the terms "local minima" and "absolute minimum" may be used to refer to number of characteristics, such as information storage, reconstruction of an original template, and energy level. Such processes of noise control are time dependent, and usually control noise by decreasing it exponentially. The method is referred to as *simulated annealing*. Kirkpatrick, Gelatt, and Becchi[93] discuss simulated annealing and apply it to several optimization problems, including the placement of computer chips on a circuit board, in which the goal is to minimize wire length and bends, and the traveling salesman problem, in which the goal is to minimize the distance traveled between cities if each city is visited only once. Simulated annealing is time dependent because it requires the noise in the system to have a decay rate, and once the noise has died out, it is necessary to introduce noise into the system again in order for it to be ready to respond to a new situation. Biological systems are generally event dependent, not time dependent. It may be difficult or impossible to determine in advance when the next challenge to survival will occur or what it will be, and when to re-inject noise into the system. Once a challenge has presented itself, there may not be enough time to adjust the rate of decay of noise.

As a step in determining how random noise might be used in adaptive systems, Burton and Mpitsos[23] devised time-independent noise algorithms (TINA) that control noise through the response of the system, as would occur in natural environments, rather than through predefined time schedules. To demonstrate the algorithm, Burton and Mpitsos used simple nonbiological neural networks that were required to learn to transmit or manipulate chaotic input signals, much as might occur if networks communicated with one another with chaotic spike trains. Networks were trained using an error-backpropagation algorithm.[166] Random noise was added to the learning-induced changes in synaptic weights and thresholds, but the level of the injected noise was adjusted on the basis of the amount of error generated each time the network responded to an input event. By such adjustments it was possible to avoid local minima and speed the process of reaching maximal levels of learning.

3.5.4 VARIATION-DEPENDENT OPTIMIZATION IN MULTIFUNCTIONAL SYSTEMS

Thus, random noise, chaos, and possibly variations arising from bifurcation conditions may provide conditions leading to two different methods of optimizations. The effect of chaotic discrete processes was shown under conditions in which chaos would act as a transmitter of information between networks, whereas the effect of noise was shown when it was added to changes in synaptic weights and thresholds during learning when the network had to respond to the chaotic signal. However, chaos is only short-term deterministic. The long-term statistics of chaotic discrete processes, as might occur in spike trains, are identical to random noise. For systems such as *Pleurobranchaea* or the mammalian olfactory bulb[172] that are multifunctional or contain multiple information within the same set of connections, variations that allow the system to search for one of many attractors or attracting states may be essential.

The three types of variation mentioned above involve different search strategies and control methods. Chaos has a deterministic search strategy and can be controlled through bifurcation parameters in membrane dynamics,[25,29] synaptic release (see the interesting suggestion in Kriebel et al.[101]) and, as we shall discuss in Section 8, in synaptic strengths. Neural systems may be able to approximate randomness simply by using weak synapses and by taking advantage of the large number of connections between cells. For example, connections between 10–100 neurons may provide sufficient degrees of freedom to approximate the high dimensionality of Gaussian noise. A number of activity-dependent changes in synaptic strengths or in the probability of transmitter release[99] might provide methods to control noise naturally and in time-independent ways. Some of the "noise" or variations that occur near bifurcation points are deterministic and self-controlled because they are transients that die out asymptotically as the activity evolves over time. Decreases in the value of Lyapunov exponents near bifurcation points would also allow random effects to become amplified, but as the system passes through bifurcation, both the transient effects and random variations diminish.

Variation, not chaos. The point, then, in thinking about adaptive mechanisms is to understand the use of a spectrum of variational types. Owing to its interesting phase-space geometry and its long-term unpredictability, chaos has received much press. The important issue, however, is not chaos, but variation and its control, and the way variation affects the ability of the system to access different dynamical states. The neural architectures that support the generation of these variabilities and ones that lead to control are unexplored. We provide suggestions in Section 8.

3.6 ERROR AS AN INTEGRATIVE PRINCIPLE

A system that has evolved to meet only one adaptive need can be highly tuned to perform that task well, but when confronted with new adaptive needs, such systems may prove extremely fragile. Alternatively, if the system is naturally variable the output may never be exactly "right" for a given task, but it may be right enough for the system to adapt successfully to different situations. Moreover, given a limited number of neurons, a greater range of outputs may be possible when the system has variable and blendable outputs than when the system contains a rigidly fixed number of output patterns.

Error may not only be a product of system dynamics, it may also be influential in the establishing the dynamics. The first indication of this was in studies of hypercycle catalytic networks originally devised to account for the first steps in chemical or prebiotic evolution.[53,106] Schnabl, Stadler, Frost, and Schuster[167] recently showed that error, expressed as mutual intermutation between reactive molecular species significantly affects the ability of a system to bifurcate into complex, chaotic oscillations. Andrade et al.[6] provide a more biologically plausible model of error utilization in catalytic networks that may be modifiable for application to studies of neural networks. In this model, error arises from faulty replication; i.e., in mutual intermutation the error is transformed into information contained in another reactant species, whereas in faulty replication, information is simply removed from the system. Although the generation of complex (chaotic) behavior in this latter model is less sensitive to changes in error than the mutual intermutation model, analysis of both models using the level of error as the bifurcation parameter shows that error plays a role in the dynamics occurring among the catalytic interaction.

3.7 DEFINITIONS: DYNAMICS, BEHAVIOR, AND MULTIFUNCTIONALITY

The above discussions provide the background for us to present several working definitions. In the most general terms, we take the term "dynamics" to imply the generation of cooperative activity among a group of interacting components of a system. There may be many different dynamical mechanisms: linear shifts in the aggregates of coactive components, bifurcations, limit-cycle and chaotic attractors, attracting states, turbulence, and self-organizing criticalities are just a few examples that we mentioned. As we shall attempt to illustrate further in Section 8, our definition of "neurocircuits" relies heavily on dynamics rather than network architecture.

In much of the preceding discussion, we have used the term "behavior" in the sense that the behaviors are distinctly different, as if feeding, regurgitation, righting, and other behaviors in the animal's repertoire, were definable. Indeed, the notion of a repertoire, seems to indicate that they are definable. However, our above discussion of "contexts" and "consensuses" shows that we do not believe that behaviors need be repeatedly the same. For example, the animal ingests food, it may regurgitate it, and it may right when inverted. Yet the animal may perform these behavioral effects in many different ways. If we are correct in our assessment of variations in neural activity and contexts, it is possible that the kinematics of the behavioral effect are always changing. Given this blurring of what the term "behavior" may mean, it is obvious that systems capable of generating many different behaviors using the same neurons must be defined in ways that include variation. Therefore, multifunctional networks to us implies patterns of activity and behavioral effects that can lead variably from one effect to another as well as the generation of distinctly different behaviors.

3.8 DEFINITION OF CONTEXTS IN GROUP ACTION: LINEAR AND NONLINEAR ORGANIZATION

To gain some perspective on our definition of contexts in group function, the above subsections provide some of the necessary background on what we mean by behavior and what we mean by nonlinear dynamics and attractors, different modes of cooperative action, and optimization and its relationship to different forms of variation as these factors play on attractors and on turbulence-like phenomena. The discussion has introduced the importance of local minima and error. The heart of all of these response phenomena lies in the anatomy of convergence and divergence. It is easy to refer to behavior, but once closely examined, we have realized that behavior may not be as definable as presumed, though we do not deny that definable behaviors do exist.

We began this section using references to studies that have considered how distributed interactions among neurons lead to behavior, and which have proposed that the appropriate behavior arises when a large number of neurons, or perhaps all or most of them, become active.[41,90,91,92,189] This is part of what we mean by "contexts" and "consensuses." Linear summations such as implied by "large number" do not address two important problems. First, if attractors or other nonlinear phenomena arise, it is not necessary for the majority, or a large number of neurons, to become active. That is, coherent activity may take place among a minority of neurons, but if the coherence is strong enough, we believe that its effect may override activity that is less strongly organized, though both coherent and noncoherent activity probably affect the actual expression of the resultant behavior. The question, then, is not how many neurons become active but how strong the coherent activity is above a "noise" level. Second, even if the interactions are linearly related, or if robust, stable attractors have not organized, adaptive responses may still take place, though the effect may not be as strong as in cases when the majority of neurons act together or when there are strong attractors.

4 BEHAVIORAL AND NEUROPHYSIOLOGICAL FINDINGS IN OTHER ANIMALS

4.1 INVERTEBRATES

4.1.1 OVERVIEW OF MULTIFUNCTIONALITY AND VARIABILITY Taking advantage of well-defined connections between four identifiable cells in the buccal ganglion of *Aplysia*, Gardner[60] has shown that synaptic effects between identified neurons vary widely from animal to animal. Drawing an analogy to connectionist neural networks, Gardner points out that the importance of a network is not so much in what its synaptic strengths are but rather in what the set of synapses together can do in expressing the information in an *algorithmic process*. The difference between biological networks and neural networks is that the temporal interrelationships in the firing of neurons may shift, and that the same network may be able to generate different patterns of activity.[138,139] Thus, in Gardner's terms, a set of connections may contain the information for many different algorithms. Our modification to this is that one must not consider the algorithm as being repeatedly the same; i.e., the algorithm is itself variably expressed.

Recent findings in the sea slug *Aplysia*[108,195] and in lobsters[26,75,76,77,108,109] are consistent with the notion that the same network can produce activity relating to different behaviors (i.e., they are multifunctional), as is the work on yet another sea slug *Tritonia*,[62,63] although only the work on *Aplysia* has taken notice of variation.[194] An important paper describes leech locomotion, and asks what it is that the "central pattern generator" really mediates since a variety of variable behaviors were observed.[8] Kien[90,91,92] has published a series of insightful papers on locust walking, and has addressed the notion of variation through observations indicating that different groups of neurons become active to produce a behavior. Variability has also been reported in walking motor patterns in cockroaches.[47]

By the late 1970s the notion that "hard wired" networks can explain behavior had received strong support form studies on genetically inherited ability to generate patterned activity in a many animals.[19] Nonetheless, ten years later, Getting[62] voiced the following interesting conclusion from his work in *Tritonia*, "*Networks with similar connections can produce dramatically different motor patterns, and, conversely, similar motor patterns can be produced by dramatically different networks*," just as one can read from the work in *Pleurobranchaea*[138] that, "*Organized activity emerges or self-organizes such that different contexts of the same coactive neurons become involved in generating the same or different motor pattern.*" Much evidence in neurobiology has shown that it is possible to ascribe particular function to identified neurons, and criteria of how to do that have been extensively discussed.[46,104,105,160] Some of the same researchers have also put forth the contrasting notion recently that conditions might exist under which it may not be possible to ascribe function to particular neurons.[102]

Thus, although the classical perspective still seems to hold, and much evidence exists to support it, there is a growing awareness of alternative possibilities. Our feeling is that it may be difficult to make direct comparisons between animals, even if there seem to be many similarities, as there are, for example, in the general neuroanatomical features of the nervous systems in snails and slugs indicating that their nervous systems contain neurons such as the BCNs in *Pleurobranchaea*. It may be, for example, that feeding systems in animals that evolved to utilize relatively stable and predictable food sources may be less variable than ones having to cope with unpredictable ones. One might envision such a comparison between certain herbivores and carnivores, though the defining experiments have not been done. What is most important in all of this is that people have begun to address the issues, and quite likely the most illuminating comparisons will be ones that involve different response dynamics. Our bias is that variation should be a common observation. In cases not exhibiting variation, the question then has to do with the mechanisms that control variation.

4.1.2 BIFURCATION AND RESPONSE MODALITY IN THE LOBSTER STOMATOGASTRIC SYSTEM The recent discovery of the ability of the stomatogastric ganglion in lobsters to generate different behaviors[75,76,77] shows clearly that one must not assume that even the simplest networks produce only single responses. The findings of Cardi et al.[26] are worth casting in our frame of reference relating to bifurcation. The stomatogastric ganglion in lobsters contains a subset of 14 neurons that comprise the pyloric network which acts as a central pattern generator. Of particular interest in this network is a further subset of three pacemaker neurons that form the oscillator. Another oscillator lying in the commissural ganglion sends projections to the stomatogastric ganglion. By using sucrose-block techniques on the nerve interconnecting the two ganglia, it was possible to reversibly interrupt the connections between the two oscillators. When the projections were blocked, systematic injection of depolarizing and hyperpolarizing current into one of the three pyloric pacemaker neurons resulted in continuous variation in the period of oscillatory bursts of activity in the pyloric rhythm. But when these projections were not interrupted, the period varied discontinuously, and, for some ranges of the injected current, two modes of oscillation emerged at a particular level of injected current. Overall the results show that the timing between the two oscillators affected the modes of integration in the pyloric network, and that the commissural projections also exerted neuromodulatory control over the pyloric network.

There are two ways to look at this data. The first is that there is some reflex circuit change that alters the oscillations in the pyloric network when the connection between the two pattern generators is intact. This seems reasonable if one considers that neuromodulation may be capable of adjusting which neurons participate in the oscillatory interactions or their interrelated timing (e.g., Marder[119,120,121]). Using John's[82] terminology, the network may use "switchboard" factors to control whether the network produces unimodal or bimodal firing in its burst patterns.

A broader perspective holds that the role of transmitters and modulator is to raise the network closer to a critical point for bifurcation. Small, systematic adjustments in the current injected into one of the three pattern-generating neurons push the system beyond the critical point allowing the network as a whole to oscillate in two modes, or to jump discontinuously from one period to another. When that transmitter (or transmitters) is not present, as when the connections between the oscillators are interrupted, the system settles into a state that is far from the bifurcation point. In this case, no amount of injected current will push the network close enough to the critical point to permit bifurcation to take place. What does happen is that the period varies continuously as a function in the strength of the injected current. This is precisely what happens when one varies the bifurcation parameter in a system that is far from a critical point (e.g., see Thompson[183] and Andrade[6]). There are two potential bifurcation parameters in the study of Cardi et al.[26] The way the experiments were conducted uses the polarization state (amount of injected current) of one of the pattern-generating neurons as the bifurcation parameter. However, if there were sufficient knowledge of the cells in the commissural ganglion that project to the pyloric ganglion, their level of firing could be used as the bifurcation parameter for each level of applied polarization in the pattern-generating neuron.

The advantage of using bifurcation analysis may not be appreciated in studies of most experimental biological systems because of their complexity and of the difficulties they pose in permitting selective control of a single parameter. The utility of the analysis becomes more obvious, however, in computer simulations. Not the least utility of bifurcation analysis is that it may provide some predictability. For example, Feigenbaum[59] observed that the succession of period-doubling bifurcations occurs in a universally predictable way. The ratio of differences in successive bifurcations is given by $\mathcal{F}_i = (\mu_i - \mu_{i+1})/(\mu_{i+1} - \mu_{i+2})$, where μ is the value of the bifurcation parameter in the sequence of bifurcations from $i = 1, \ldots, \infty$. For many bifurcation maps, $\mathcal{F}_i$ quickly converges to 4.6692 to the fourth decimal place. The pyloric network may be small enough to permit the use of computational methods. The major task will be to determine what parameter to control, though information from neurohumoral experiments may point to candidate factors. Different bifurcation states may use the underlying network architecture in different ways. The way the network expresses the various firing patterns among its constituent neurons is not predictable from knowledge of the bifurcation parameter itself nor of the anatomy of the neuronal connections. Predictability of these functional or emergent networks is even more difficult in large networks or if variability is a factor. If there are many weak synapses, there may be insufficient synaptic power to control how the activity traverses the connections among the neurons. Previous activity in the network may alter how the neurons participate in the future to produce similar overall patterns of activity. Both factors have been observed in *Pleurobranchaea*,[139] and may affect how the network responds during bifurcation.

4.2 MAMMALS

The importance of variation in brain function was, to our knowledge, noted first in mammalian studies. The work of Adey and coworkers (see summary in Adey[3]), done over twenty years ago, on the chimpanzee and human electroencephalogram (EEG), and on firing of cortical neurons in cats, clearly expressed the need to consider that noise may have a crucial role in the organization of brain function. Adey noted that while information must be contained in structure, the way the information is expressed quite likely is not obtainable from knowing the connections of structure itself. At about the same time, John[82] discussed the problem of considering cortical structure as statistical rather than as "switchboard" circuits that can be deciphered simply by examining the connections. The ideas expressed by Adey and John were seminal in solidifying reservations in our own laboratory about the viability of ascribing whole-animal behavioral phenomena to simple neurocircuits.[143] Wetzel and

Stuart[189] clearly favored a variable neuronal group hypothesis to account for vertebrate walking. More recently, Braitenberg[21] examined the connectivity of visual cortex and suggested that activity flowing through it may resemble a random walk. Rapp et al.[158] analyzed spontaneous firing in cortical neurons and suggested that the variations observed in cortical may not be random, but rather may arise from deterministic low-dimensional mechanisms such as chaos. Variation appears to be an important avenue for self-organization of cooperative activity occurring simultaneously over the entire surface of olfactory bulb. Freeman and Skarda[172] have proposed that the dynamical state of the bulb shifts from chaotic baseline variations into memory-specific limit cycles that are evoked when the animal inhales odors.

All of these findings are consistent with our own findings in *Pleurobranchaea*, and, in turn, our findings suggest that the different variational types may provide for response optimization into different attractors. Although the work in *Pleurobranchaea* represents the first demonstration that chaotic activity underlies adaptive responses in animals, it is necessary to take the evidence extremely cautiously, as has been pointed out.[133,137,144] However, to the extent that chaos does hold to be the case in *Pleurobranchaea*, and in the various observations described above in mammals, it may prove a general principle to pursue further that the variations may not only convey information for a behavior but also may provide for one of the methods for response optimization discussed in Section 3.

4.3 DIVISIONS OF THE MAMMALIAN MOTOR SYSTEM: RELATIONSHIP TO DIVERGENCE AND CONVERGENCE

Mammalian motor behavior may be classified as involving the pyramidal system (PS) or the extrapyramidal system (EPS). According to the classical view, execution of all voluntary movement in mammals is initiated by motor cortex acting through the PS, which constitutes a two-neuron chain. The upper motor neuron descends from the cortex and synapses in the spinal cord with the lower motor neuron, which innervates the muscle. Going backwards, each muscle fiber is innervated by a single lower motor neuron, which is contacted by only a few, perhaps a single upper motor neuron. So, each skeletal muscle of the body has a topical representation in a specific zone of the motor cortex. Stimulation of a specific region results in a stereotype response, which, if the stimulus is focal, includes one muscle fiber only. A given cortical neuron can act in two different states depending on the context defined by preceding impulses from the associative cortex.[181] This seems much like a switchboard, showing a precise structure-function correspondence. It can function as such, but the result is not the kind of movement we would like to perform. We get an idea of what kind of movements the PS can produce by itself by watching patients with dysfunction of the cerebellum or the basal ganglia, as in the case of Parkinson's disease. Their movements are coarse, as if the limb-moving is not quite sure of the goal. They have often heavy tremor, suggesting an imbalance of muscular tone at rest. Similar imbalance during movement is indicated by rigidity, suggesting that processing of the sensory information about continuously altered position is not occurring fast enough or precisely enough. We might say that the PS does not tolerate nearly as much error as the EPS. It is interesting to emphasize that in cases of cerebellar infarcts or in Parkinson's disease, the spinal cord with all its reflexes is supposed to be intact and functioning the best it can perform. Therefore, the PS may exhibit considerably less convergence of overlapping information and less distributed action. The one-to-one mapping allows the PS to execute precise control of movement but may make it extremely error prone should a particular line fail, whereas the EPS may exhibit less precise control yet may be less error prone when its components fail.

Although the physiological finding that given muscular responses can only be obtained by stimulation of certain cortical neurons indicates that there is little convergence, histochemical data suggest that multiple transmitter systems, presumably from the EPS and spinal cord, converge onto the lower motor neuron. The substances involved include dopamine, noradrenaline, serotonin, histamine, substance P, and thyrotropin releasing hormone (TRH).[80] The upper motor neuron shows some degree of divergence, since its collaterals contact with EPS neurons and spinal cord interneurons before synapsing with the lower motor neuron.

Classically, anything regulating motor functions other than the PS is defined collectively as the EPS. It includes the basal ganglia, the vestibular system, and the cerebellum, and it is thought to be responsible for coordination of movements. Its components connect indirectly with the PS both at cortical and spinal cord levels. The components of EPS are highly interconnected, although the precise circuitry is incompletely known, a high degree of convergence and divergence are likely to occur in the EPS, as suggested by the morphology of, e.g., the cerebellar Purkinje cells and basket cells. By contrast, the PS has significantly fewer connections among its constituent neurons.

This distinction between PS and EPS, however, may not be immutable, as indicated by motor learning. Consider a musician learning a new piece or a jongleur learning a new number. Initially, the motor pattern is established under cortical control. This always happens relatively slowly and, once it gets fast enough, the cortex cannot handle it and may even inhibit the pattern. Where is the pattern transferred to? It must be some subcortical level that takes over the pattern. All we know is that the control levels must be above the lower motor neuron, which is the final common pathway and that the pattern must be processed by the EPS. Control can be switched back and forth between the different levels, but the PS and EPS seem almost to have switched their functional categorization. To be sure, learning may model EPS to conform to convergence architectures that exhibit less convergence and variation, as discussed below in relation to Figure 4.

The diffuse reticular activating system (RAS) is perhaps most apropos to discussions of convergence and divergence, and adds a control factor that must be considered with all somatic motor functions. We know from everyday experience that rather sophisticated motor activity can take place at the lowest states of activation (sleepwalking) or rather gross errors may occur, if the state of activation is overly high. The structure classically thought to be related to the state of activation is the RAS of the brain stem. Interestingly, this is not really a structure in the sense of the nuclei or the cortex. Rather, its neurons are diffusely spread over a large proportion of the brain stem. Considering the anatomical fact that most of the vital regulation centers are located in that region over a very small space, RAS must be in contact with just about everything. It has been thought that RAS controls mainly autonomic vital functions. However, it has turned out that a reticular system is found all over the spinal cord as well. So it is reasonable to expect that RAS is intimately involved with motor functions too. (Our guess is that the RAS extends over all the cortex as well, if we only had markers to identify the cell types.) Thus, a better understanding of differences in the connectivity and function of the PS, EPS, and RAS, and their interactions, may shed some light on the functional significance of convergence and divergence.

5. NEUROMODULATION

5.1 CONVERGENCE AND DIVERGENCE OF NEUROTRANSMITTER SYSTEMS

5.1.1 INVERTEBRATES In the classic view, experimental manipulation of individual neuromodulators often generates predictable effects, as has long been demonstrated in other animals.[116,118,119,149] Our own work began with a similar intention: to identify behavior-specific neurotransmitter evidence relating to associative learning. There is good pharmacological evidence for the classically defined type of cholinergic muscarinic receptors (and of a new form) in *Pleurobranchaea*.[150] Behavioral evidence shows that muscarinic receptors have a role in associative learning.[147] Development of immunofluorescence methods for detecting the transmitter for these receptors, acetylcholine (ACH), has allowed us to identify the location of presynaptic cholinergic neurons.[174,175] Using complete serial histological sections to examine the full extent of the projections led us to the finding that we should have expected from our physiological work, but, interestingly, we did not. The histology showed that a relatively few cells diverge perfusely throughout the nervous system, hardly leaving any portion of the neuropil untouched.

This led us to examine the distribution of over a dozen putative neurotransmitters in complete serial sections of all ganglia in both *Aplysia* and *Pleurobranchaea*.[174,175,176] Examples of these findings are shown in Figure 2 (A-F) for *Aplysia* and in Figure 2 (G-I) for *Pleurobranchaea*. Each transmitter we examined involved a few neurons that diverged and converged extensively over the same target areas of the neuropil, and on individual neurons. The alternative possibility that neurotransmitters projected selectively onto different areas was seldom seen. Our present working hypothesis, which is being examined physiologically, is that there may be little motor specificity in the projection of neuromodulators, though there may be differences in their actions. Recent physiological findings in *Aplysia*[130] support this hypothesis since individual bath-applied transmitters and neuromodulators appear to affect all motor systems examined.

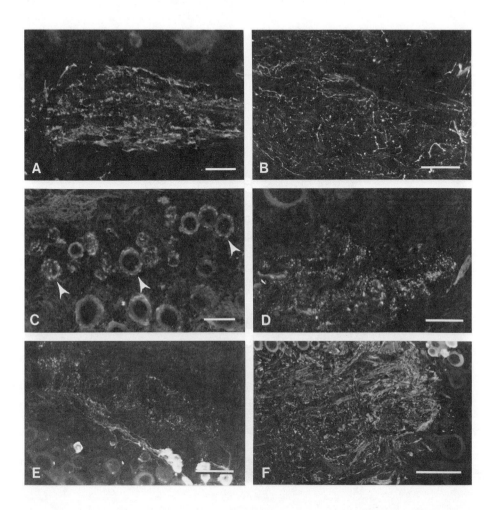

FIGURE 2 A-F: Photomicrographs of the neuropil region of *Aplysia* buccal ganglion showing immunoreactivity for (A) histamine, (B) serotonin, (C) ACH, (D) GABA (gamma-aminobutyric acid), (E) VIP (vasoactive intestinal peptide), (F) FMRFamide (Phe-Met-Arg-Phe-NH$_2$), cross in (C) indicates immunoreactive neuropil, and the arrowhead shows immunoreactive terminals around nonreactive neurons. Bar = 100 μm (A,D,E,F) or 50 μm (B,C). (G)-(I) (now labeled (A)-(C); will be changed): Photomicrographs of the neuropil region of *Pleurobranchaea* buccal ganglion showing immunoreactivity for (G) histamine, (H) GABA, (I) FMRFamide. Bar = 100μm. Note the extensiveness of the immunoreactive coverage throughout the neuropil in all tissues from both animals. Positive immunoreactivity is indicated by the white profiles that are extensively (cont'd.)

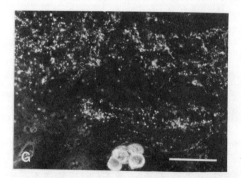
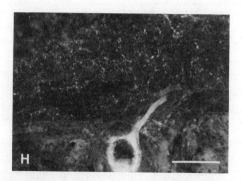
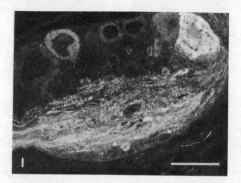

FIGURE 2 (cont'd.) distributed over the black nonreactive areas. For reference, in Figure 2(I), FMRF-amide covers the entire neuropil of the buccal ganglion. The large cell at the right is the buccal giant, and the commissure leading to the left half of the buccal ganglion is at the left margin. The anterior margin of the ganglion is delineated by the row of dimly stained cells at the top of the micrograph, and the posterior margin is shown at the bottom edge of the neuropil. The area between the neuropil and the row of dimly stained cells contains cell bodies which are not seen because they contain no immunoreactivity. Reprinted with permission from *Biol. Bull.* **181** (1991): 484–499.

Given the physiological finding of the extensive convergence and divergence in *Pleurobranchaea*,[139] and the corollary finding in *Aplysia* that sensory stimulation activates perhaps the majority of neurons in a ganglion,[195] the interesting possibility arises that conditions may often arise when many or possibly all neurotransmitters may become active at the same time. In this case, the classic view of neuromodulation that has been generated using selective applications of single transmitters may not provide adequate insight into the physiological effects produced under normal behavioral conditions. The classic view comes, we believe, dangerously close to making an unstated assumption that the effects of the individual transmitters on common target neurons sum linearly. But if conditions arise when the interactions are nonlinear, the classic experimental approach provides us with little insight into how neuromodulation acts to control network function in normally behaving animals.

5.1.2 VERTEBRATES As in the above discussion, we provide only selected examples here. Extensive innervation by nerve fibers staining for a large number of transmitters, such as ACH, dopamine, serotonin, histamine, GABA, taurine, glutamate, enkephalin, angiotensin, cholecystokinin, TRH, and vasoactive intestinal polypeptide, has been described in the mammalian striatum.[70] Likewise, multiple transmitters (ACH, serotonin, noradrenaline, glutamate, GABA) have been localized throughout the cerebellar cortex.[169] The wulst ("bulge") is a structure in the avian brain that resembles the mammalian neocortex. It is bipartite and runs the length of the dorsomedial portion of the hemisphere. A medial portion is similar to the mammalian hippocampus (wulst regio hippcampalis, Wrh), and a lateral portion is similar to regions of the somatosensory neocortex (wulst regio hyperstriatica, Whs). Both structures are laminated, permitting experiments that can determine whether neurotransmitters are differentially distributed between and within laminae. Shimizu and Karten[171] examined the immunohistochemical location of cell bodies and fibers containing serotonin, ACH (through localization of choline acetyltransferase, ChAT, and nicotinic ACH receptors, nAChR), catecholamine (through localization of the enzyme tyrosine hydroxylase), GABA (through localization of the enzyme glutamic acid decarboxylase, GAD, and the $GABA_A$ receptor), and the neuropeptides substance-P (SP), leucine-enkephalin (L-ENK), neuropeptide Y (NPY), neurotensin (NT), somatostatin releasing-inhibiting factor (SRIF), corticotropin releasing-factor (CRF), vasoactive intestinal polypeptide (VIP), and cholecystokinin (CCK). Although these substances exhibited laminar specificity, evidence was obtained showing that many regions of the Whs contained overlapping transmitters and neuromodulators. For example, in some portions of a large region, the hyperstriaticum accessorium, evidence was obtained for all substances except CCK, though the density of distribution for each substance was different.

An ideal structure to use for such purposes in vertebrate animals is the retina because of its well-known function and neuroarchitecture, and the ease with which its various cell types can be identified.[50,188] Present findings indicate that many neurotransmitters and neuromodulators are located in the various cells of the retina,[85] but the methods do not show clearly enough how much divergence and convergence among the cells in the retina or wulst, and how much occurs from the retinal ganglion cells onto other brain areas. A better method of analysis is to use evidence from the location and distribution of transmitter receptors. Progress in the laboratory of Professor Harvey J. Karten[86] at the Department of Neuroscience, University of California at San Diego, indicates that individual retinal cells contain receptors for many different neurohumoral factors, and that many cells stain for the same receptors, indicating that there is extensive convergence and divergence of neurotransmission and neuromodulation. Because of its experimental approachability and well-known function, the retina may provide a rich experimental source for understanding how multiple converging factors interact to control neuronal function.

In human physiology, Parkinson's disease is probably the best-known example of a transmitter-specific defect in human motor function. Its cause is considered to be a decrease in the activity of the dopaminergic nigrostriatal tract. Clinical neurology has established that when the amount of dopamine is too low, the action of the dopamine antagonist, the cholinergic system of the basal ganglia, becomes too strong. The treatment, l-dopa, increases dopamine levels to retain the balance between the two systems. However, there is nothing in here to prove that the action of the dopamine-ACH system is necessarily based on fixed circuits and that it acts individually in normal brain function. Although dopamine is found in a specific tract, we do not know how much divergence or convergence is involved in that system, and what the effects may be when many neurons and transmitters act together.

Although the pituitary is not a classically definable motor organ, it provides an excellent example of multi-humoral control. The intermediate lobe is a morphologically homogeneous group of cells that all contain the same hormones, melanocyte-stimulating hormone and beta-endorphin. The question is why are so many different transmitters needed for the simple regulation of inhibition-excitation. Stimulatory (serotonin, ACH) and inhibitory (dopamine, opioids, probably GABA) actions have been described for one substance at a time, but we have no idea how these substances act together. Since the output is so simple and easily measurable (hormone secretion), this tissue may provide a model to study the implications of divergence and convergence of multiple neurotransmitter inputs.

Figure 3 summarizes some of our findings in rat pituitary. The data clearly support the possibility of high convergence onto the same target areas, but since there is presently no morphometric evidence of how many neurons provide the innervation, we cannot presently provide an estimate of the ratios of convergence and divergence. The pituitary is particularly interesting since the output of the system in response to converging actions is neurohumoral rather than electrical.

In conclusion, we suggest that the properties of nonlinearity, distributed function, variability, multifunctionality, convergence/divergence, and the likelihood that the system is error-prone, all of which we have attributed to the electrical neurocircuit, may also be ascribable to neuromodulation. It may be possible to obtain repeatable effects when controlling certain transmitters, but what the effects may be or how to conceptualize the interaction of many transmitters (acting at very low concentrations) is presently unclear. If the dynamics of target processes (electrical or chemical) are far from bifurcation points, the nonlinearities (or any effect) may not be observable. But given that the bifurcation points are accessible, the number of possible effects arising from electrical nonlinearities and from the effects of transmitters, cotransmitters, and neurohormones become enormous. If we are

In Search of a Unified Theory of Biological Organization

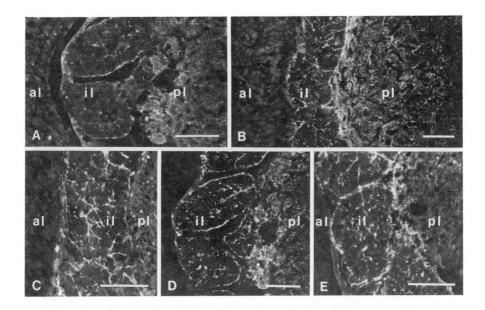

FIGURE 3 Photomicrograph of rat pituitary. al: Anterior lobe. il: intermediate lobe. pl: posterior lobe. (A) Acetylcholine. (B) MEAGL (Met^5-enkephalin-ARG^6-GLY^7-LEU^8. (C) Serotonin. (D) GABA. (E) Tyrosine hydroxylase, the dopamine-synthesizine enzyme. Note convergence of these substances onto similar areas of the intermediate and posterior lobes, as shown in Figure 2 for neural tissues of *Aplysia* and *Pleurobranchaea*.

to believe that neurohumoral agents act variably and in concert, then we must envision further that the subcellular mechanisms that each of these receptors and channels activates, may lead to converging and diverging nonlinear actions within the cell itself. Thus, it is conceivable that the clarity of the mechanisms presented for a single neurotransmitter or a single second-messenger system may be somewhat misleading. The point that needs to be examined further is that there may be many different sites of converging interactions in biological systems that process the same information in parallel, and perhaps in different ways, but may be capable of sharing the results of such processing. Thus, systems may exist in which it may not be possible to ascribe unique function to any motor, cellular, or subcellular process.

6. REDUCTION AND EMERGENCE IN CONTROL MECHANISMS

How are these widely distributed physiological and neurohumoral processes controlled? We suggest that many are not, at least not explicitly. It would be too costly, for the same reasons that it would be too costly to devise neurocircuits for each behavior. It seems better to allow the system to be error-prone. As discussed in studies on *Pleurobranchaea*,[133,138,139] some looseness may actually be beneficial since systems needing to be highly tuned to specific tasks may prove to be brittle in variable, unpredictable environments. Put differently, it seems better to allow the interaction between the organism and the environment to determine the behavior than to "hard wire" encode all of the behaviors that an animal can perform.

6.1 TRANSMITTERS CONTROL NETWORK FUNCTION AND ARCHITECTURE

There are, of course, demonstrable control mechanisms that we need to remember that show hard-wiring. For example, as we have mentioned previously, it has been shown that selective application of neurotransmitters evokes different patterns of activity in simple ganglia,[48,111,119] just as there is a vast textbook literature showing evidence of the classical "neurocircuit."[83] Most published evidence weighs heavily in this direction. Thus, good evidence exists to show that *"Each neurotransmitter or neurotransmitter system may...be able to elicit, from the same neuronal circuit, a characteristic and different 'operational state.' In this way it would be possible to obtain a wide range of stable neuronal outputs from a single circuit."*[121]

A remarkable series of experiments by Kater and coworkers (e.g., Kater and Mills[87] and Lipton and Kater[111]), begun initially in the fresh water snail *Helisoma* and now extended to mammalian neural tissues, shows the ability of transmitter receptors to control neuronal growth, plasticity, and even survival of neurons. The work has examined a spectrum of neurotransmitters and neuromodulators, including ACH, GABA, dopamine, glutamate, norepinepherin, serotonin, somatostatin, and VIP. Taking advantage of cell culture of identified neurons, the work has been able to provide a strong basis of control experiments. As one example in *Helisoma*, serotonin, an excitatory transmitter in this system, retards neurite outgrowth whereas the addition of ACH, an inhibitory transmitter, prevents the serotonin-induced inhibition. The transmitters work through the depolarization state of the cell. For example, presenting an excitatory transmitter alone retards the normal neurite outgrowth, but superimposing hyperpolarizing current on transmitter-induced excitation allows the neurite to resume its normal growth rate. The transmitters may act either through voltage- or receptor-activated channels on a common intracellular messenger, calcium. As Lipton and Kater[111] summarize, neuronal architectures (and therefore neurocircuits) are determined by a fine *balance* in the activation of these two types of channels through an interplay of excitatory and inhibitory transmitters (though different mechanisms may be used in other neural systems; see Garyantes[61]).

The term "balance" clearly indicates that Lipton and Kater are aware that control in natural biological systems may be high-dimensional since neural tissues are known to contain many transmitters. The problem, then, is to determine how the high dimensionality is expressed. One possibility is that there is simple linear summation of the effects produced by the various transmitter. However, it is well known that the electrogenic properties of the postsynaptic cell can easily change a simple synaptic input into a nonlinear response. Twenty years ago, Wilson and Cowan[191] conducted computer simulations on a population model to illustrate that groups of cells intercommunicating through excitatory and inhibitory connections exhibit damped oscillations, multiple stable states, and, under certain constraints, stable limit-cycle oscillations in the number of excitatory and inhibitory neurons firing per unit time. A rather interesting feature of the model is that local interactions were essentially random, yet the long-range effects were quite organized. Another interesting feature of the model that is pertinent to the present discussion is that the population of excitatory and inhibitory cells were homogeneous; differences arose statistically through use and refractory period. In even simpler networks involving one-shot activation between converging inputs to a common neuron can lead to linear and nonlinear effects in the postsynaptic cell.[5,98] In single neurons, it may be possible to generate many different periodic and aperiodic firing patterns by means of fine adjustments to a single ion channel.[29] This latter study also showed that intracellular calcium concentration may fluctuate differentially and nonlinearly in each dynamical state. Therefore, the controlling balance between converging transmitters and neuromodulators that affect neuronal structure need not be a simple linear affair. What may seem a linear balance, under some parameter ranges of the neurohumoral state, can easily switch to drastically different conditions at critical bifurcation conditions.

The dynamics of interactions arising in population of cells need not employ the full high-dimensional space. Going back to our notion of attractors, the different dynamics that a network will allow determine the characteristics of temporal visitation of activity at any given neuron in the coactive group; i.e., a set of connections will be activated differently by the types of attractors that it can sustain. Although a developing network at some primitive state may exhibit different dynamical capabilities than a finely tuned, mature one, the same questions of nonlinear conditions arise in both. Finally, if attractors arise either in the responses of single neurons or in networks of them, the high-dimensionality we see in the number of transmitters present may not necessarily be expressed as a high-dimensional process. It is an interesting possibility, raised by numerical studies, that coordinated activity in potentially high-dimensional systems often results in low-dimensional attractors.[144,172] From a simple listing of the number of transmitter resulting from experiments in which transmitters are applied one at time or in pairs, it is not evident how the system dynamically collapses into low-dimensional control, and which of the transmitters become involved. Even in small model networks in which all of the driving differential equations are known, it is not obvious from the equations themselves, nor presently from the connectivity, how it is that a lower dimensionality arises from a larger possible set of available variables unless the system is examined after activating it.[6]

Given a linear system, it may be possible to say that neurotransmitters are architects of neural structure. But, as we shall discuss later in Section 8 when dealing with bifurcation in minimal networks, conditions may arise when the activity itself is what fine tunes a network, and in turn, the network redefines the type of activity that can emerge. There is a dialectical interplay between the two elements, and this dialect, we believe, can act as an architect of neurons and circuits. The chain of events that we might envision of the events that control cell structure is as follows: The dynamics of firing in individual neurons and in networks of them acts on structure through transmitters; the transmitters act on the cell through calcium. The dynamics of changes in intracellular calcium sets up a chain of events that affect cell growth. But cell growth redetermines what the dynamics will be, and so forth recursively. Other factors may contribute, such as synaptic competition. If the notion that many neurons act in close temporal association, or in coordination, is correct, we must then add the complication that the system as a whole is extremely high dimensional and that many types of nonlinearities may occur. As we shall speak below of the locus of learning, there may be no *sine qua non* balance of neurohumoral agents for a given architecture to appear. Although there may be many systems in which there is always a precise connection between a balance between a particular set of chemical elements and structure, understanding these systems gives little insight into others in which variability is an issue.

Thus, while the scientific method at our disposal provides elegant connections between cause and effect, much as Descartes and Euclid would like us to believe, the possibility of high-dimensional space, of nonlinearities, and of the dialectic between structure and dynamics indicate that our view of complex systems may be too simple. However, the scientific methods, as they are, are nonetheless the only ones we have. Therefore, our concern is not that the methods and conclusions are simplistic but rather it is that they do not address fundamental questions that need to be asked. Moreover, the clarity of some of these reductionistic methods and the importance of the resulting findings have overshadowed the need to go beyond them and to develop methods of data collection that may be useful in taking that step.

6.2 CONTROL OF WHOLE-ANIMAL BEHAVIOR: CRITIQUE OF REDUCTIONIST EXPLANATION OF LEARNING IN *APLYSIA*

6.2.1 SYNAPSE-SPECIFIC CONTROL OF BEHAVIOR A tradition in invertebrate neurobiology holds that an advantage of using invertebrate animals is that once a behavior is identified with a particular motor pattern, the same behavior can then be studied neurophysiologically in the motor patterns of isolated nervous systems. As discussed briefly in subsection 3.3, this is quite difficult to do in *Pleurobranchaea*.[138] However, the most elegant example of such reductionist approaches has been the identification of site-specific learning in the gill-withdrawal response in *Aplysia*.[27,58,84,146] A long series of studies have attempted to show how changes at monosynaptic sites between sensory neurons and motor neurons can explain whole-animal phenomena such as sensitization, dishabituation, and associative learning. The mechanism involves serotonin as a neurotransmitter in the reinforcing pathway. The original series of experiments showed that activation of serotonin receptors on sensory neurons leads to a chain of events involving adenosin 3′,5′-monophosphate (cyclic AMP) that depress a potassium current when the cell fires. This exposes an inward calcium current that broadens the action potential, and, owing to the increase in intracellular calcium, leads to increased transmitter release onto the follower motor neuron. A group of sensory cells, referred to as the LE-neurons, which are usually activated electrically in isolated ganglia, provides the input to identified motor neurons of which neuron L_7 is perhaps the most important in terms of its effect on the movement of the gill. A group of cells, referred to as L_{29}, provides the serotonergic input.

6.2.2 COMPLICATIONS A number of important extensions and problems have arisen that both greatly illuminate and complicate this simple model system. We cite only a few examples:

1. *Peripheral nervous system.* From the beginning of work in the late 1960s, evidence has existed indicating that emergent effects may involve the peripheral nervous system which is distributed within the gill itself. Indeed, in many cases the abdominal ganglion seems not to be necessary for generating robust gill withdrawal responses and simple forms of learning.[146]
2. *Complex behavior.* The once-presumed simple withdrawal reflex has turned out not to be so simple, and consists of several different types of movements.[108]
3. *Neuronal function.* Some of the major identifiable motor neurons have variable function within the same experimental preparation within the same behavior.[109] This raises strong questions in *Aplysia* as to the veracity of assuming that identified neurons have consistently the same role in a given behavior, much as Mpitsos and Cohan[139] have raised regarding the function of neurons in *Pleurobranchaea*.
4. *Complex network.* Small, well-localized sensory taps activate perhaps half of the cells in the abdominal ganglion, showing that there is extensive divergence of sensory and possibly other effects.[195]
5. *Non constant activity.* Cells partaking in successive taps are variable,[194] suggesting that localization of the network may be difficult or impossible.
6. *Source of serotonergic control is unidentified.* Activating L_{29} produces enhanced transmitter release. Serotonin applied experimentally produces same effect. But L_{29}, which was thought to provide the serotonergic enhancement, apparently does not contain serotonin.[94,153]
7. *Multiple neurohumoral factors enhance synaptic release.* We now know that at least two other transmitters, small cardioactive peptide A and B (SCP_A, SCP_B), broaden action potentials in LE cells and produce synaptic facilitation on their follower motor neurons,[2] but apparently they are not located in L_{29}.[103]

Interestingly, SCP_B produces spike broadening but not facilitation of transmitter release in depressed sensory neurons,[156] which may relate to mobilization of transmitter.

8. *Multiple subcellular processes.* There may be diverging cyclic AMP-dependent processes in different forms of synaptic facilitation.[68] Conversely, in both the gill-withdrawal system and the analogous tail-withdrawal system, cyclic AMP-dependent and cyclic AMP-independent subcellular processes may converge onto the same spike-broadening mechanisms in both the gill-[95] and tail-sensory neurons.[179]

9. *More than one group of sensory inputs.* The possibility has been raised that under some conditions, novel sensory neurons may be involved in modification of a siphon withdrawal response whose behavioral modification has been thought to be controlled by changes in the LE sensory neurons.[193]

10. *LE cell activity lacks timing to be primary site of learning.* Most importantly, it now appears that there is a second group of sensory cells that have lower thresholds than the LE cells,[36] and are probably more likely to activate than the LE cells during training of the gill withdrawal response itself. It has now been reported[36] that the latency of responses in mechanoactivated LE cells in all of the 32 preparations that were tested always occurred *after* the initiation of the discharge in the motor neurons. Their timing in the behavioral reflex has been difficult to determine.[24] The problem, then, is if the *cellular basis of behavior* relies on the LE cells as the site of facilitated transmitter release, the responses of the LE cells must occur *before* the initiation of motor output for that behavior, but the recent findings show clearly that they do not.

6.3 EMERGENT CONTROL OF *APLYSIA* BEHAVIOR: PARALLEL DISTRIBUTED PROCESSING

6.3.1 DON'T WORRY, BE HAPPY: NEW SYNTHESIS It might be tempting to some interpreters of the above-mentioned complications in *Aplysia* to disparage the original conclusions about site-specific learning. We believe, however, that that would be a mistake. To dismiss the original conclusions would be to fall to the temptation that has faced previous work on learning in *Aplysia*, and of most such attempts in other animals, that there is, in fact, some other reducible locus of learning, or some reducibly identifiable neurocircuit as the generator of behavior. But by making the dismissal, one would miss the more important issue that emerges from the findings, namely, that the data may be influential in redirecting the focus from reductionism to a higher level of analysis. It is not just that behavior may be different on different occasions. A general scheme appears to have emerged in all of the work on *Aplysia* that is not inconsistent with the findings we have obtained in our attempts to understand the integrative processes that generate behaviors in *Pleurobranchaea*. This scheme relates to our discussion above of parallel processing arising from the extensive distribution and sharing of information, as we summarize below in subsections 5.3.2 and 5.3.3.

6.3.2 THE LOCUS OF LEARNING MAY NOT BE AT A UNIQUE CELLULAR SITE The evidence cited in the above list of complications may be reinterpreted as in the following general scheme: Different sites in the nervous system are capable of generating similar components of the same behavior, and each site is capable of affecting the other; i.e., there is apparently extensive convergence and divergence between different sensory and motor centers. Within a given sensory-motor system, divergence is an inherent effect of even small, highly localized stimulation. At the same time, different sensory pathways converge on the same motor neurons. Similar convergence occurs among neurohumoral systems and their subcellular effects.[174] Thus, mounting evidence indicates a cascade of diverging and converging chemical interactions that distribute sensory and motor effects widely.

Evidence exists that supports these possibilities. For example, we know that weak, highly localized tactile stimulations, as used in training experiments to show learning, activates large numbers of neurons,[195] i.e., that divergence distributes information over many cellular loci. We also know that learning occurs in both the peripheral and central components of the nervous system of *Aplysia* (see review in Mpitsos and Lukowiak[146]). We also know from studies in isolated nervous systems and from more intact preparations that conditioning-related changes occur on LE sensory neurons that synapse on different gill motor neurons. Training-induced changes may occur at the neuromuscular junction.[81] Additionally, changes may occur during training that follow all of the criteria established for associative learning but which do not take place between the sensory neurons and their follower neurons. For example, Lukowiak and Colebrook[115] have obtained evidence of associative conditioning that excludes the major gill motor neurons. The conditioned stimulus (CS) consisted of weak tactile stimulation of the siphon skin. The unconditioned stimulus (UCS), in one set of experiments, consisted of strong electrical stimulation of the pedal nerve which connects the brain with the foot, and in another set of experiments, it consisted of strong tactile stimuli to the gill itself. During training, dual intracellular recordings were made from sensory neurons and major identifiable gill motor neurons (L_7, LDG_1, LDG_2, L_9). The movement of the gill itself was also monitored. In the course of training, the CS produced gill-withdrawal movements that increased as a function of the number of training trials, and the efficiency of the sensory-to-motor neuron synapses increased. Appropriate control experiments showed that the effects were consistent with associative conditioning. However, the number of action potentials produced in the motor neuron in response to the CS correlated well with the actual movement of the gill only during the initial stages of training. But most of the amplitude changes in the gill-withdrawal response was not correlated with any changes in the number of action potentials generated in the motor neurons. In another set of experiments, designed to mimic associative learning observed whole-animal studies, evidence was obtained for associative learning in a significant number of reduced preparations in which there was an increase in the number of action potentials produced in the motor neurons, but there was no change in the amplitude of the gill-withdrawal response.

Findings such as these show that associative learning, and simpler forms of learning such as sensitization and habituation, may take place at many different loci. Thus, as regards complication 10 noted above, it is not too big a jump to realize that learning could also happen in classes of sensory neurons other than the LE cells, and eventually to discover that learning-related physiological changes may also be shown postsynaptically in the motor neurons themselves, not just presynaptically in the sensory neurons. Additionally, as Mpitsos et al.[143] have pointed out in detailed control studies of associative learning in *Pleurobranchaea*, let us not be wedded dogmatically to a definition of associative learning that forces physiology to comply with a particular protocol of stimulus presentations applied by the experimenter to whole animals. Single-trial training in this study showed, for short intervals between CS and the UCS, that backward conditioning produced almost as strong conditioning as forward conditioning. Mpitsos et al. pointed out that what may be temporally controllable experimentally in the application of sensory inputs may not hold physiologically. The same set of subcellular mechanisms producing learning-related changes in forward between the CS and UCS (which is required by the definition of associative learning) may exist to some extent when the stimuli are presented in close temporal pairing but in reverse order. To us, changes arising from both the forward and backward temporal relationships between the CS and UCS can represent associative learning (though this does not exclude arguments for different mechanisms, should they occur, to account for backward conditioning). For these reasons, it also may not be too big a jump to accept the fact that learning may still take place in the LE neurons of *Aplysia*, even if their responses arising from stimulation of sensory skin do not occur until after the motor neurons are activated by other sensory neurons.

Thus, while it is possible that a unique "locus of learning," the engram in *Aplysia*, might still be found, the data indicate strongly that *the system seems to consist of many parallel, redundant, and possibly interacting components, none of which may be the sine qua non element in the learning process or in the generation of the motor responses, irrespective of whether or not they involves learning.*

6.3.3 THE NEUROCIRCUIT MAY NOT BE DEFINABLE Another tradition of reductionism in neurobiology, particularly in studies of invertebrate studies, has been the notion that cells and their function are repeatedly identifiable. We have already mentioned some of the problems in identifying function in *Aplysia*.[60,109] The recent computer simulations of simple neural networks relating to the feeding system of *Aplysia* have led to a similar conclusion that, "...*tests done on individual neurons can provide misleading information on the actual role of the neuron in generating behavior.*"[102] Compare this quote with one from Mpitsos and Cohan,[139] p. 538: "...*these findings indicate that the classic technique of driving a particular neuron in order to assess its effect in evoking activity or a behavior may be an insufficient criterion for identifying its functional role.*" That is, a given neuron's function depends on the context of activity in which it takes part. But, given variability in the activity in the firing patterns within such contexts or "mobile consensuses," even this might be an insufficient definition.[32,133,139]

The neurocircuit for a behavior is misrepresented by even the most complete mappings of identified neurons that we see in publications. Studies using voltage-sensitive dyes show that weak, localized stimulation of sensory skin of the siphon produces massive and variable activation of neurons in the abdominal ganglion of *Aplysia*.[194,195] As we have discussed of the simplified networks shown in Figure 1 for *Pleurobranchaea*, the connectivity of the actual circuit of interacting neurons is quite large. The larger the overall pool, and the greater the number of weak synapses that exist, the greater will be the possibility that the actual network generating a behavior will be variable and undefinable.

6.3.4 DIFFERENT LEVELS OF LEARNING WITHIN DEFINABLE SETS OF SYNAPSES.
Let us assume for the moment that a small group of neurons can be isolated functionally from the effects of other groups of cells. Can we then obtain sufficient information about the network to define it completely by looking at the network and knowing all of the connection parameters? We think not. Consider just one example relating only to the strength of synapses. In our own neural network simulations, the data indicate that synapses contain different forms of information.[23,134] One form of information ("knowledge") is task-specific relating to the computations of one or more functions that network must perform. Another form ("metaknowledge") has to do with the process by which that task was learned; it does not affect the network performance on the specific tasks, but only becomes evident when the network is confronted with new tasks. These conclusions were drawn from experiments that compared learning performance in networks that used random noise to optimize changes in synaptic weights against networks that were not exposed to noise. Both types of networks were allowed to reach the same level of learning on a given task, but the noise-exposed networks learned a subsequent task faster, even when noise was not included during training of the second task, than networks that did not use noise. Starting networks at different initial synaptic strengths at the beginning of a training session yields different final synaptic settings, but all final networks perform the same learned task equally well. Because of this, Burton and Mpitsos initialized networks using different synaptic strengths and thresholds. Examination of a large number of networks at the end of the first training session revealed that the two types of training methods did not generate statistically significant differences in the means and standard deviations of the synaptic weight settings. Both types of networks contained the same information for generating equally accurate computations relating to the first task, but networks that were exposed to noise contained further information that permitted them to perform well on a second task. Each task has a particular error landscape associated with it (see Figure 8 in Burton and Mpitsos[23] and Figure 13 in Mpitsos[136] for examples of error landscapes and volumes). Burton and Mpitsos suggest that noise-exposed networks sample these error-structures more completely than networks that were not exposed to noise. Thus, when confronted with new tasks having any similarity in their error structures as the first task, the synaptic settings of networks exposed to noise already contain information about the new task and are able to navigate its error fields rapidly. By contrast, since networks that are not exposed to noise

contain less of such information, they are not able to navigate as rapidly through the new error structure.

The implication of these findings for the present discussions is that one may look for changes relating to a given task, but depending on the conditions under which that task has been learned, the aggregate of synapses within a pool of neurons may contain different types of information, where one type pertains specifically to one or more tasks that have been learned, and the second type pertains to more general conditions that do not affect the accuracy of the first, but nonetheless may camouflage the results that the experimenter is seeking to identify. The rabbit olfactory bulb[172] may be a useful example to contrast our findings. In this structure, odor-specific information is stored spatio-temporally, but apparently all neurons take part in expressing the code for each odor. Our simulation networks can also be constructed to encode information relating to multiple tasks,[136] but the noise-induced changes in the network represent an informational abstraction that goes beyond the information need specifically to perform well on previously learned tasks. Therefore, if our computer simulations of connectionist neural networks have analogs in biological systems, the understanding of synaptic modification and the information that the synapses contain cannot be deciphered simply by examining the synapses themselves as they relate to only one task. In their studies of Mauthner neurons, Faber, Korn, and Lin[57] raise the related caveat, but for different reasons, that "...although it is possible to derive generalized rules of the operation of synapses, their variants may exert a major role in shaping the behavior of complex circuits."

Analogous problems as those described above and in the preceding two subsections may have beset Lashley[107] whose unsuccessful attempts to identify the locus of stored memories (engrams) in the cortex have been more inspiring and illuminating, at least to us, than were he to have found them. It is interesting that much of neuroscience has followed the same course as Lashley. But now the search has been on the cellular level in attempting to identify behavioral phenomena in terms of single synapses and single neurons. It is also interesting that Pavlov, before Lashley, was apparently discontent with the possibility that learning could be localized to particular areas of the cortex since learning persisted in his animals even after they had suffered brain damage (see Boakes,[20] pp. 127–128).

6.4 "FUZZY" CONTROL

Thus, the "control" we seek to define for the physiological and neurohumoral aspects of the nervous system is oblique and emergent rather than being crisply Euclidean in postulating particular causes and effects as would be expected of reflexes. One feature of such emergence is that there may be many ways to do the same thing, and even gradations between these ways. We know, for example, that under some conditions, removal of a neuron from acting in a motor pattern can be compensated by shifts in the activity of other neurons.[139] Redundancy, arising from information sharing among convergent pathways, compensates for error or failure in some of its components, even if these components originally generated strong control over the other members of the coactive group. Are neurohumoral systems equally redundant, or does each of the ever-growing number neurotransmitters being identified daily have a unique task? Our own work leans heavily toward the first of these possibilities.[174] In the same sense that there may be "lazy" synapses in neural networks,[136] whose presence is required only under some conditions, are there "lazy" or even unnecessary transmitters? Some of what we see in a given system may represent baggage of evolutionary or developmental processes. This, however, provides for yet another form of variation that permits possible adventitious incorporation into further evolution or behavior.

6.5 IS OUR VIEW HOLISTIC?

No. Being concerned with mechanisms that generate global behavior is not necessarily being holistic. In our approach, global behavior depends on local rules followed by individuals acting within a large group. It is these rules that we seek to identify, though there may be different rules that relate to global behavior directly. Even in simple processes such as building of sand-grain mounds[13] and affine transformations,[16] the global consequences of local behavior are not predictable. Nevertheless, emergent function need not be a property of large groups of neurons.

It is interesting, however, that one of the best examples of work in artificial intelligence in many decades employed a top-down analysis in which a principle obtained from studies on the behavior of whole animals was used to gain insight into how that behavior might have emerged from individual neuronal units. The work we refer to is Klopf's[97] drive-reinforcement model of associative learning, which extends Hebb's[74] rule to account for Pavlovian conditioning. Hebb's rule states that, *"When an axon of cell A is near enough to excite cell B and repeatedly and persistently takes part in firing it, some growth or metabolic change takes place in one or both cells such that A's efficiency as one of the cells firing B is increased."* Before Klopf's model, computer simulations of Hebb's rule in simple networks were not successful in demonstrating learning that mimicked findings in biological systems.

Hebb's rule may be interpreted as a three-cell network,[143] one input cell for the CS and one input cell for the UCS, both of which synapse on a common follower cell (cell B). Klopf[97] made the following crucial modifications to the rule to make it work in such a simple system: (1) Temporal delay was added between the onset of the CS and UCS. (2) Synaptic modification was made proportional to the rate of change in the CS and UCS. (3) The follower cell (B) itself expressed a form of behavior analogous to tendencies that may be observed in whole animals: Whole animals seek to optimize some quality of their environment, such as avoiding pain and enhancing pleasure. Klopf[96] made the simple, but crucial analogous assumption that cells tend to optimize excitation and reduce inhibition. Additionally, to account for excitation and inhibition, the follower cell received excitatory and inhibitory terminals in its CS input pathway.

The methodology for training the network is the same as for training the whole animal. In each training trial, a pulse is presented to the CS input, which initially produces little effect, and after a short delay, a pulse is presented to the UCS input. The only parameter that is arbitrarily set in the model is the constant for the rate of learning. Amazingly, training-induced changes in the synaptic effect of the CS input on the follower cell reproduced all of the known Pavlovian conditioning phenomena in experimental animals and in humans (e.g., backward conditioning, CS alone, UCS alone, trace conditioning, second-order conditioning, foreshadowing, blocking, conditioned inhibition, etc.).

The model has now been extended to account for instrumental conditioning.[131] The work also made progress in resolving the long-standing debate relating to the theoretical relationship between Pavlovian and instrumental conditioning since the instrumental conditioning effects in the model emerge from Pavlovian conditioning. Thus, computational methods may have resolved what psychological debate and experimentation in biological systems have not been able to do. The studies discussed in Section 8 pursue the same rationale of using simple rules to lead to understanding of global effects.

7. DOES A THEORY EXIST?

At least three important principles have emerged from dynamical systems studies that are important to biologists: (1) The notion that distributed networks can generate attractors. (2) A considerable amount of information about a system can be gained from bifurcation analysis. And (3) an understanding of the dynamics of a system can be obtained from the phase-space geometry of such attractors. By these methods, it is possible to discover much about a system without having to resort to the difficult if not impossible task of uncovering the sets of equations that actually run the system.

A long history of work has developed these ideas, from Poincaré to Lorenz, Crutchfield, Farmer, Packard, Rössler, Ruelle, Takens, Swinney, Shaw, Yorke, and others of the many recent contributors to the knowledge of nonlinear dynamics.[1,183] There are many theorems in the field of nonlinear dynamics, and there are many discussions of how to handle the nonlinearities,[71,72,170] beautiful demonstrations of attractor topologies, bifurcations, and stability analyses, when these are in fact available. As important as these are, they do not constitute a unified theory, at least not as it might apply to brain function, though Bak and coworkers suggest that their mathematics or models of self-organizing criticalities,[10,11,12,13,15,31,187] which apparently account well for many physical and biological phenomena, may provide an encompassing dynamical theory.

One way to get around the theoretical problems, as is often suggested by physiologists and non-physiologists alike, is to perform computer simulations on systems whose state space is completely defined and parameterized, that is, to determine all of the connections between neurons, membrane properties, neurotransmitters, firing thresholds, and the like. However, one look at the complexity of the connections and at the wide divergence and convergence occurring in even "simple" systems should provide convincing evidence that this approach is hopeless.[35,139,174,195] Moreover, as discussed above, the reductionist neurocircuits that have been developed over the years to account for behaviors are but a caricature of the actual "network" that generate the behaviors in intact animals.

The possibility might also be suggested that insight into the integrative principles might be obtained from the mathematics describing the biological systems. This also seems an unlikely possibility at present, even in relatively small systems. Even in well-defined experimental systems, the first evidence of dynamical states and their bifurcations came from direct observations. One such example is the Belousov-Zhabotinsky reaction which consists of about 30 chemical constituents in which malonic acid is oxidized in an acidic bromate solution.[163,164] While it may be possible to define the various reactant species and list the reactions, it has not been possible, to our knowledge, to predict the dynamics of the system using the mathematics of the reactions. Another example is the demonstration of different dynamical states in yeast glycolysis.[122] As yet another example, near the turn of the century, Duffing extensively studied damped-driven oscillators, yet the full force of the dynamics in his simple model system was not uncovered until recently using computer simulations.[183,184] Lorenz's landmark paper[112] showing the first instance of persistent chaos in a simple mathematical model of fluid convection was found accidentally in computer simulations, not theory.

Finally, even the application of extant dynamical systems tools to time series of experimental data provides little recourse.[133] These tools have largely been developed using simple models whose responses can be generated sufficiently long to obtain an indication of their dynamics. Biological responses, by contrast, are often extremely short lived. For example, chewing and swallowing behaviors in humans as in *Pleurobranchaea* may be generated by robust attractors, but so few cycles are generated that characterization of their dynamics, whether they be limit-cycle or chaotic attractors, is not possible. Even in ideal systems, a certain amount of guess-work needs to be done. For example, the Grassberger-Procaccia algorithm can significantly overestimate the attractor dimension of limit cycles and underestimate it for chaotic systems, particularly as the dimension increases, even for model systems such as the Rössler hyperchaos.[161]

The positive side of all of these problems is that biology stands on an exciting albeit difficult threshold of growth in theories and concepts. And it is biology that will force further development of dynamical tools. The work of Ellner and coworkers on nonparametric methods to calculate Lyapunov exponents is an example.[54,55,56]

8. COMPUTER SIMULATIONS: MINIMAL MULTIFUNCTIONAL NETWORKS

Computational analogies may provide insight where theory is lacking. Lorenz's[112] work on convection provides an excellent example of how computer simulations may spark insight into new methods for handling complex systems. The work of Klopf and coworkers,[97,131] which was discussed above under Reductionism, is another example in which computational methods have proved decisive in addressing an important problem in the theory of learning. In Lorenz's case, the outcome was unexpected. In Klopf's case, the outcome was planned because of the equivalence of the statement of drive reinforcement at both the unit and global levels. Both of these examples show that certain statements or assumptions about interacting systems can be used to address complex behavior through computational methods without having first to develop a proved theory about the global system. Put differently, given certain assumptions about local events, it may be possible to allow the system to generate itself. In the same way, we discuss here four topics that may be addressable computationally and which may eventually prove beneficial in understanding some of the complexities of biological organization.

8.1 NONLINEARITIES AND BIFURCATIONS IN SIMPLE NETWORK ARCHITECTURES

As we have referred to repeatedly above, we do not yet understand the functional meaning of convergence and divergence beyond the notion of reflexes,[136,139,174] or as Sperry put it,[178] of the "three-bodies problem." In studies of associative learning and motor pattern generator, there is as much need now for a new language to handle the emergent properties arising from convergence as there was fifteen years ago.[143] But we can point at least to two small interrelated advancements: identification of the nonlinear interactions that arise from network architectures, and the identification of architectures that permit bifurcations to arise from such interactions. The discussion below uses several model systems to clarify what we mean, and to inquire into the problem of continuous versus discrete processes in neuronal activity.

8.1.1 NONLINEARITY AND BIFURCATION IN MODEL SYSTEMS *Rössler and logistic*.
Nonlinearities are easy to see in simple models such as the Rössler system[162] of coupled ordinary differential equations that generate complex chaotic dynamics:

$$\frac{dx}{dt} = -y - z \qquad \frac{dy}{dt} = x + ay \qquad \frac{dz}{dt} = b_z x - cz,$$

where $a, b,$ and c are constants. Here X is a function of Y and Z, Y is a function of X and itself, and Z is a nonlinear function of itself and X. Each of these variables is expressed nonlinearly through the others. The logistic equation, $X_{n+1} = R(1 -$

$X_n)X_n$, is an even simpler example, where the new value on the left is generated by the nonlinear drive of the previous value on the right (initialized between 0 and 1), and is then reintroduced into the system to generate the subsequent number. For values of the constant R between 0 and about 3.55, the process of nonlinear action followed by recursive folding back into the equation produces periodic sequences of numbers, but for R greater than 3.55, the system generates chaotic sequences.[123] Successive, linear adjustments to a constant such as R may produce only minor changes in the system over a large portion of R's allowable range. But at critical points, very small alterations in R produce nonlinear shifts (bifurcations) in the sequence of numbers. At low R-scale resolutions, regions are observed at which only chaos appears to occur. By expanding the R-scale, one observes that chaotic regions contain periodic regimes.

Bifurcation in Hodgkin-Huxley membrane. Teresa Chay's[29] seminal paper examined a three-variable Hodgkin-Huxley membrane precisely in this way. The time variation of voltage in the model is given by

$$\frac{dV}{dt} = g_I^* m_\infty^3 h_\infty (V_I - V) + g_{K,V}^* n^4 (V_K - V) + g_{K,C}^* \frac{C}{1+C}(V_K - V) + g_L^*(V_L - V).$$

I: mixed inward currents (sodium, calcium). K,V: voltage-sensitive potassium channel. C: internal calcium concentration. K,C: calcium-sensitive potassium current. L: leakage. n: probability of opening K,V. m,h: probabilities of activation, inhibition. g*: maximal conductance divided by capacitance.

The three variables in the system are (1) membrane potential (V); (2) n, the probability of opening the voltage-dependent potassium channel; and (3) intracellular concentration of calcium (C). Intracellular calcium is voltage-dependent, as are sodium, one of the potassium channels, n, m, and h. It can be easily seen

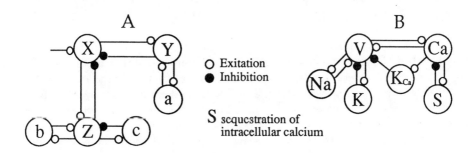

FIGURE 4 Cartoon of "minimal" neurocircuit transpositions of the three-variable Rössler system of coupled differential equations (A) and of the Chay's three-variable Hodgkin-Huxley membrane (B). See text.

mathematically that all of these variables affect one another through voltage (as a consequence of their effects on currents), and that the system of such interactions is highly nonlinear, although examination of the equations would not necessarily give immediate insight into which parameters to use to control bifurcations. The bifurcation parameter is the calcium-dependent potassium conductance $g_{K,C}$, and, as described above for the logistic equation, the membrane produces many different firing patterns when this conductance was systematically changed.

8.1.2 RELATIONSHIP BETWEEN BIFURCATION DYNAMICS AND NETWORK ARCHITECTURES. To illustrate the difficulties encountered in attempting to understand the dynamical capabilities of network architectures, and the direction we have taken in some of our computer studies, consider the (overly) simplified cartoons in Figure 4 that transpose the Rössler system and the Chay membrane into "realistic" analogs of neuronal networks. "Realistic" might include voltage-sensitive ion channels, calcium-dependent ones, transmitter release dynamics, transmitter re-uptake, and second messenger systems, and other processes one might want to include in an experimental system.

Given tonic excitatory input to X in Figure 4(a), and making X capable of post-inhibitory rebound, it may be possible for X and Y, and X and Z, to oscillate in opposition if there is sufficient accommodation in the firing of Z and/or Y. Figure 4(b) shows a network cartoon of a subset of the variables in the Chay membrane. Given Chay's simulations, it might be predicted that the synapse of K_{Ca} onto V would provide access to bifurcation dynamics. The nonlinearities in the Rössler and Chay systems are easily identifiable in the differential equations that compose them. And it is possible to see how the calcium-dependent potassium conductance can influence the dynamics of the Chay model. But it is considerably more difficult to identify analogous nonlinearities and bifurcation conditions in neuronal networks. It has long been established that synaptic activation of neurons leads to nonlinear responses because of the firing threshold in the driven neuron. It is also known how to simulate individual synapses using digital integration, by describing the kinetics mathematically, or by examining nonlinear interactions between different types of synapses.[98] But the dynamical implications of different network architectures and of the synapse characteristics that affect the dynamics of regenerative electrical activity of neurons in these networks are problems that remain largely untapped.

Along this line, present efforts in our laboratory are aimed at understanding what types of converging and diverging centers in minimal networks are required for bifurcations to occur. In the same way as Chay used the calcium-dependent potassium conductance to control the bifurcations, our efforts are to determine whether synaptic strengths can also be used as bifurcation parameters. The problem facing us in dealing with the biological system is much more difficult than that which faced Chay because: (1) our system has many more degrees of freedom. (2) Our system is not as smoothly continuous as the Hodgkin-Huxley membrane; i.e., the membrane responses may seem continuous, but cells usually receive information in short pulses or bursts. (3) There are no previous network examples for us to follow in which bifurcation have been demonstrated. Interestingly, the types of convergence centers that have proved capable of bifurcating into variable activity in our preliminary computer simulations, are ones having similar structures as the one shown in Figure 4(b).

As our knowledge grows of the connectivity among the BCNs and of their connections with other neuronal groups, we shall construct computer simulations of networks having increasing sizes. We shall then progressively introduce the effects of the many converging neurotransmitter systems. Additionally, by implementing early behavioral evidence of synaptic competition during learning in *Pleurobranchaea*,[143] and the evidence for synaptic competition in mammalian cortex,[129] we expect to see our networks remodel their connections overtime. Interactive groups may actually grow or shrink in time; large populations may split into subsets; the spatial boundaries between coactive groups may move in time; and network architectures may emerge that affect the amount of variation occurring in the network.

8.1.3 CONTINUOUS VS. DISCRETE PROCESSES. The Rössler and Chay model are both three-variable systems, as required of any continuous bounded system that is capable of generating chaos. We summarized the reasons behind the need for three variables using mixing of trajectories in three-space and an examination of Lyapunov exponents in subsection 3.5.2. By contrast, discrete processes can generate chaos in one dimension, as in the case of the logistic equation, and coupled discrete processes can generate chaos in two-space, as shown by the Hénon system, where $X_{n+1} = 1aX_n^2 + Y_n$ and $Y_{n+1} = bX_n$.[78] Recall also that the issue is not whether a system generates chaos, but its ability to exhibit both simple and complex behaviors, depending on its bifurcations conditions arising from simple quantitative alterations rather than from qualitative changes in network structure. Moreover, if the bifurcation parameter is the driving frequency of an input signal, it is not necessary even for quantitative changes to occur in the network for simple and complex dynamics to appear.

The difference between continuous and discrete processes is of significance to neurobiologists. The neural networks studies of Mpitsos and Burton[136] indicate that when signals between networks are chaotic discrete processes, simple networks are able to perform difficult tasks on these signals that would otherwise require more complex networks to perform if the mode of transmission used continuous periodic or continuous chaotic processes. Continuous processes are used in neural integration,[188] but the usual mode of information transfer is through trains of action potentials. Trains of action potentials in pacemaker firing cells are generated by continuous fluctuations in membrane potentials and in the dynamics of ionic species. Examples may be found in computer simulations of the parabolic burster neuron R_{15} in *Aplysia*,[25] and in the Chay model described above. The information in these spike trains, though generated by continuous processes, is in a pulse code. Therefore, there are a number of questions that need examination. For example, is there an informational difference between the dynamics of spike trains by comparison to the information contained in the continuous membrane processes that generate them? What happens in postsynaptic cells when they receive such spike trains, and when are we to consider the dynamics in the postsynaptic cells as continuous processes or analogs of discrete processes? The membrane potentials of these follower cells may appear continuous, but they are driven by discontinuous input events.

The differences between discrete and continuous processes pose problems in numerical analyses. Experimental data usually consists of the time series of one or several dependent variables, but the methods provide little knowledge of the number of dependent variables that actually drive the system. Numerical methods provide some help. For example, it is possible to conduct phase-space analyses that give information about the topological dimension of attractors and about the number of dependent variables (embedding space) that may be involved in generating the attractors.[137,144] The evidence provides some justification supporting chaotic attractors and low-dimensional embedding space.

However, some of the calculated attractor dimensions were lower than two, posing some difficulties in interpretation of what the dynamics is. Continuous systems must have at least three Lyapunov exponents; there must be at least two non-negative ones, one being positive, as required for chaos, and one having zero value, as required by Haken's theorem (subsection 3.3.2). Given two non-negative exponents, calculations using the Kaplan-Yorke conjecture should be expected that the lowest attractor dimension for continuous chaotic systems be greater than two (examples are given in Andrade et al.[6]; Wolf[192]). One-variable discrete processes, such as the logistic equation, have dimensions less than 1. Two-variable discrete processes have dimensions between one and two; our own estimate of the Henn system gives dimension of about 1.36. Knowing the mathematical representation of a system allows one to place such numbers in appropriate context, but experimental data leaves numerical results ambiguous. Do we assume that attractor dimensions less then two are coupled discrete processes or is it a problem with the analytical methods? Of the latter possibility, the available tools, whether using time series of a single variable or all variables, calculation of attractor dimensions are difficult to obtain even for model systems.[6]

Answers to questions as the one given above are necessary because they provide an indication about how information is processed and encoded. We are presently addressing them using numerical analyses of data from computer simulations of membrane patches and of responses of cells in networks where we have access to all parameters and variables of the system. Comparison of analyses on the data from measurements of continuous variables and from spike trains may yield some insight into implications relating to continuous and discrete processes.

8.2 RESPONSE OPTIMIZATION, ENERGY GRADIENTS, AND ATTRACTORS IN BIOLOGICAL NETWORKS

8.2.1 ATTRACTORS, FROM SEA SLUGS TO BEES
Real[159] has shown recently that bees are able to adjust their behavior so as to optimize the use of food resources. Whether or not this involves gradients and attractors has not been addressed. The idea is consistent with the possibility that biological networks (and biological systems generally) may exhibit behavior that tends to minimize some gradient factor

(as error or energy) through the ability of attractors to dissipate energy.[134,135] Attractors (see subsection 3.4) pull in any phase-space trajectory that falls within their basin of attraction. Thus, for example, in limit cycles, externally applied perturbations move the trajectory of the system away from the limit set, but if the state of the trajectory remains within the attractor's basin of insets, the trajectory will fall asymptotically back into the limit set. Chaotic attractors also attract nearby states but dissipate perturbations over their entire surface. We might say that attractors minimize energy or error.[135] Put differently, attractors optimize the match between their attracting set and activity that falls near it. In either case, the action may be consider a minimization process. On the behavioral level, bees are able to control their foraging techniques so as to optimize the use of food resources.[159]

8.2.2 COMPARISON THROUGH ANALOGY IN PRINCIPLES, NOT IN IDENTITY OF MECHANISMS The potential consequences of the identity between attractors and optimization are rather interesting. Consider the following situations. In attempting to simplify computer simulations, it is often difficult to determine exactly where to limit the characterization of the biology. For example, the connectionist methods of error back-propagation are usually faulted because of their obvious non-biological nature. But the answers that come from the use of such networks depends on the principles that are actually being simulated. The major driving element of error back-propagation is that the system must follow a negative error gradient between a teacher function and the output of the system.[165] If the question being addressed has to do with the principle of error reduction, rather than, say, what second messengers might be involved in a cellular process, or how feedback actually occurs in a real nervous system, the back-propagation method might give some insight into how gradient-seeking systems store information in their distributed elements.

Response thresholds. Following this rationale, Mpitsos and Burton[136] obtained a number of results that might have relevance to biological systems. They found, for example, that the computational capabilities of networks are severely limited when only trainable synaptic strengths are used. Adding trainable thresholds significantly expands the computational power of the networks. In invertebrate learning studies, thresholds (as might be inferred from membrane changes in postsynaptic cells) have either not been observed at the cellular level or have not been generally attended to.[146] Studies on long-term potentiation (LTP) in rats have, however, provided evidence implicating response thresholds through changes in synaptically induced changes in the ratio of excitation and inhibition rather than changes in membrane impedance.[17,28] Heretofore, the methods used to test LTP have not focused on assessing the computational implications of threshold adjustments, nor the technical conditions to extend the findings, but it would be extremely interesting to determine whether adjustments in the ratio of excitation to inhibition were set differently for each cell, as might occur in gradient descent adjustments in thresholds during learning in neural networks.

Network size may be self-limiting. An unexpected finding in the studies of Mpitsos and Burton[136] was that increasing the number of neurons in a hidden layer or interneuronal layer beyond a certain point slows and eventually causes the system to cease learning; i.e., group size may be self-limiting. Limitation of group size has been enforced algorithmically in simulations of mammalian cortex through synaptic competition and inhibitory synapses.[52,155] It is also conceivable, however, that group size may be additionally limited by the gradient tendencies of attractors. If the findings of Mpitsos and Burton hold biologically, the slower organizational times of large networks may be superseded by smaller subsets of neurons as they form attractors. Once sufficiently formed, the attractors themselves may restrict group size, partly by their gradient processes, and partly by learning-related synaptic competition. To our knowledge, the network-forming aspects of synaptic competition have been viewed only at the level of neuronal trophic factors and whether or not activity occurs. What we are attempting to point out here is that the network not only generates activity, but that the dynamics of this activity may affect the characteristics of the network architecture.

A similar distinction between activity and dynamics may be raised in studies of motor pattern switching. In a traditional sense, switching between patterns of activity require some network change or the introduction of activity in a controlling neuron.[39] We do not deny this possibility, but add that the notion of bifurcation raises the discussion from the level of activity alone to a level involving dynamical processes. Using John's terminology,[82] the former is a "switchboard" effect relating to particular neuron(s), whereas the latter is an abstraction of the self-organizing activity in neurons, and quite likely may not be identifiable in network structure,[139] although some identifiable structural indices may be obtainable as discussed for the studies of Figure 4.

Metaknowledge and lazy synapses. Metaknowledge represents that ability of networks to store different forms of information.[23] We discussed it above in dealing with reductionism (subsection 6.3.4), and we believe that it may be a consequence of gradient tendencies. Our computational studies also found that although networks set their synaptic weights and thresholds at optimum levels, many of the synaptic weights produce little effect when removed from the network; i.e., they are "lazy." Mpitsos and Burton[136] discuss a number of uses for such synapses. One of the most interesting possibilities comes from somewhat different studies by Warren[186] who showed that certain synapses may be deleted after training without significantly affecting network performance on a previously learned task, but networks were unable to learn the task if they started with the reduced number of synapses in the first place. This poses interesting problems to biologists since weak connections are often observed between the interactive components of their experimental systems. The tendency in the past has been to dismiss such connections, or to presume that they would be "pruned" away if not used. Our findings along with Warren's indicate that these synapses may be crucial for learning new tasks. By analogy to computers, they might be considered as temporary registers that permit gradient descent, but once gradient descent has been reached, they are no longer needed for that task.

8.3 LOCAL ERROR MINIMA IN BIOLOGICAL ADAPTATION

The idea that a system tends to optimize its behavior has a somewhat different expression in biological systems than it might have in computer simulations of connectionist neural networks. With enough time and stable environmental conditions, we can envision that evolutionary competition between organisms will produce changes that best adapt the species to the environment. One might think of the process as reaching an absolute error minimum between the response of the organism and the best possible response under the imposed conditions. Any response that is not optimal represents a local minimum. In neural networks, methods have been developed (see subsection 3.5.3) to avoid local minima using, for example, simulated annealing[93] and time-invariant noise algorithms (TINA).[23] Simulated annealing usually involves exponential decay of noise over time. TINA adjusts noise as a function of the amount of error that is produced when a system responds to its input stimuli. This method, however, was chosen only as a vehicle to demonstrate the idea of TINA. Other methods, not necessarily directly related to error feedback, may also be used that retain time invariance. For example, our present attempts to implement TINA in networks consisting of neurons having biologically realistic characteristics is to adjust the probabilistic release of transmitter[99] or to use short-term activity-dependent learning rules such as sensitization[146] to maintain the flow in a given part of the network. Our goal is to assign certain facilitatory responses to classes of neurons, and then to allow the actual pathway to emerge dynamically. Low-error would be represented by activity recurring through a particular part of the network. As error increases, diffusely distributed feedback onto the network would disrupt such preferentially frequented pathways, permitting others to emerge. If these new pathways lead to low error, feedback decreases, allowing the flow through the pathway to continue. If attractors self-organize, the preferential pathways would then be further entrenched, because, as discussed above, the basin of insets to the attractor itself may represent an energy or error-minimizing process.

This process does not require that the tendency to follow a gradient actually reach an optimal minimum, or, equivalently, that the attractor be spatio-temporally a robust, stable structure. Biologically, in both the daily behavior of organisms and in their evolutionary succession, local minima are extremely important in generating adaptive responses. Whatever works is sufficient, whether the response is optimal or not. Thus, our notion of an adaptive system is one that can generate different minima that can be addressed rapidly, and exited rapidly if they do not meet the need. Indeed, we believe that it is from the ability to generate many local minima that multibehavioral networks may have evolved.

Part of the understanding about the generation of local minima will be to see how multibehavioral networks generate different attractors in computer simulations. Transitions between different attractors may yield labile intermediate forms that only partially resemble more stable ones. The most difficult problem that we face here is to determine how best to visualize temporal activity graphically for spike trains.[137,144] Continuous non-spiking processes pose less of a problem.[6] Part of the answer may also come from an understanding of spatio-temporal dynamics.

8.4 VISUALIZATION OF SPATIO-TEMPORAL DYNAMICS

The more we study biology, the more it seems that we must somehow leave it to gain a feel for what may be happening there. Put simply, biological systems are too complex and uncontrollable even to perform experiments as those represented by Figure 4. We must imbue these simulation networks with as much biological information as needed to obtain activity that somehow resembles the activity of the biological system. But complete state-space parameterization of the biological system is beyond hope, as one glimpse of the complexity in Figures 2 and 3 will show. At the level at which we can attribute realistic biological characteristics to a network, the system becomes intractable even for simple analyses of steady states (see example analysis of a simple model system in Andrade et al.[6]).

Given the growing power of computer graphics and the increasingly easier access to supercomputers, the recourse for biologists interested in the emergence of group dynamics is to conduct the type of experiments shown in Figure 4, and, especially, to visualize the spatio-temporal flow of activity in large-scale simulations involving many interacting units. An understanding of such spatio-temporal flows is, we believe, one of the central questions facing neuroscience. Walter Freeman and coworkers were perhaps the first to begin a detailed account of spatially distributed recordings in their studies of rabbit olfactory bulb (e.g., see review in Skarda and Freeman[172]). But even in these studies, the analysis of the temporal flow is of the time series of single recording sites. Perhaps the major lesson in dynamical systems work over the decade has been the fact that much can be learned about the activity of a system by the analysis of its phase-space geometry. Up to four variables can be analyzed simultaneously using time series analysis (e.g., see Figures 8-11 in Andrade et al.,[6] and Figure 13 in Mpitsos and Burton[136]). We need to do the same for many variables, both spatially and temporally.

By such methods it may be possible to examine the possibility of limit cycles, chaotic attractors, SOCs and turbulence, the coexistence of multiple attractors, movement of these attractors spatially, and possibly even their blending into one another. It may also be possible to determine how particular circuit structures emerge, how variability appears controlled by particular circuit characteristics. In the long term it will be important to ask how such structures are affected by system-wide factors. If we are to believe our neurochemical findings, it is quite likely that bifurcation parameters may be more accurately defined as being distributed over a large number of cells rather than, for example, in the conductance modification of a single cell. The first possibility may explain the fact that some systems are relatively insensitive to changes in only a few of their components.

9 CONCLUSION

In answer to the title of this paper, we have actually said little about what sea slugs can tell us explicitly about the neurointegration of specific human movement.

But we believe that the findings tell us considerably about what must be addressed in order to gain a unified perspective of biological integration that might eventually affect how we view human movement. We understand that much has been said appropriately by others about coordination of limbs in invertebrates and vertebrates, the rightful importance of FAPs, and selective control of individual neurotransmitters on pattern generation and in the formation of network structure, and that such findings may be applicable to human motor behavior. Perhaps most of the time all of these studies provide the best answers, as most of the time Newtonian physics provides the right answers in daily engineering problems. Perhaps also, the neurointegrative processes in *Pleurobranchaea* and *Aplysia* follow the same predictabilities most of the time.

The instances that are not explainable by traditional neurocircuit perspectives might be dismissed as biological aberrance. Alternatively, owing to the fact that the animal seems to function well enough with them, they may be pursued as being of adaptive significance. We have followed the latter route, and have been forced into a perspective that is more statistical mechanical and dynamical than classically "switchboard." Lorenz[113] voiced the long-held view that all biological information is stored in structure. We hardly disagree with that. But the question is, how do we read that information, and is much of it redundant and even of nonsense or accidental value? The latter possibilities may actually provide certain adaptive value adventitiously in ever changing and unpredictable environments. In reaching a new theoretical perspective that addresses these issues, our view is that there are two levels of solution: the special case, relating to the switchboard neurocircuit, and the general solution, that must be reducible to the special case but must also provide a general theoretical foundation that is extensible to many other cases.

The shift to dynamics, or at least away from answering all questions by using reflexes, marks a shift away from mechanism to organization. Although each biological level of organization may express the dynamics in its own processes, the dynamical principles may be applicable to all levels of organization. The central question in all of these systems is *"How does the individual influence the group, and, in turn, how does the group influence the actions of the individual?"* We have tried as much as possible to couch our ideas on biological findings, though much more data needs to be gathered (and re-gathered) before we feel more comfortable. If what we have discussed is accurate, then, as Barbara McClintock envisioned, "We are going to have a new realization of the relationship of things to each other."[88]

ACKNOWLEDGMENTS

This work was supported by AFOSR 89-0262 to George J. Mpitsos and by a grant from the Finnish Cultural Foundation to Seppo Soinila. The authors wish to thank Dr. Janet Leonard for her critical reading of a previous version of this manuscript, and Professor Lavern Weber, Director of the Mark O. Hatfield Marine Science Center, for making space available to us and for his continuing encouragement.

REFERENCES

1. Abraham, R. H., and C. D. Shaw. *Dynamics–The Geometry of Behavior, Parts 1-4*. Santa Cruz: Aerial Press, 1983.
2. Abrams, T. W., V. F. Castellucci, J. S. Camardo, E. R. Kandel, and P. E. Lloyd. "Two Endogenous Neuropeptides Modulate the Gill and Siphon Withdrawal Reflex in *Aplysia* by Presynaptic Facilitation Involving cAMP-Dependent Closure of a Serotonin-Sensitive Potassium Channel." *Proc. Natl. Acad. Sci. USA* **81** (1984): 7956–7960.
3. Adey, R. W. "Organization of Brain Tissue; Is the Brain a Noisy Processor?" *Int. J. Neurosci.* **3** (1972): 271–284.
4. Albano, A. M., A. I. Mees, J. Muench, P. E. Rapp, and C. Schwartz. "Singular-Value Decomposition and the Grassberger-Procaccia Algorithm." *Phys. Rev. A.* **38(6)** (1988): 3017–3026.
5. Andersen, P. O. "Properties of Hippocampal Synapses of Importance for Integration and Memory." In *Synaptic Function*, edited by G. Edelman, W. E. Gall, and W. M. Cowan, 403–429. New York: Wiley & Sons, 1987.
6. Andrade, M. A., J. C. Nuño, F. Moran, F. Montero, and G. J. Mpitsos. "Complex Dynamics of a Catalytic Network Having Faulty Replication into an Error Species." 1992, submitted.
7. Aristotle. *Physica*. First ed., Vol. I(9), 192b,6. Athens: Tsipouro Press of Mpitsopoulos & Sons, Scribners Attendee at the Lyceum, 330BC.
8. Ayers, J., G. Carpenter, S. Currie, and J. Kinch. "Which Behavior Does the Lamprey Central Motor Program Mediate?" *Science* **221** (1983): 1312–1315.
9. Babloyantz, A., J. M. Salazar, and C. Nicolis. "Evidence of Chaotic Dynamics of Brain Activity During the Sleep Cycle." *Phys. Lett.* **111A** (1985): 152–155.
10. Bak, P. "Is the World at the Border of Chaos?" *Ann. New York Acad. Sci.* **581** (1990): 110–118.
11. Bak, P. "Self-Organized Criticality." *Physica A* **163** (1990): 403–409.
12. Bak, P. "Simulation of Self-Organized Criticality." *Physica Scripta* **T33** (1990): 9–10.
13. Bak, P., and K. Chen. "Self-Organized Criticality." *Sci. Am.* **264** (1991): 46–53.
14. Bak, P., K. Chen, and C. Tang. "A Forest-Fire Model and Some Thoughts on Turbulence." *Phys. Lett. A* **147** (1990): 297–300.
15. Bak, P., and C. Tang. "Self-Organized Criticality." *Phys. Rev. A* **38** (1988): 364–374.
16. Barnsley, M. *Fractals Everywhere*. San Diego: Academic Press, 1988.

17. Barrionuevo, G., S. R. Kelso, D. Johnston, and T. H. Brown. "Conductance Mechanism Responsible for Long-Term Potentiation in Monosynaptic and Isolated Excitatory Synaptic Inputs to Hippocampus." *J. Neurophysiol.* **55** (1986): 540–550.
18. Bellman, K. L. "The Conflict Behavior of the Lizard, *Sceloporus Occidentalis*, and Its Implication for the Organization of Motor Behavior." Ph.D. Thesis, University of California, San Diego, 1979.
19. Bentley, D., and M. Konishi. "Neural Control of Behavior." *Ann. Rev. Neurosci.* **1** (1978): 35–59.
20. Boakes, R. *From Darwin to Behaviorism*. Cambridge: Cambridge University Press, 1984.
21. Braitenberg, V. "Some Arguments for a Theory of Cell Assemblies in the Cerebral Cortex." In *Neural Connections, Mental Computations*, edited by L. Nadel, L. A. Cooper, P. Culicover, and R. M. Harnish, 137–145. Cambridge, MA: MIT Press, 1989.
22. Bullock, T. H. "Comparative Neuroscience Holds Promise for Quiet Revolution." *Science* **222** (1984): 473–478.
23. Burton, R. M., and G. J. Mpitsos. "Event-Dependent Control of Noise Enhances Learning in Neural Networks." *Neural Networks*, in press.
24. Byrne, J. H., V. F. Castellucci, and E. R. Kandel. "Contribution of Individual Mechanoreceptor Sensory Neurons to Defensive Gill-Withdrawal Reflex in *Aplysia*." *J. Neurophysiol.* **41** (1978): 418–431.
25. Canavier, C., J. W. Clark, and J. H. Byrne. "Routes to Chaos in a Model of a Bursting Neuron." *Biophys J* **57** (1990): 1245–1251.
26. Cardi, P., F. Nagy, J.-R. Cazalets, and M. Moulins. "Multimodal Distribution of Discontinuous Variation in Period of Interacting Oscillators in the Crustacean Stomatogastric Nervous System." *J. Comp. Physiol. [A]* **167** (1990): 23–41.
27. Carew, T. J., and C. L. Sahley. "Invertebrate Learning and Memory: From Behavior to Molecules." *Ann. Rev. Neurosci.* **9** (1986): 435–487.
28. Chavez-Noriega, L. E., J. V. Halliwell, and T. V. Bliss. "A Decrease in Firing Threshold Observed after Induction of EPSP-Spike (E-S) Component in Rat Hippocampal Slices." *Exp. Brain Res.* **79** (1990): 633–641.
29. Chay, T. R. "Chaos in a Three-Variable Model of an Excitable Cell." *Physica* **16D** (1985): 233–242.
30. Chay, T. R., and J. Rinzel. "Bursting, Beating, and Chaos in an Excitable Membrane Model." *Biophys. J.* **47** (1985): 357–366.
31. Chen, K., and P. Bak. "Is the Universe Operating at a Self-Organized Critical State?" *Phys. Lett. A* **140** (1989): 299–302.
32. Cohan, C. S. "Centralized Control of Distributed Motor Networks in *Pleurobranchaea Californica*." Ph.D. thesis, Case Western Reserve University, 1980.
33. Cohan, C. S., and G. J. Mpitsos. "The Generation of Rhythmic Activity in a Distributed Motor System." *J. Exp. Biol.* **102** (1983): 25–42.

34. Cohan, C. S., and G. J. Mpitsos. "Selective Recruitment of Interganglionic Interneurons During Different Motor Patterns in *Pleurobranchaea*." *J. Exp. Biol.* **102** (1983): 43–58.
35. Cohen, L., H. P. Hopp, J. Y. Wu, C. Xaio, and J. London. "Optical Measurement of Action Potential Activity in Invertebrate Ganglia." *Ann Rev Physiol.* **51** (1989): 527–541.
36. Cohen, T. E., V. Henzi, E. R. Kandel, and R. D. Hawkins. "Further Behavioral and Cellular Studies of Dishabituation and Sensitization in *Aplysia*." *Soc. Neurosci. Abstr.* **17** (1991): 1302.
37. Conrad, M. "What is The Use of Chaos?" In *Chaos*, edited by A. V. Holden, 3–14. Princeton: Princeton University Press, 1986.
38. Croll, R. P., and W. J. Davis. "Motor Program Switching in *Pleurobranchaea*. I. Behavioral and Electromyographic Study of Ingestion and Egestion in Intact Specimens." *J. Comp. Physiol.* **145** (1981): 277–287.
39. Croll, R. P., and W. J. Davis. "Motor Program Switching in *Pleurobranchaea*. II. Ingestion and Egestion in the Reduced Preparation." *J. Comp. Physiol.* **147** (1982): 143–153.
40. Darwin, C. *The Origin of Species*. London: John Murray, 1859.
41. Davis, W. J. "Organizational Concepts in the Central Motor Networks of Invertebrates." In *Neural Control of Locomotion*, edited by R. M. Herman, S. Grillner, P. S. G. Stein, and D. G. Stuart, 265–292. New York: Plenum, 1976.
42. Davis, W. J., and D. Kennedy. "Command Interneurons Controlling Swimmeret Movements in the Lobster. I. Types of Effects on Motoneurons." *J Neurophysiol.* **35** (1972): 1–12.
43. Davis, W. J., and D. Kennedy. "Command Interneurons Controlling Swimmeret Movements in the Lobster. II. Interaction of Effects on Motoneurons." *J. Neurophysiol.* **35** (1972): 13–19.
44. Davis, W. J., and G. J. Mpitsos. "Behavioral Choice and Habituation in the Marine Mollusk *Pleurobranchaea californica*." *Z. Vergl. Physiol.* **75** (1971): 207–232.
45. Davis, W. J., G. J. Mpitsos, and J. M. Pinneo. "The Behavioral Hierarchy of the Mollusc *Pleurobranchaea*. I. The Dominant Position of the Feeding Behavior." *J. Comp. Physiol.* **90** (1974): 207–224.
46. Delcomyn, F. "Neural Control of Movement." *Science* **210** (1980): 492–498.
47. Delcomyn, F., and J. H. Cocatre-Zilgien. "Individual Differences and Variability in the Timing of Motor Activity During Walking in Insects." *Biol Cybern.* **59** (1988): 379–384.
48. Dickinson, P. S., C. Mecsas, and E. Marder. "Neuropeptide Fusion of 2 Motor-Pattern Generator Circuits." *Nature* **344** (1990): 155–158.
49. Doty, P., J. Marmur, J. Eigner, and C. Schildkraut. "Strand Separation and Specific Recombination in Deoxyribonucleic Acids: Physical Chemical Studies." *PNAS USA* **46** (1960): 461–476.
50. Dowling, J. E. *The Retina: An Approachable Part of the Brain*. Cambridge, MA: Harvard University Press, 1987.

51. Edelman, G. M. "Group Selection and Phasic Reentry Signalling: A Theory of Higher Brain Function." In *The Mindful Brain*, edited by G. M. Edelman and V. B. Mountcastle, 55–110. Cambridge, MA: MIT Press, 1978.
52. Edelman, G. M. *Neural Darwinism: The Theory of Neuronal Group Selection*. New York: Basic Books, 1987.
53. Eigen, M., and P. Schuster. *The Hypercycle. A Principle of Natural Self-Organization*. New York: Springer-Verlag, 1979.
54. Ellner, S. "Estimating Attractor Dimensions from Limited Data: A New Method with Error-Estimates." *Phys. Lett. A* **133** (1988): 128–133.
55. Ellner, S. "Detecting Low-Dimensional Chaos in Population Dynamics Data: A Critical Review." In *Does Chaos Exist in Ecological Systems*, edited by J. Logan and F. Hain. Charlottesville: University of Virginia Press, 1991.
56. Ellner, S., A. R. Gallant, D. McCaffery, and D. Nychka. "Convergence Rats and Data Requirements for Jacobian-Based Estimates of Lyapunov Exponents From Data." *Phys. Lett.* (1991): in review.
57. Faber, D. S., H. Korn, and J.-W. Lin. "Role of Medullary Networks and Postsynaptic Membrane Properties Regulating Mauthner Cell Responsiveness to Sensory Excitation." *Brain, Behav. Evol.* **37** (1991): 286–297.
58. Farley, J., and D. L. Alkon. "Cellular Analysis of Gastropod Learning." In *Cell Receptors and Cell Communication in Learning*, edited by A. J. Greenberg, 220–266. Basel: S. Karger, 1986.
59. Feigenbaum, M. J. "Universal Behavior in Nonlinear Systems." *Physica* **7D** (1983): 16–39.
60. Gardner, D. "Paired Individual and Mean Postsynaptic Currents Recorded in 4-Cell Networks of *Aplysia*." *J. Neurophysiol.* 63 (1990): 1226–1240.
61. Garyantes, T. K. and W. G. Regehr. "Electrical Activity Increases Growth Cone Calcium but Fails to Inhibit Neurite Outgrowth From Rat Sympathetic Neurons." *J. Neurosci.* **12** (1992): 96–103.
62. Getting, P. A. "Emerging Principles Governing the Operation of Neural Networks." *Ann. Rev. Neurosci.* **12** (1989): 185–204.
63. Getting, P. A., and M. S. Dekin. "*Tritonia* Swimming: A Model System for Integration within Rhythmic Motor Systems." In *Model Neural Networks and Behavior*, edited by A. I. Selverston, 3–20. New York: Plenum, 1985.
64. Gillette, R. "Command Neurons-FAP." *Behav. Brain Sci.* **9** (1986): 727–729.
65. Gillette, R., M. Kovac, and W. J. Davis. "Command Neurons in *Pleurobranchaea* Receive Synaptic Feedback from the Motor Network They Excite." *Science* **199** (1978): 798–801.
66. Gillette, R., M. Kovac, and W. J. Davis. "Control of Feeding Motor Output by Para-Cerebral Neurons in the Brain of *Pleurobranchaea*." *J. Neurophysiol.* **47** (1982): 885–908.
67. Glanzman, D., and G. Mpitsos. Unpublished.
68. Goldsmith, B. A., and T. W. Abrams. "Role of Adenylate Cyclase in Several Forms of Synaptic Facilitation in *Aplysia* Sensory Neurons." *Soc. Neurosci. Abstr.* **15** (1989): 1624.

69. Grassberger, P., and I. Procaccia. "Characterization of Strange Attractors." *Phys. Rev. Lett.* **50** (1983): 346-349.
70. Graybiel, A. M., and C. W. Ragsdale. "Biochemical Anatomy of the Striatum." In *Chemical Neuroanatomy*, edited by P. C. Emson, 427-504. New York: Raven Press, 1983.
71. Grossberg, S. *Studies of Mind and Brain*. Boston: Reidel, 1980.
72. Grossberg, S., and M. Kuperstein. *Neural Dynamics of Adaptive Sensory-Motor Control*. Amsterdam: North-Holland, 1986.
73. Haken, H. "At Least One Lyapunov Exponent Vanishes if the Trajectory of an Attractor Does Not Contain a Fixed Point." *Phys Lett.* **94A** (1983): 71-72.
74. Hebb, D. O. *Organization of Behavior*. New York: Wiley, 1949.
75. Heinzel, H. G. "Gastric Mill Activity in the Lobster. 1. Spontaneous Modes of Chewing." *J. Neurophysiol.* **59** (1988): 528-550.
76. Heinzel, H. G. "Gastric Mill Activity in the Lobster. 2. Proctolin and Octopamine Initiate and Modulate Chewing." *J. Neurophysiol.* **59** (1988): 551-565.
77. Heinzel, H. G., and A. I. Selverston. "Gastric Mill Activity in the Lobster. 3. Effects of Proctolin on Isolated Central Pattern Generator." *J Neurophysiol.* **59** (1988): 565-585.
78. Hénon, M., and Y. Pomeau. "Two Strange Attractors with a Simple Structure." In *Turbulence and Navier-Stokes Equations*, Springer Lecture Notes in Mathematics, Vol. 688. New York: Springer-Verlag, 1975.
79. Hofbauer, J., and K. Sigmund. *The Theory of Evolution and Dynamical Systems*. Cambridge, MA: Cambridge University Press, 1988.
80. Hunt, S. P. "Cytochemistry of the Spinal Cord." In *Chemical Neuroanatomy*, edited by P. C. Emson, 53-84. New York: Raven Press, 1983.
81. Jacklet, J. W., and J. Rine. "Facilitation at the Neuromuscular Junction: Contribution to Habituation and Dishabituation of the *Aplysia* Gill Withdrawal Reflex." *Proc. Natl. Acad. Sci. USA* **74** (1977): 1267-1271.
82. John, E. R. "Switchboard Versus Statistical Theories of Learning and Memory." *Science* **177** (1972): 850-864.
83. Kandel, E. R. *Behavioral Biology of Aplysia*. San Francisco: Freeman, 1979.
84. Kandel, E. R., and J. H. Schwartz. "Molecular Biology of Learning: Modulation of Transmitter Release." *Science* **218** (1982): 433-443.
85. Karten, H. J., K. T. Keyser, and N. C. Brecha. "Biochemical and Morphological Heterogeneity of Retinal Ganglion Cells." In *Vision and the Brain*, edited by B. Cohen and I. Bodis-Woliner. New York: Raven Press, 1990.
86. Karten, H. J. Personal communication.
87. Kater, S. B., and L. R. Mills. "Neurotransmitter Activation of Second Messenger Pathways for the Control of Growth Cone Behaviors." In *Molecular Aspects of Development and Aging of the Nervous System*, edited by J. M. Lauder, 217-225. New York: Plenum Press, 1990.
88. Keller, E. F. *A Feeling for the Organism: The Life and Work of Barbara McClintock*. New York: W. H. Freeman, 1983.

89. Kelso, J. A. S., J. P. Scholz, and G. Schoner. "Nonequilibrium Phase Transitions in Coordinated Biological Motion: Critical Fluctuations." *Phys. Lett.* **118A** (1986): 279–284.
90. Kien, J. "The Initiation and Maintenance of Walking in the Locust: An Alternative to the Command Hypothesis." *Proc. Roy. Soc. Lond. B* **219** (1983): 137–174.
91. Kien, J. "Neuronal Activity During Spontaneous Walking. I. Starting and Stopping." *Comp. Biochem. Physiol.* **95A** (1990): 607–621.
92. Kien, J. "Neuronal Activity During Spontaneous Walking: II. Correlation With Stepping." *Comp. Biochem. Physiol.* **95A** (1990): 623–638.
93. Kirkpatrick, S., C. D. Gelatt, and M. P. Becchi. "Optimization by Simulated Annealing." *Science* **220** (1983): 671–680.
94. Kistler, H. B., R. D. Hawkins, H. W. Koester, W. M. Steinbusch, E. R. Kandel, and J. H. Schwartz. "Distribution of Serotonin: Immunoreactive Cell Bodies and Processes in the Abdominal Ganglion of Mature *Aplysia*." *J. Neurosci.* **5** (1985): 72–80.
95. Klein, M., O. Bratha, N. Dale, and E. R. Kandel. "Analysis of a Newly Described Cellular Process Contributing to Facilitation at Depressed Neuron Synapses." *Soc. Neurosci. Abstr.* **15** (1989): 1264.
96. Klopf, A. H. *The Hedonistic Neuron: A Theory of Memory, Learning, and Intelligence*. New York: Hemisphere, 1982.
97. Klopf, A. H. "A Neuronal Model of Classical Conditioning." *Psychobiol.* **16** (1988): 85–125.
98. Koch, C., and T. Poggio. "Biophysics of Computation: Neurons, Synapses, and Membranes." In *Synaptic Function*, edited by G. Edelman, W. E. Gall, and W. M. Cowan, 637–697. New York: Wiley & Sons, 1987.
99. Korn, H., and D. S. Faber. "Regulation and Significance of Probabilistic Release Mechanisms at Central Synapses." In *Synaptic Function*, edited by G. M. Edelman, E. W. Gall, and W. M. Cowan, 57–108. New York: John Wiley & Sons, 1987.
100. Kovac, M. P., and W. J. Davis. "Reciprocal Inhibition Between Feeding and Withdrawal Behaviors in *Pleurobrachaea*." *J. Comp. Physiol.* **139** (1980): 77–86.
101. Kriebel, M. E., J. Vautrin, and J. Holsapple. "Transmitter Release: Prepackaging and Random Mechanism or Dynamic and Deterministic Process?" *Brain Res. Rev.* **15** (1990): 167–178.
102. Kupferman, I., D. Deodhar, and K. R. Weiss. "Simple Neural Network Models Provide Heuristic Tools for Understanding the Possible Role of Command-Like Neurons Controlling Behaviors in *Aplysia*." *Soc. Neurosci. Abstr.* **17** (1991): 1591.
103. Kupferman, I., A. Mahon, R. Scheller, K. R. Weiss, and P. E. Lloyd. "Immunocytochemical Study of the Distribution of Small Cardioactive Peptide (SCP_b) in *Aplysia*." *Soc. Neurosci. Abstr.* **10** (1984): 153.
104. Kupferman, I., and K. R. Weiss. "The Command Neuron Concept." *Behav. Brain Sci.* **1** (1978): 3–39.

105. Kupferman, I., and K. R. Weiss. "Command Performance." *Behav. Brain Sci.* **9** (1986): 736–739.
106. Küppers, B. O. *Molecular Theory of Evolution*. Berlin: Springer-Verlag, 1983.
107. Lashley, K. "In Search of the Engram." *Symp. Soc. Exp. Biol.* **4** (1950): 454–482.
108. Leonard, J. L., J. Edstrom, and K. Lukowiak. "A Re-Examination of the 'Gill Withdrawal Reflex' of *Aplysia Californica* Cooper (Gastropoda; Opisthobranchia)." *Behav Neurosci.* **103** (1989): 585–604.
109. Leonard, J. L., M. Martinez-Padron, J. P. Edstrom, and K. Lukowiak. "Does Altering Identified Gill Motor Neuron Activity Alter Gill Behavior in *Aplysia*." In *Molluscan Neurobiology*, edited by K. S. Kits, H. H. Boer, and J. Joose, 30–37. In press. Amsterdam: North Holland, 1990.
110. Ling, G., and R. W. Gerard. "The Normal Membrane Potential of Frog Sartorius Fibers." *J. Cell Comp. Physiol.* **34** (1949): 383–396.
111. Lipton, S. A., and S. B. Kater. "Neurotransmitter Regulation of Neuronal Outgrowth, Plasticity, and Survival." *TINS* **12** (1989): 265–270.
112. Lorenz, E. N. "Deterministic Non-Periodic Flows." *J. Atmos. Sci.* **20** (1963): 130–141.
113. Lorenz, K. Z. "Analogy as a Source of Knowledge." *Science* **185** (1974): 229–234.
114. Lorenz, K. Z. *The Foundations of Ethology*. New York: Simon and Schuster, 1981.
115. Lukowiak, K., and E. Colebrook. "Classical Conditioning of *in vitro Aplysia* Preparations: Multiple Sites of Neuronal Changes." In *Neurobiology of Molluscan Models*, edited by H. H. Boer, W. P. M. Geraerts, and J. Joose, 320–325. New York: North-Holland, 1986.
116. Lukowiak, K., J. Goldberg, W. F. Colmers, and J. P. Edstrom. "Peptide Modulation of Neuronal Activity and Behavior in *Aplysia*." In *CRC Handbook of Comparative Opioid and Related Neuropeptide Mechanisms*, edited by G. B. Stephano. Boca Raton: CRC Press, 1986.
117. Marcus, P. S. "Numerical Simulation of Jupiter's Great Red Spot." *Nature* **331** (1988): 693–696.
118. Marder, E. "Mechanisms Underlying Neurotransmitter Modulation of a Neuronal Circuit." *TINS* **7** (1984): 48–53.
119. Marder, E. "Pattern Generators: Modulating a Neural Network." *Nature* **335** (1988): 296–297.
120. Marder, E., and S. L. Hooper. "Neurotransmitter Modulation of the Stomatogastric Ganglion of Decapod Crustaceans." In *Model Neural Networks and Behavior*, edited by A. I. Selverston, 319–338. New York: Plenum Press, 1985.
121. Marder, E. E., S. L. Hooper, and J. S. Eisen. "Multiple Neurotransmitters Provide a Mechanism for the Production of Multiple Outputs from a Single Neuronal Circuit." In *Synaptic Function*, edited by G. Edelman, W. E. Gall, and W. M. Cowan, 305–327. New York: Wiley & Sons, 1987.

122. Markus, M., D. Kuschmitz, and B. Hess. "Properties of Strange Attractors in Yeast Glycolysis." *Biophys. Chem.* **22** (1985): 95–105.
123. May, R. M. "Simple Mathematical Models with Very Complicated Dynamics." *Nature* **261** (1976): 459–467.
124. McClellan, A. D. "Feeding and Rejection in *Pleurobranchaea*: Comparison of Two Behaviors Using Some of the Same Musculature." *Neurosci. Abstr.* **4** (1978): 201.
125. McClellan, A. D. "Swallowing and Regurgitation in the Isolated Nervous System of *Pleurobranchaea*: Distinguishing Features and Higher Order Control." *Neurosci. Abstr.* **5** (1979): 253.
126. McClellan, A. D. "Feeding and Regurgitation in *Pleurobranchaea californica*: Multibehavioral Organization of Pattern Generation and Higher Order Control." Ph.D., Case Western Reserve University, 1980.
127. McClellan, A. D. "Movements and Motor Patterns of the Buccal Mass of *Pleurobranchaea* During Feeding, Regurgitation, and Rejection." *J. Exp. Biol.* **98** (1982): 195–211.
128. McClellan, A. D. "Re-Examination of Presumed Feeding Motor Activity in the Isolated Nervous System of *Pleurobranchaea*." *J. Exp. Biol.* **98** (1982): 212–228.
129. Merzenich, M. M., J. H. Kaas, J. T. Wall, R. J. Nelson, M. Sur, and D. J. Felleman. "Topographic Reorganization of Somatosensory Cortical Areas 3b and 1 in Adult Monkeys Following Restricted Deafferentation." *Neurosci.* **8** (1983): 33–55.
130. Morgan, J. L. M. "Peptidergic Regulation of Visceral Motor Circuits in the Sea Hare, *Aplysia Californica*." Ph.D. thesis, Oregon State University, 1991.
131. Morgan, J. S., E. C. Patterson, and A. H. Klopf. "A Drive-Reinforcement Model of Simple Instrumental Conditioning." *Proc. IJCNN* **2** (1990): 227–232.
132. Mpitsos, G. J. "Physiology of Vision in the File Clam *Lima Scabra*." *J. Neurophysiol.* **367** (1973): 371–383.
133. Mpitsos, G. J. "Chaos in Brain Function and the Problem of Nonstationarity: A Commentary." In *Dynamics of Sensory and Cognitive Processing by the Brain*, edited by E. Basar and T. H. Bullock, 521–535. New York: Springer-Verlag, 1989.
134. Mpitsos, G. J. "Neural Network Error Surfaces: Limitation of Network Size, Input Signal Dynamics, and Metaknowledge in Memory Storage." *Soc. Neurosci. Abstr.* **17** (1991): 484.
135. Mpitsos, G. J. "Attractors Provide Mechanism for Gradient Descent in Biological Organization: Effect of Chaos and Noise." In *Behavioral Mechanisms in Evolution Ecology*, edited by L. Real. Chicago: University of Chicago Press, 1992.
136. Mpitsos, G. J., and R. M. Burton. "Convergence and Divergence in Neural Networks: Processing of Chaos and Biological Analogy." *Neural Networks* (1992), in press.

137. Mpitsos, G. J., R. M. Burton, H. C. Creech, and S. O. Soinila. "Evidence for Chaos in Spike Trains of Neurons that Generate Rhythmic Motor Patterns." *Brain Res. Bull.* **21** (1988): 529–538.
138. Mpitsos, G. J., and C. S. Cohan. "Comparison of Differential Pavlovian Conditioning in Whole Animals and Physiological Preparations of *Pleurobranchaea*: Implications of Motor Pattern Variability." *J. Neurobiol.* **17** (1986): 498–516.
139. Mpitsos, G. J., and C. S. Cohan. "Convergence in a Distributed Motor System: Parallel Processing and Self-Organization." *J. Neurobiol.* **17** (1986): 517–545.
140. Mpitsos, G. J., and C. S. Cohan. "Differential Pavlovian Conditioning in the Mollusk *Pleurobranchaea.*" *J. Neurobiol.* **17** (1986): 487–497.
141. Mpitsos, G. J., and C. S. Cohan. "Discriminative Behavior and Pavlovian Conditioning in the Mollusc *Pleurobranchaea.*" *J. Neurobiol.* **17** (1986): 469–486.
142. Mpitsos, G. J., and S. D. Collins. "Learning: Rapid Aversive Conditioning in the Gastropod Mollusc *Pleurobranchaea.*" *Science* **188** (1975): 954–957.
143. Mpitsos, G. J., S. D. Collins, and A. D. McClellan. "Learning: A Model System for Physiological Studies." *Science* **199** (1978): 497–506.
144. Mpitsos, G. J., H. C. Creech, C. S. Cohan, and M. Mendelson. "Variability and Chaos: Neurointegrative Principles in Self-Organization of Motor Patterns." In *Dynamic Patterns in Complex Systems*, edited by J. A. S. Kelso, A. J. Mandell, and M. F. Shlesinger, 162–190. Singapore: World Scientific, 1988.
145. Mpitsos, G. J., and W. J. Davis. "Learning: Classical and Avoidance Conditioning in the Mollusk *Pleurobranchaea.*" *Science* **180** (1973): 317–320.
146. Mpitsos, G. J., and K. Lukowiak. "Learning in Gastropod Molluscs." In *The Mollusca*, edited by A. O. D. Willows, 95–267.8. New York: Academic Press, 1985.
147. Mpitsos, G. J., T. F. Murray, H. C. Creech, and D. L. Barker. "Muscarinic Antagonist Enhances One-Trial Food-Aversion Learning in *Pleurobranchaea.*" *Brain Res. Bull.* **21** (1988): 169–179.
148. Mpitsos, G. J. Unpublished observations.
149. Murphy, A. D., K. Lukowiak, and W. K. Stell. "Peptidergic Modulation of Patterned Motor Activity in Identified Neurons of *Helisoma.*" *Proc. Natl. Acad. Sci. USA* **82** (1985): 7140–7144.
150. Murray, T. F., and G. J. Mpitsos. "Evidence for Heterogeneity of Muscarinic Receptors in the Mollusc *Pleurobranchaea.*" *Brain Res. Bull.* **21** (1988): 181–190.
151. Murray, T. F., G. J. Mpitsos, J. F. Siebenaller, and D. L. Barker. "Stereoselective L-[^{3}H] Quinuclidinyl Benzilate-Binding Sites in Nervous Tissue of *Aplysia Californica*: Evidence for Muscarinic Receptors." *J. Neurosci.* **5(12)** (1985): 3184–3188.
152. Nadel, L., L. A. Cooper, P. Culicover, and R. M. Harnish, ed. *Neural Connections, Mental Computation*. Cambridge MA: MIT Press, 1989.

153. Ono, J., and R. E. McCaman. *Neuroscience* **11** (1984): 549.
154. Osovets, S. M., D.-A. Ginzburg, V. S. Gurfinkel, L. P. Zenkov, L. P. Latash, V. B. Malkin, P. V. Mel'nichuk, and E. B. Pasternak. "Electrical Activity of the Brain: Mechanisms and Interpretation." *Sov. Phys. Usp.* **26** (1984): 801–828.
155. Pearson, J. C., L. H. Finkel, and G. M. Edelman. "Plasticity in the Organization of Adult Cerebral Cortical Maps: A Computer Simulation Based on Neuronal Group Selection." *J. Neurosci.* **7** (1987): 4209–4223.
156. Pieroni, J. P., and J. H. Byrne. "Differential Effects of Serotonin, SCP_B, and FMRFamide on Processes Contributing to Presynaptic Facilitation in Sensory Neurons of *Aplysia*." *Soc. Neurosci. Abstr.* **15** (1989): 1284.
157. Pribram, C. *Languages of the Brain*. Englewood Cliffs: Prentice Hall, 1971.
158. Rapp, P. E., I. D. Zimmerman, A. M. Albano, G. C. Deguzman, and N. N. Greenbaun. "Dynamics of Spontaneous Neural Activity in the Simian Motor Cortex: The Dimension of Chaotic Neurons." *Phys. Lett.* **110A** (1985): 335–338.
159. Real, L. A. "Animal Choice Behavior and the Evolution of Cognitive Architecture." *Science* **253** (1991): 980–986.
160. Rosen, S. C., T. Teyke, M. W. Miller, K. R. Weiss, and I. Kupferman. "Identification and Characterization of Cerebral-to-Buccal Interneurons Implicated in the Control of Motor Programs Associated with Feeding in *Aplysia*." *J. Neurosci.* **11** (1991): 3630–3655.
161. Rössler, O. "An Equation for Hyperchaos." *Phys. Lett.* **71A** (1979): 155.
162. Rössler, O. E. "An Equation for Continuous Chaos." *Phys. Lett.* **57A** (1976): 397–398.
163. Roux, J.-C. "Experimental Studies of Bifurcations Leading to Chaos in the Belousov-Zhabotinsky Reaction." *Physica* **7D** (1983): 57–68.
164. Roux, J. C., R. H. Simoyi, and H. L. Swinney. "Observation of a Strange Attractor." *Physica* **8D** (1983): 257–266.
165. Rumelhart, D. E., G. E. Hinton, and R. J. Williams. "Learning Internal Representations by Error Propagation." In *Parallel Distributed Processing: Explorations in the Microstructure of Cognition, Vol 1. Foundations*, edited by D. E. Rumelhart and J. L. McClelland, 318–362. Cambridge: MIT Press, 1986.
166. Rumelhart, D. E., J. L. McClelland, and PDP Group, ed. *Parallel Distributed Processing: Explorations in the Microstructure of Cognition, Vol 1. Foundations*. Cambridge: MIT Press, 1986.
167. Schnabl, W., P. F. Stadler, C. Frost, and P. Schuster. "Full Characterization of a Strange Attractor: Chaotic Dynamics in Low-Dimensional Replicator Systems." *Physica D* **48** (1991): 65–90.
168. Schroeder, M. *Fractals, Chaos, Power Laws—Minutes from an Infinite Paradise*. New York: W. H. Freeman, 1991.
169. Schulman, J. A. "Chemical Neuroanatomy of the Cerebellar Cortex." In *Chemical Neuroanatomy*, edited by P. C. Emson, 209–228. New York: Raven Press, 1983.

170. Seydel, R. *From Equilibrium to Chaos. Practical Bifurcation and Stability Analysis.* New York: Elsevier, 1988.
171. Shimizu, T., and H. J. Karten. "Immunohistochemical Analysis of the Visual Wulst of the Pigeon (*Columba livia*)." *J. Comp. Neurol.* **300** (1991): 346–369.
172. Skarda, C. A., and W. J. Freeman. "How Brains Make Chaos in Order to Make Sense of the World." *Behav. Brain Sci.* **10** (1987): 161–195.
173. Skinner, J. E., M. Mitra, and K. W. Fulton. "Low-Dimensional Chaos in a Simple Biological Model of Neocortex: Implications for Cardiovascular and Cognitive Disorders." In *An International Perspective on Self-Regulation and Health*, edited by J. G. Carlson and A. R. Seifer. New York: Plenum, 1989.
174. Soinila, S., and G. J. Mpitsos. "Immunohistochemistry of Diverging and Converging Neurotransmitter Systems in Molluscs." *Biol. Bull.* **181** (1991): 484–499.
175. Soinila, S., and G. J. Mpitsos. "Distribution of Acetylcholine in the Nervous System of *Aplysia* and *Pleurobranchaea*." 1992, in preparation.
176. Soinila, S., G. J. Mpitsos, and P. Panula. "Comparative Study of Histamine Immunoreactivity in Nervous Systems of *Aplysia and Pleurobranchaea*." *J. Comp Neurol.* **298** (1990): 83–96.
177. Sommeria, J., S. D. Meyers, and H. L. Swinney. "Laboratory Simulation of Jupiter's Great Red Spot." *Nature* **331** (1988): 689–693.
178. Sperry, R. W. "Changing Priorities." *Ann. Rev. Neurosci.* **4** (1981): 1–15.
179. Sugita, S., D. A. Baxter, and J. H. Byrne. "Serotonin- and PKC-Induced Spike Broadening in Tail Sensory Neurons of *Aplysia*." *Soc. Neurosci. Abstr.* **17** (1991): 1590.
180. Swinney, H. Personal communication.
181. Tanji, J., and E. V. Evarts. "Anticipatory Activity of Motor Cortex Neurons in Relation to Direction of an Intended Movement." *J. Neurophysiol.* **39** (1976): 1062–1068.
182. Thom, R. *Semio Physics: A Sketch of Aristotelian Physics and Catastrophe Theory.* Redwood City, CA: Addison-Wesley, 1990.
183. Thompson, J. M. T., and H. B. Stewart. *Nonlinear Dynamics and Chaos.* New York: John Wiley & Sons, 1986.
184. Ueda, Y. "Steady Motions Exhibited by Duffing's Equation: A Picture Book of Regular and Chaotic Motion." In *New Approaches to Nonlinear Problems in Dynamics*, edited by P. J. Holmes, 311–322. Philadelphia: SIAM, 1980.
185. Vemuri, V. *Artificial Neural Networks: Theoretical Concepts.* Neural Networks, Washington, DC: Computer Society Press of the IEEE, 1988.
186. Warren, A. H. "An Investigation in Size Reduction in Neural Networks." Masters thesis, Oregon State University, 1989.
187. Weissenfeld, K., C. Tang, and P. Bak. "A Physicist's Sandbox." *J. Stat. Phys.* **54** (1989): 1441–1458.
188. Werblin, F. S., and J. E. Dowling. "Organization of the Retina of the Mud Puppy, *Necturus Maculosus*." *J. Neurophysiol.* **32** (1969): 339–355.

189. Wetzel, M. C., and D. G. Stuart. "Ensemble Characteristics of Cat Locomotion and Its Neuronal Control." *Progr. Neurobiol.* **7** (1976): 1–98.
190. Wilson, D. M. "The Central Nervous Control of Flight in a Locust." *J. Exp. Biol.* **38** (1961): 471–490.
191. Wilson, H. R., and J. D. Cowan. "Excitatory and Inhibitory Interactions in Localized Populations of Model Neurons." *Biophysical J.* **12** (1972): 1–24.
192. Wolf, A., J. B. Swift, H. L. Swinney, and J. A. Vastano. "Determining Lyapunov Exponents from a Time Series." *Physica* **16D** (1985): 285–317.
193. Wright, W. G., E. A. Marcus, and T. J. Carew. "Dissociation of Monosynaptic and Polysynaptic Contributions to Dishabituation, Sensitization, and Inhibition in *Aplysia*." *Soc. Neurosci. Abstr.* **15** (1989): 1265.
194. Wu, J., C. X. Falk, H. Höpp, and L. B. Cohen. "Trial-to-Trial Variability in the Neuronal Response to Siphon Touch in the *Aplysia* Abdominal Ganglion." *Soc. Neurosci. Abstr.* **15** (1989): 1264.
195. Zečević, D., J. Wu, L. B. Cohen, J. A. London, H. Höpp, and C. X. Falk. "Hundreds of Neurons in the *Aplysia* Abdominal Ganglion are Active During the Gill-Withdrawal Reflex." *J. Neurosci.* **9** (1989): 3681–3689.

Timothy R. Thomas,† George J. Papcun,† and Barry N. Guinn‡
†Computing and Communications Division, Mail Stop B-265, Los Alamos National Laboratory, Los Alamos, NM 87545, and ‡Department of Computer Science, Indiana University, Bloomington, IN 47405

Mapping from Speech Acoustics to Tongue Dorsum Movement: An Application of a Multilayer Perceptron

INTRODUCTION

In this lecture we will illustrate the application of nonlinear neural network techniques to a large-scale problem of some importance. While the strategies and methods we adopted may not be exactly the correct approach for other situations, we suggest that a detailed examination of how one complex, nonlinear problem was solved will facilitate the search for good solutions to other similar problems. The focus will be on how we managed to reach a satisfactory result with the technique, rather than on the theoretical benefits of the method.

The problem explored in this paper is that of finding a map from speech acoustics to the movement of the speech articulators. If we were successful in this task, we could then produce a simulated X-ray of the movements of the tongue during speech. This would be a substantial aid to the deaf since such a display would permit a kind of "lip reading" of the tongue. It would also benefit the field of speech therapy, where knowledge of tongue movements would help to monitor and guide treatment. As an instantiation of the problem, we have chosen to map to the vertical movement of the tongue dorsum, which we define as a point on the midline upper surface of the tongue 30 mm behind the tip—a point which generally touches the roof of the mouth near the posterior boundary of the hard palate when a speaker forms consonants like /k/ and /g/.

We also view our research as a first step toward a device that will recognize natural, continuous speech produced at different rates by different speakers. Our approach is based on a substantial body of theoretical[5,6] and experimental[3,10,11,21] evidence that suggests people perceive speech by extracting from the acoustic signal information regarding the vocal tract gestures that were used to produce the speech. According to this approach all sounds in a particular language can be represented by a unique pattern of gestural units. For example, the difference between /k/ and /g/ is that the velar closing gesture is accompanied by a glottal adduction in /g/ but not in /k/. If we could detect the motion of the tongue dorsum during speech, we would be able to identify the occurrence of a velar closure gesture, and would have developed a system that might be extended to other gestures. This in turn would eventually enable us to detect the unique combination of gestures that specify each utterance in the language. One advantage of gestural units for speech recognition is that gestures are relatively invariant across speaking rate.[23] For a more detailed discussion of the mapping problem, as well as its implications for recognizing speech gestures, see Papcun et al.[16]

We chose to attack the problem with neural networks because we expected the relationship between acoustics and tongue positions to be extremely complex, and because previous attempts to map from speech acoustics to vocal tract position by analytic[1] or other techniques[2,25] have not been particularly successful. Also, multilayer perceptrons (MLP) have been used successfully to map from speech acoustics to phonemes[4,24] and to words.[7,17] Therefore, we elected to use a feed-forward, fully interconnected MLP to attempt to map from speech acoustics to the underlying motion of the articulators that produced that speech.

INPUT AND OUTPUT

Data were collected at the University of Wisconsin's Waisman Center's X-ray microbeam facility. The microbeam tracked the movement of a 2.5 mm gold pellet attached to the tongue dorsum during natural speech. The acoustic signal was also simultaneously recorded. We recorded three male speakers repeating monosyllabic words three times each in lists of eight words. Each list formed a record and took 20 sec to complete. The words contained one of five vowels: /U/ as in *hud*, /ae/ as in *had*, /I/ as in *hid*, /i/ as in *he*, or /o/ as in *hoe*. These were preceded either by a glottal fricative /h/, an unvoiced velar stop /k/, or an unvoiced alveolar stop /t/. After the vowel, the words were ended either by an unvoiced velar /k/, a /ks/, or by a voiced alveolar /d/. Words beginning with /h/ had no final consonant. Examples of the words are *he, cud, toke, tucks,* and *tax*. These words were selected because they featured velar stops in a variety of vowel contexts.

As in any use of neural network techniques, a crucial decision was how to represent the input. A good representation contains the necessary information for a successful map and eliminates unnecessary, confusing, or irrelevant information. The speech signal contains large amounts of this latter type of information, such as that related to the loudness of the speech signal, and the sex and identity of the speaker. However, as usual in the application of neural networks, we do not know exactly what aspect of the input will contribute to a successful map—if we did have this exact knowledge, we would probably not have selected a neural network approach, but rather would have used a more structured artificial intelligence technique.

The representation problem can be summarized as a choice between two different philosophical approaches: (1) the know-nothing approach and (2) the know-it-all approach. If we had followed the first, we would have presented the speech signal in something close to its raw form and claimed that the network would adjust its weights to ignore irrelevant information; if we had adopted the second approach, we would have thrown out everything judged to be irrelevant and let the network examine only a highly filtered signal that highlighted the supposedly critical features. The advantage of the first approach is that we are fairly certain to have included the necessary information, while the disadvantage is that all the excess input will slow the finding of a solution or will permit a solution that is idiosyncratic to the training set. This will cause a failure to generalize successfully. The advantages of the second approach are that it is likely the features we have highlighted will actually be used by the net, thus aiding generalization. Moreover, we could expect the reduced size and complexity of the net to result in faster training. The disadvantages are that faulty filtering of the speech will exclude key information and the net will fail to learn or will learn a solution that will fail to generalize. Our strategy was to lean to the know-nothing approach, but to filter the speech in ways that are well known to be done by the peripheral auditory system—i.e., transform short time segments to the frequency domain, and represent the spectral energy in log scales in both power (db scale) and frequency (bark scale).

The acoustic signal was sampled at 10000 Hz, low pass filtered at 5000 Hz, and adjusted to a mean of 0 to remove any dc bias. It was then segmented into overlapping 12.8 msec sections, passed through a Welch window,[18] and transformed to frequency domain by an FFT. The resulting power spectrum was converted to a db scale and then redistributed into 19 bark-scale categories to more nearly represent the frequency resolution of the human auditory system. The first two bark bins (0–200 Hz) were deleted because they primarily contain information about the pitch of the speaker's voice. This helped to reduce the dimension of the input vectors, thus simplifying the network by reducing the number of weights. After this step, each record was represented by sequential 6.4 msec frames, each containing 17 numbers representing the power spectrum.

The next step was to normalize both the articulatory data and the acoustic data processed as described above. Normalization can enhance the effectiveness of the network by eliminating unwanted sources of variation, emphasizing crucial features of the data and pre-adjusting the data to the characteristics of the nonlinear function used in the network. We have generally found that normalizing our input/output values to within the 0–1 range provides a consistent setup for a variety of network problems and thus helps to transfer insights gained in one problem to other situations. For some problems we have scaled the input to a wider range to take advantage of the interaction of the size of the input values with the initial weights, biases, and dynamic range of the nonlinear sigmoid function. Output values are also typically normalized to within the 0–1 range to permit a choice of making the final network output either a simple sum of the input to the final units, or of making the output units like all other units by passing the sum through the sigmoid function.

For the problem at hand, we used normalization to accomplish five goals: (1) to eliminate spectral tilt (the trend toward lower energy values at higher frequencies) and thus help to equate differing recording environments; (2) to eliminate recording glitches (where sporadic very large or small values were anomalously recorded); (3) to insure that each bark in the spectrum was given equal initial weighting; (4) to equalize the loudness for different acoustic recordings of different subjects; and (5) to make the range of tongue movements for different vocal tracts equivalent. For these purposes the acoustic representation was normalized within barks, with the highest 0.1% of the values in each bark of each record assigned a value of 1, and the lowest 0.1% a value of 0. Other inputs were given proportionate values. The tongue position values were normalized in the same fashion, but in this case the limits were set so that the maximum value of both dimension of each pellet were assigned a value of 0.9 and a minimum of 0.1. These values were chosen because 1 and 0 are the limits of the nonlinear sigmoid function applied to the final network output, and so are not realizable. Also, the values of 0.9 and 0.1 permit the output for extreme patterns to over or undershoot the target, thus facilitating overall reduction of the error.

THE CONTEXT WINDOW CONCEPT AND THE PROBLEM OF TIME

A critical characteristic of speech that must be accounted for by any successful application of network technology to speech problems is that of the temporal distribution of the speech signal. The meaning of an acoustic event at any one time depends heavily on the acoustic signal which surrounds it. Perhaps the simplest technique for capturing this temporal dependency is to cut out from the signal a section of time that is presumed to contain the category of interest, and present it to the net as a static representation. This is appropriate if the goal is to identify isolated phonemes or words, and has been used by Burr[7] and Waibel.[24] While this solves the context problem, it creates serious problems of segmentation and alignment.[7] The segmentation problem is particularly critical in the selection of prototypes that are subword units, since phonemic information is overlapped in the speech signal[11] and therefore any particular short-term segment contains information about several phonemes. For word or syllable-length prototypes, segmentation can be artificially accomplished by having the speaker produce the tokens in isolation. Unfortunately, this requires that the user of the trained network also separate his words by brief pauses, vastly reducing the speech recognizer's utility. Furthermore, this form of segmentation creates a critical alignment problem in that the test tokens must now be aligned in the input window in the same way as the training tokens. How this is to be done automatically is not clear, though Burr[7] has successfully used a modification of dynamic time warping, and many researchers simply have aligned by hand,[26] leaving the problem for future research.

In any case, since our problem is to map from a vector-valued time series to a continuous curve, segmentation is not a viable option and alternative methods for capturing temporal dependencies are needed. One option would be to introduce recurrences into the network architecture so as to provide the network with a kind of memory for past events. This method was applied by Watrous et al.[26] to the identification of phonemes; by Robinson and Fallside[19] to the speech coding problem; and by Laszlo and Zahorian[13] to speaker identification. However, since this technique involves complexities that we poorly understand, and presents problems of how to prepare the input, we elected to use another technique—that of a *context window*. In this technique a time window of n frames is passed, one frame at a time, over the acoustic input, producing an $n \times Number\_of\_Frequency\_Bins$ input matrix associated with each desired output value. Thus, each of the tongue's particular momentary positions is associated with an $n \times Time\_Slice\_Length$ segment of speech acoustics. Whether this segment should be arranged symmetrically around the moment at which a particular tongue position is serving as the desired network output, or be arranged so that most or all of it is in the future or past relative to the desired output, must be decided by empirical exploration or phonetic analysis. A somewhat similar technique, called a time delay neural network, was used by Waibel et al.[24] and Hampshire and Waibel.[12] In their technique, each hidden unit in the network received input from a context window involving only the past, and each hidden unit was aligned to a time step in the acoustic input. However, the input patterns were static in that they were segmented out of speech and temporally aligned by hand, so this technique is not compatible with the present objective. Nonetheless, the idea of a time delay unit on the second hidden layer may contribute to further development of the context window technique employed here.

THE NEURAL NETWORK

Having already decided to use a MLP to learn the mapping, we first faced the problem of how to implement the network. Since many authors had warned that a major practical disadvantage of traditional back-propagation implementations was that training could take a very long time,[8,24] we early decided to implement the network on a CRAY supercomputer. This led to a decision to write the code in Fortran ourselves, rather than modifying existing general purpose programs. This proved to be a fortunate decision, because by using the program optimization utilities of the CRAY, we were able to spot ways to maximize vectorization for this particular problem, and to then easily modify the code for faster execution. The most important change in this respect was to read all the acoustic data into the program once as a $Number\_of\_Frequency\_Bins \times Number\_of\_Frames$ matrix. Then when selecting a particular context window for input, we would point to the first frame in that window and, by reading beyond the row boundary of the data matrix, implicitly provide a one-dimensional vector of length $Number\_of\_Frequency\_Bins \times Context\_Window\_Size$ as an input vector for the net. If this pointer technique had not been used, it would have been necessary to store each input/output pair separately; since each pair in a typical run contained about 850 floating point numbers, this would have greatly increased the amount of memory required. Such tricks can make an enormous difference in the time and memory required for training, and illustrate that successful use of neural networks may require tailoring the code to suit your problem, as well as tailoring the input representation to suit your code.

THE TRAINING PROBLEM—FINDING A GOOD SET OF WEIGHTS

Another set of decisions involved the selection of a training method. Training is here defined as finding a good set of weights for the network—ones that will map speech acoustics to tongue dorsum position for words not in the training set. While training time itself does not define success, it should be short enough to permit reasonable exploration of the network parameter space and to allow research into the best form and size of the training set. Thus, achieving rapid training, while not crucial to successful application of back propagation, is a sensible goal. If we could handcraft weights, and thus avoid the training problem entirely, we would do so—but in that case we could have probably solved the problem in some more efficient way and would not be using neural networks at all.

We considered three approaches to training: gradient descent plus momentum, implemented as a back-propagation algorithm[20]; Quickprop, a modification in which parabolic projections on the weight changes are used[9]; and conjugate gradient optimization, in which the entire network is treated as a vector function to be minimized.[18] All three techniques, when applied to our task, were able to find a set of weights that would map the training input to the desired output within a small RMSE. However, back propagation produced the best overall results both in terms of time to train and success of generalization.[22] Within the back-propagation technique, we used three methods to reduce training time: (1) use the best hardware, (2) optimize the code for that hardware, and (3) find good combinations of network parameters. Of these methods, 1 and 2 do not affect generalization; however, we have found that good training parameters are generally good settings for successful generalization.

Having decided to use a gradient descent method, the question arises as to how large a step to take down the gradient at each weight update. This is the problem of selecting a proper learning rate. In general, as is illustrated in Figure 1, there is a U-shaped function that relates $Learning\_Rate$ to the time required to reach the RMSE error level set for termination. In the example illustrated, and in others we have explored, there is a broad range of $Learning\_Rate$ values that produce approximately equivalent results, with rapid deterioration of performance outside this range. The strategy we adopted for choosing the size of the step in weight space was to move down the gradient as far as possible consistent with a rough overall decrease in the total error. Since we are only interested in finding a set of weights that will produce a map with the required accuracy and not in how we found it, we accept fairly erratic iteration-to-iteration movement of the RMSE in the hope that one large step will move quickly to an adequate position. This strategy will be more successful if there are large, easy-to-find regions of weight space that produce adequate maps. If the solution set is a small portion of the space, then this sort of erratic search will be proportionately less successful.

A second parameter closely related to learning rate is the *Momentum_Term*, which determines the proportion of the last weight change to be added to the current weight change. Again, we expect a U-shaped function relating the size of the momentum term and the time required to reach termination criterion. Figure 2 shows this function for the case where *Learning_Rate* = 1.0. Peeling and Moore[17] found that for a static digit recognition task, the appropriate value of the *Momentum_Term* depended on an interaction with the *Learning_Rate* value. We would certainly expect this to hold here, too, but also expect that the shape of the error surface in weight space produced by the particular problem will have an even greater effect on the choice of a *Momentum_Term*. A large momentum term would likely be helpful if the error surface were relatively flat and smooth, while harmful if the surface were irregular with narrow valleys. At any rate, our tactic has generally

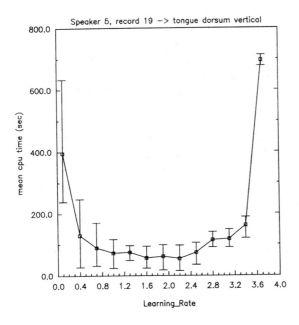

FIGURE 1 Mean cpu time to reach termination criterion as a function of the *Learning_Rate*. Error bars represent the range of five runs with different initial weights.

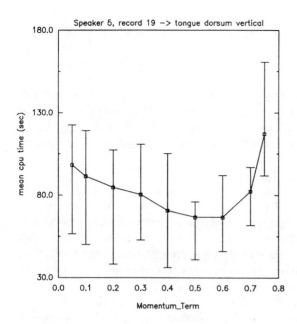

FIGURE 2 Mean cpu time to reach termination criterion as a function of *Momentum_Term*. Error bars represent the range of five runs.

been to initially set the *Momentum_Term* to a relatively low value, then find a good *Learning_Rate*, and finally do a rough search for a better *Momentum_Term*.

We would not expect that the choice of *Learning_Rate* or *Momentum_Term* would have any significant effect on generalization success, providing the net was trained to the same final RMSE. Figure 3 show the correlation between the obtained and desired output for a test set as a function of the *Momentum_Term*. As expected, no substantial effects were found for this parameter, nor were they found when *Learning_Rate* was varied. Thus, unless one needs very rapid training, such as might be required in an on-line training situation, a sensible strategy might be to guess values for *Learning_Rate* and *Momentum_Term* and do a crude search only if the time to train interferes with the progress of the research.

An issue related to *Learning_Rate* is that of the proper learning termination criterion. The goal is to find a level of training that can be achieved in an affordable period of time, which will also map both the training and test sets to the correct output within a tolerable error. The tendency is to assume that better performance on the test set will necessarily mean better performance in generalization testing, and therefore train to the lowest level attainable. We have found that this strategy generally produces an effect we call *overtraining,* in which the network weights come to encode specific characteristics of the training exemplars which are not present in the test patterns. This causes poorer performance during testing. To avoid this, one can experiment with different termination levels or increase the size and generality of the training set, thus making overtraining more difficult.

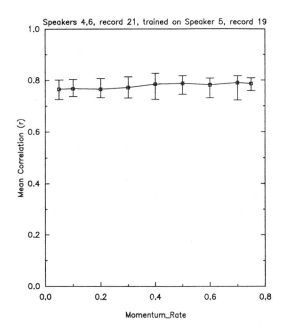

FIGURE 3 Mean correlation between the obtained and desired output during the generalization test as a function of the $Momentum\_Term$. Error bars represent the range of five runs.

For training regimes such as the present one in which there are a relatively small fixed number of input/output pairs, one can update the network weights after each pair (pattern training), or after the entire set has been presented (epoch training). The advantage of epoch training is that the measure of total error, $\sum_{i=1}^{n}(obtained_i - desired_i)^2$, where n is the total number of patterns, is guaranteed to decline with training, providing the size of the step down the gradient is sufficiently small. With pattern training no such guarantee exists, and any gain accomplished by moving down the gradient computed for one pattern may be lost when moving in the direction specified by another pattern. Nonetheless, we have found pattern training to produce much more rapid learning, with no loss in quality of the solution obtained.[22] Of course, pattern training must be accompanied by random presentations of the patterns in the training set to avoid a limit cycle where early patterns move the weights in one general direction, later patterns in another, and then the next presentations of the early patterns move the weights back again. Apparently, pattern training combined with relatively large steps (high $Learning\_Rate$) finds weights which reduce the total error because the majority of the patterns produce gradients which approximate the total gradient, and after those majority gradients become flatter, the unusual patterns (those with grossly different gradients relative to the weight position obtained) are then resolved. But in any case, as a practical matter, pattern training has been vastly superior to epoch training for this project.

THE SIZE AND CONFIGURATION OF THE NETWORK

One way to make training relatively easy is by increasing the number of weights in the net, so that there are more free parameters to adjust. However, if the number of weights begins to approach, or even exceed, the number of input/output pairs in the training set, the weights can potentially encode a sort of look-up table or memorization of the training set and thus fail to extract features that are generally characteristic of the input class. This may cause the network to fail to correctly classify new exemplars when they are presented during the generalization phase. However, just because it is possible for the net to reach a poor solution, does not mean that it will actually do so in practice. It is quite possible, especially if many of the input/output pairs are relatively similar to each other, that the solution found will be a general one that will successfully map new input to the desired output.

The easiest way to increase the number of free parameters in a fully connected net is to increase the number of hidden units in the first layer. The motivation for doing so is to provide enough units for relatively rapid learning, while still forcing the extraction of general features. Figure 4 shows the time to reach termination criterion as a function of the number of hidden units in the first layer. As the net size increases, each pass through the net takes more time because there are more connections to evaluate. However, fewer iterations through the training set are required because a solution should be easier to find. For this problem we did not find that the number of hidden units affected the success of generalization. This result was unexpected because we had felt that fewer hidden units would force more abstract representations of the input and, thus, produce a better generalization to input not in the training set. We expected to pay for this benefit by longer training times and poorer performance on the training set. However, the expected result was not obtained, perhaps because the few units (> 2) required for successful training encode features that are equally prominent in the test set.

For some problems it has been shown that adding a second hidden layer to the network can increase the success of generalization[14] since the second hidden layer can form a more abstract representation of the input pattern. While this may be true for many problems, for this problem we did not find any advantage to adding a second layer. For example, a net with one layer of 13 units generalizes from training on one recording to a second recording from two other speakers with a correlation between the obtained and desired tongue dorsum trajectories of $r = 0.76$. Adding a second layer of five units produces a nearly identical correlation of $r = 0.77$.

Of the parameters that define the state of the network, two are directly concerned with the representation of speech: the *Context_Window_Size*, and *Percent_Future*—the proportion of the context window which is future or past relative to the momentary tongue dorsum position serving as target. Since it is reasonable to presume that values which are useful for one speech problem will likely also be useful for other similar problems, we first varied the value of *Context_Window_Size* from 64 msec to 768 msec, trained on one record from one

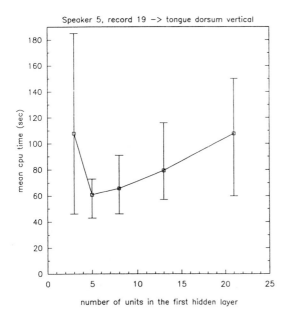

FIGURE 4 Mean cpu time to reach termination criterion of as a function of the number of units in the first hidden layer. Error bars represent the range of five runs.

speaker, and tested on different records from two other speakers. Figure 5 shows the mean CPU time required to reach a RMSE of 0.06, while Figure 6 shows the mean correlation between obtained and desired during the testing phase. Fortunately, the two criteria for good parameter settings provide similar answers. A good *Context_Window_Size* for both training time and generalization success is from 256 to 384 msec. This period is about the duration of a spoken syllable, which might be assumed to be the maximum size that has clear acoustic information about the phonemes being formed and consequently the tongue motions producing them. Figures 7 and 8 show the same measures as a function of the proportion of the context window that is in the past relative to the target. Examination of these results show that a good value for this parameter is around 0.4, suggesting that information from both the past and future is useful, while elimination of all future acoustic information makes the problem much more difficult.

THE SIZE OF THE TRAINING SET

One of the most obvious ways to attempt to increase the success of generalization is to increase the size of the training set. The argument is that since a single set of weights is now required to map many more examples to the appropriate output, the net will be forced to use features common to all members of each class in order to successfully learn the mapping. What can happen, however, is that large training sets may make it more likely that the net will reach some local minimum

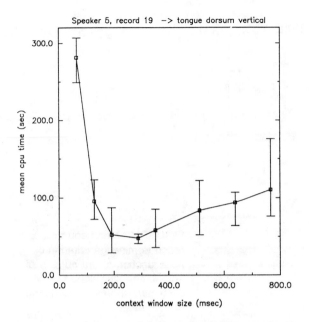

FIGURE 5 Mean cpu time to reach termination criterion as a function of the context window size. Error bars represent the range of five runs.

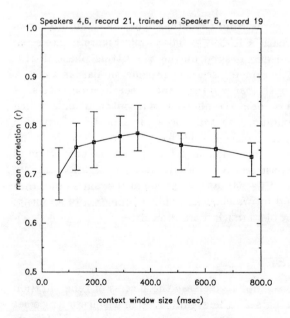

FIGURE 6 Mean correlation between obtained and desired output during the generalization test as a function of the context window size. Error bars represent the range of five runs.

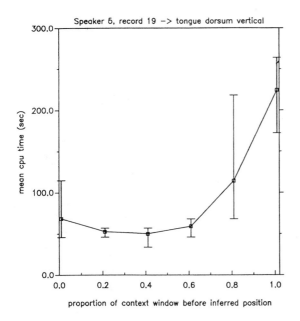

FIGURE 7 Mean cpu time to reach termination criterion as a function of the proportion of context window that is in the past relative to the momentary position of the tongue serving as the target. Error bars represent the range of five runs.

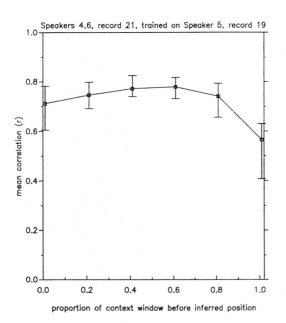

FIGURE 8 Mean correlation between the desired and obtained output during the generalization test as a function of the proportion of the context window that is in the past relative to the momentary position of the tongue serving as the target. Error bars represent the range of five runs.

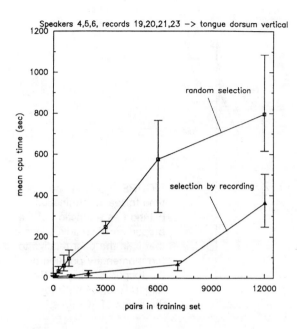

FIGURE 9 Mean cpu time to reach termination criterion as a function of the number of input/output pairs in the training set. The upper curve represents the results when the pairs presented to the net were selected at random from the entire training set, while the lower curve shows the results when the pairs were selected sequentially. Error bars represent the mean of five runs.

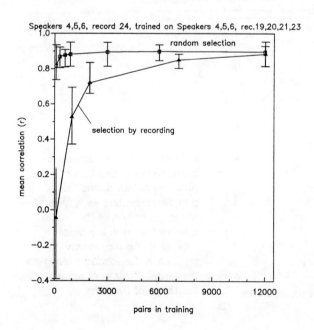

FIGURE 10 Mean correlation between desired and obtained output during the generalization test as a function of the number of input/output pairs in the training set. The upper curve represents the results when the pairs presented to the net during training were selected at random from the entire set, while the lower curve shows the results when the pairs were selected sequentially. Error bars represent the mean of five runs.

or that training time will become too long for practical application.[14] Furthermore, if the extra training vectors are not fairly representative of the distribution of the set of potential input vectors, then the net may overlearn one class of patterns—since correctly mapping that class will have a disproportionate effect on reducing the total error, while incorrectly mapping the poorly represented classes will have a small effect on overall error. Thus, increasing the training set size should not automatically be considered a wise strategy, though should clearly be considered if generalization success is below the required level.

Figure 9 shows the time to train as a function of training set size. Two techniques were used to select the members of the training set. In the first technique, members of the training set were randomly chosen from the 26,912 pairs available from four recordings from three different speakers, with the restriction that equal numbers of pairs were taken from each speaker. A fifth recording of each speaker was reserved for testing. In the second technique, the pairs were selected sequentially from the recordings, beginning with the first recording from the first speaker and continuing until all pairs in that record were exhausted, and then more pairs were included by adding all the pairs in additional recordings. This sequential technique produced considerably faster training than did random selection because the training sets contained many vectors that were similar to each other. This was expected because a recording contains one speaker repeating eight words three times. While this rapid training might appear to be an advantage, the sequential selection technique produced problems during generalization. New input vectors that are dissimilar to those that were well represented in the training set are likely to be poorly mapped. This effect is illustrated in Figure 10, where the success during testing of the two selection techniques are plotted against training set size. There the advantage of the random selection method is clearly shown, since very satisfactory results were obtained after selection of about 100 vectors from each speaker.

RESULTS

The purpose of an application of back propagation is generally not to understand how the network responds to varying its parameters, but rather to obtain a system that accurately maps new input to the correct output. The degree of accuracy required generally depends crucially on the uses to which the output will be applied. Figure 11 shows the actual and inferred tongue dorsum vertical position for one of the three speakers for a segment of speech not included in the training set. The correlation between the two curves was $r = 0.94$, showing that the general shape of the curves was highly similar, while the magnitudes of the two curves were somewhat different (RMSE = .09). However, for the purpose of identifying when the velar closure occurred, or for providing a visual display of the tongue dorsum's movement, the inferred plot is clearly adequate.

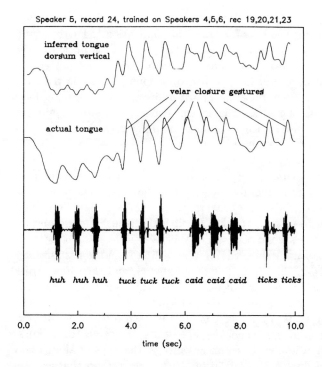

FIGURE 11 The actual and inferred tongue dorsum vertical movements during the speech segment shown in the acoustic record at the bottom of the figure. The location of the actual velar closures corresponding the /k/ phonemes are also shown.

These results were produced by a one-layer network with eight hidden units and one output unit. Training was done with a *Learning_Rate* of 1.0 and a *Momentum_Term* of 0.5, and was terminated when RMSE reached 0.08. The context window size was 320 msec with 60% of the window in the future, and the training set consisted of 6000 randomly selected pairs from four different recordings.

Obviously, since an exhaustive search of the network parameter space was not completed, some other network configuration might have produced better results. This is not of particular concern here, however, because we have achieved satisfactory results with the current training set. With other problems, where good results have not been achieved, the possibility of improvement by manipulation of the network parameters presents a problem—how to decide if further tweaking of the parameters will be beneficial or a waste of computer cycles. In general, our solution would be to try large variations in the parameters to get a rough idea of the useful range of values, and then select some moderate value and not attempt excessive fine tuning. We have found that potent effects on results are more easily obtained by changing the input representation or the training set than by making minor adjustments in network parameters.

In some respects, the results we have achieved are suspiciously good—particularly the success obtained after training on only 100 randomly selected 320 msec bits of speech from each speaker. Perhaps we have overestimated the difficulty of the problem. Since the speech signal is obviously caused by movements of the vocal apparatus, it should not be surprising that a network can easily find a functional relationship between the two domains. On the other hand, perhaps our success is the result of choosing training and test sets that are so closely matched that they are a poor representation for any realistic application.

Clearly, we now need to extend the current results to more difficult problems involving other speech articulators, differing speaking rates, more general speech samples, and a greater variety of speakers. Our optimistic hypothesis is that the experience gained here will transfer to new situations and permit rapid resolution of new problems as they develop. If this hypothesis is correct, back-propagation techniques will surely find increasingly widespread application to currently intractable problems.

ACKNOWLEDGMENTS

This work was partially supported by DOE contract W-7405-ENG-36. The DOE's Science and Engineering Research Semester provided financial support for Mr. Guinn. The Waisman Center's X-ray microbeam facility was operated under grant 5-P50-DC00162 from the NIDCD. We would like to thank Bob Nadler and Carl Johnson for crucial help during the data collection phase of this work. We would also like to thank the Computing and Communications Division and the Center for Nonlinear Studies at Los Alamos National Laboratory for providing a stimulating environment. We are indebted to Kathy Berkbigler for teaching us many valuable techniques for optimizing our code.

REFERENCES

1. Atal, B. S. "Determination of the Vocal-Tract Shape Directly from the Speech Wave." *J. Acoust. Soc. Am.* **47** (1969): 65.
2. Atal, B. S., M. V. Chang, and J. W. Tukey. "Inversion of Articulatory-to-Acoustic Transformation in the Vocal Tract by a Computer-Sorting Technique." *J. Acoust. Soc. Am.* **63** (1978): 1535–1555.
3. Best, C. T., B. Morrongiello, and R. Robson. "Perceptual Equivalence of Acoustic Cues in Speech and Nonspeech Perception." *Perception & Psychophysics* **29** (1981): 191–211.
4. Bourland, H., and C. J. Wellekens. "Speech Pattern Discrimination and Multilayer Perceptrons." *Computer Speech & Language* **3** (1989): 1–19.
5. Browman, C. P., and L. M. Goldstein. "Towards an Articulatory Phonology." *Phonology Yearbook* **3** (1986): 219–252.
6. Browman, C. P., and L. M. Goldstein. "Articulatory Gestures as Phonological Units." *Phonology* **6** (1989): 201–251.
7. Burr, D. J. "Speech Recognition Experiments with Perceptrons." In *AIP Conference Proceeding, Neural Information Processing Systems*, Denver, Colorado, 1987.
8. Elman, J. L., and K. Zipser. "Learning the Hidden Structure of Speech." *J. Acoust. Soc. Am.* **83** (1988): 1615–1626.
9. Fahlman, S. "An Empirical Study of Learning Speed in Back-Propagation Networks." CS-88-162, Carnegie-Melon University, 1988, 1–17.
10. Fitch, H. L., T. Halwes, D. M. Erickson, and A. M. Liberman. "Perceptual Equivalence of Two Acoustic Cues for Stop-Consonant Manner." *Perception & Psychophysics* **27** (1980): 343–350.
11. Fowler, C. A. "Segmentation of Coarticulated Speech in Perception." *Perception & Psychophysics* **36** (1984): 359–368.
12. Hampshire, J. B., and A. H. Waibel. "A Novel Objective Function for Improved Phoneme Recognition Using Time Delay Neural Networks." *IEEE Trans. on Neural Networks* **1** (1990): 216–228.
13. Laszlo, R., and S. A. Zahorian,. "Text-Independent Talker Identification Using Recurrent Neural Networks." *J. Acoust. Soc. Am.* **87** (1990): S104.
14. Martin, G. L., J. Lovgren, and J. Pittman. "Improving Generalization in Neural Nets." Technical Report No. ACA-HI-362-88, MCC, 1988, 1–39.
15. Martin, G. L., and J. A. Pittman. "The Value of a Second Hidden Layer in Improving Generalization of Neural Net Learning." Technical Report No. ACA-HI, MCC, 1988, 1–8.
16. Papcun, G. J., J. Hochberg, T. R. Thomas, F. Laroche, and J. Zacks. "Inferring Articulation and Recognizing Gestures from Acoustics with a Neural Network Trained on X-ray Microbeam Data." *J. Acoust. Soc. Am.*, in press.
17. Peeling, S. M., and R. K. Moore. "Isolated Digit Recognition Experiments using the Multi-Layer Perceptron." *Speech Comm.* **7** (1988): 403–409.

18. Press, W. H., B. A. Flannery, S. A., Teukolsky, and W. T. Vetterling. *Numerical Recipes.* New York: Cambridge University Press, 1987.
19. Robinson and Fallside. "The Utility Driven Dynamic Error Propagation Network." Cambridge University Technical Report CUED/F-INFENG/TR.1, Cambridge, England, 1987.
20. Rumelhart, D. E., G. E. Hinton, and R. J. Williams. "Learning Internal Representations by Error Propagation." In *Parallel Distributed Processing*, edited by D. E. Rumelhart & J. L. McClelland, 318–362. Cambridge, MA: MIT Press, 1986.
21. Spoehr, K. T., and W. J. Corin. "The Stimulus Suffix Effect as a Memory Coding Phenomenon." *Memory & Cognition* **6** (1978): 583–589.
22. Thomas, T. R., and T. L. Brewster. "Experiments in Finding Neural Network Weights." Technical Report LA-11772-MS, Los Alamos National Laboratory, 1990.
23. Thomas, T. R., and M. T. Vo. "Control of Tongue Movement Dynamics." In *1990 Lectures in Complex Systems*, edited by L. Nadel and D. Stein, 253–268. Santa Fe Institute Studies in the Sciences of Complexity, Lect. Vol. III. Redwood City, CA: Addison-Wesley, 1991.
24. Waibel, A., T. Hanazawa, G. Hinton, K. Shikano, and K. J. Lang. "Phoneme Recognition Using Time-Delay Neural Networks." *IEEE Trans. on Acoustic, Speech, & Signal Process.* **37** (1989): 328–339.
25. Wakita, H. "Estimation of Vocal-Tract Shapes from Acoustical Analysis of the Speech Wave: The State of the Art." *IEEE Trans. Acoustics, Speech, & Signal Process.* **27** (1979): 281–285.
26. Watrous, R. L., B. Ladendorf, and G. Kuhn. "Complete Gradient Optimization of a Recurrent Network Applied to /b/,/d/,/g/ Discrimination." *J. Acoust. Soc. Am.* **87** (1990): 1301–1309.

H. F. Nijhout
Department of Zoology, Duke University, Durham, NC 27706

Pattern Formation in Biological Systems

Pattern formation refers to the processes in development by which ordered structures arise within an initially homogeneous or unstructured system. Understanding these processes is absolutely essential for understanding regulatory mechanism in development. It is also essential for understanding the developmental origin of biological form, and ultimately, for understanding morphological evolution. In practice, pattern formation refers to things like the processes in embryos that determine where gastrulation will occur, or the processes that define where bones will condense in the mesenchyme of a developing limb, how many there will be, their shape, and their positional relation to each other. Or in plants, where leaves will form on the stem of a plant, and what shape those leaves will have.

Here we will be particularly concerned with processes of pattern formation that occur quite late in animal development, in particular, the development of pigment patterns. Pigment patterns have several advantages as model systems in which to study the principles of pattern formation. First, color patterns are almost always two dimensional, so they can be studied on the plane without having to use projections or collapse dimensions. This makes them far easier to deal with than three-dimensional processes in development, and makes color patterns particularly attractive for computer modeling, because the whole pattern can be represented on the two-dimensional computer screen. Second, since they develop relatively late, the processes that give rise to the pattern occur in a system that is usually macroscopic and, therefore, more easily manipulated experimentally than are early embryos. Third, since pattern is manifested as the local synthesis of pigment, it is easy to detect. Fourth, since the chemical nature and biosynthetic pathways of most pigments are known, it is in principle possible to fully understand all control pathways in the system at the chemical and molecular level. Finally, it is in systems like color patterns, where all the molecular and biochemical steps are in principle knowable and understandable, that we have the best chance of uncovering the full sequence of events that links genotype and phenotype, something that has yet to be done for any morphological event.

THREE MECHANISMS

The processes that result in local specialization of structure and function can be formally subdivided into two distinctive kinds: those that involve cell migration and mechanical interactions among cells (such as traction and differential adhesivity), and those that involve chemical pre-patterning.[14,19] Murray[14] points out that the two mechanisms are quite different because in chemical pre-patterning the chemical pattern precedes morphogenesis, while in patterning by mechanochemical cell-cell interactions, morphogenesis is the immediate consequence of the patterning process. There are a few examples of patterning systems that are purely one or the other (the formation of butterfly wing patterns, which we will deal with below, is one of them), but in most cases both mechanisms seem to operate, such as when a chemical gradient allows migrating cells to aggregate and interact.

Among the best studied examples of cell movement-mediated patterning are aggregation and fruiting body formation in the slime mold *Dictyostelium*, and the formation of bones in the developing limbs of vertebrates. In *Dictyostelium* we have one of the very few cases in which we actually know what the chemical morphogen is whose gradient stimulates the initial aggregation. Here the aggregation signal is cyclic AMP (cAMP), which is secreted by isolated cells when they run out of food. When other cells perceive this signal, they are attracted to its source and migrate up the cAMP gradient. Such migrating cells also begin to secrete cAMP themselves, and a complex set of interactions ensues that transiently gives rise to interesting cell aggregation patterns and eventually results in the clumped aggregates. While aggregating, the population of cells does exhibit spatial patterns of spiral waves very similar to those seen in the Belousov-Zhabotinski reaction, and in many models using cellular automata (see figures in Winfree,[32] Tomchik and Devreotes,[27] and Murray[14]).

Patterned bone formation in the mesenchyme of developing vertebrate limbs has been studied in a variety of contexts. Perturbation experiments have revealed complex interactions that have been modeled conceptually as the well-known clock face model of French et al.[8] and Bryant et al.,[2] and mechanistically as a traction-aggregation mechanism by Oster et al.[20] Evolutionary morphologists have been particularly interested in the development of bone patterns in vertebrate limbs because the well-established homologies among the bones, and the extensive historical pattern of their transformation preserved in the fossil record, makes this one of the most attractive (and tractable) systems in which to study the interplay of developmental and evolutionary processes in the shaping of biological form.[9,21]

Pigment patterns arise by the same two mechanisms, of cell movement and chemical pre-patterning and, in mollusk shells, by a third distinctive mechanism that involves a complex interplay between the tissue of the mantle and the shell as it is secreted. In vertebrates, the pigment pattern of the skin is produced by melanophores, which are cells that produce the black/brown pigment, melanin. Melanophores arise from the neural crest (along the dorsal midline) early in embryonic development and from there migrate across the body surface.[5,29,30] The color patterns of fish, frogs, zebras, giraffes, and leopards are therefore the consequence of the migration and patterned accumulation of pigment-producing cells.

In insect color patterns, the mechanism is quite different. The insect epidermis is only one cell layer thick, and the cells are attached to the overlying cuticle most of the time. Cell migration and cell rearrangement are therefore generally impossible. All patterning in the epidermis must therefore take place by mechanisms of cell-to-cell communication. The cells of the insect epidermis are interconnected by gap junctions and are thus coupled electrically and are potentially coupled by diffusion. Signals can thus be transmitted across substantial distances and control over this communication can be exercised by modulating the number and distribution of gap junctions that are open at any one time. Pigment patterns are thus the result of local cell differentiation in a static monolayer of cells. Formation of the pattern does not involve cell migration, nor is the pattern subsequently modified by cell rearrangement.

In the shells of gastropods (snails) and bivalves (clams), the color pattern is laid down as the shell is secreted. The pigment of the pattern resides not in cells but in the non-living shell. In contrast to the two previous cases, pattern formation in shells is essentially a one-dimensional process. During growth the mantle adds material to the leading edge of the shell and at the same time secretes pigments at appropriate locations to produce species-characteristic color patterns (stripes, spots, zig-zags, etc.). The mantle is a motile organ and moves frequently relative to the margin of the shell as the animal locomotes, rests, and hides. Consequently, shell deposition is not continuous but shows both regular and erratic periods of growth and rest. The mantle ultimately controls where and when the shell will grow, and also where exactly pigment will be deposited. The pigment pattern is thus the result of the behavior of the whole mantle and of the way it interacts with the growing edge of the shell.

The color patterns of vertebrates, of insects, and of mollusk shells thus come about by three fundamentally different mechanisms. Theoretical work has shown, however, that the essence of these three pattern-forming processes can captured by very similar sets of mathematical equations.[7,14,19] This suggests that the principles involved in each process could be fundamentally similar even though the actual mechanisms are not. In almost all cases, lateral inhibition (short-range activation coupled with long-range inhibition of a particular event) provides the organizing mechanism, and, while systems may differ in the exact means by which effective activation and inhibition is achieved (e.g. cooperativity, autocatalysis, or positive feedback, versus catabolism, interference, or competition), the final spatial results of the process are similar if not identical.

In the sections that follow we will assume, for the sake of simplicity, that chemical pre-patterning is the process at work because such a process can be modeled without having to take account of the movement of cells relative to one another. In addition, we will assume a perfectly two-dimensional system. Thus, what follows will apply, strictly speaking, only to pattern formation in the insect integument. As we will see, these assumptions produce a rich, complex, and largely non-intuitive world of patterns, that begs further exploration, both experimental and theoretical.

DIFFUSION

In biological systems, convection (usually via a circulatory system) and diffusion (within cells and, via gap junctions, between cells) provide the most common means of chemical communication within and among cells and tissues. Convection is generally used for long-range transport and appears to play no role in any of the pattern formation systems that have been studied so far. Thus, to understand patterning, we need to understand the mechanism and consequences of diffusion.

Diffusion comes about by the random movement of particles produced by thermal agitation. The mathematics of diffusion has been widely studied, and the reader is referred to the text by Crank[4] for the fundamentals, and to Carslaw and Jaeger[3] for a more elaborate treatment of special cases. In one spatial dimension, the diffusion equation is usually written as:

$$\frac{\partial \mathbf{c}}{\partial t} = D \nabla^2 \mathbf{c} \qquad (1)$$

where $\mathbf{c}$ is the concentration of the diffusion substance, and D is the diffusion coefficient. On macroscopic scales diffusion is a slow process. The dimension of the diffusion coefficient, D, can be used to get an idea of the rate of diffusion. If the diffusion coefficient is expressed in the units cm^2/sec, then the average time (in seconds) it takes for a particle to diffuse through a given distance, d (in centimeters), is approximately d^2/D.[6] Moderately large biochemical molecules diffuse through the cytoplasm of a cell with $D = 10^{-7}$. Such a molecule would take an average of $(10^{-4})^2/10^{-7} = 10$ seconds, to diffuse across the diameter of a typical 10 micron cell. The average distance over which diffusion acts within a given period of time is proportional to $(Dt)^{1/2}$.[6] Even though diffusion is an inherently slow process, it does clearly provide a relatively effective means of communication over the small distances (usually 1 mm or less) and time periods (hours to days) that are relevant to most developmental systems.

Diffusion-dependent processes can also exert their effect rapidly and over much larger distances if they are coupled to some amplifying machinery. The diffusion of a large charged molecule (say, $D = 10^{-7}$ cm2/sec) across a cell membrane can rapidly change the local balance of charge, and cause the diffusion of small ions towards or away from the area. If the small ions have a diffusion coefficient of, say, 10^{-5}, and act as intermediate messengers, then the rate of "signal" propagation caused by diffusion of the large molecule would have been amplified 100-fold. The propagation of an action potential, which is basically a cytochemical cascade mechanism, is a well-known example of the amplification of a diffusible signal.

EVOKING AN EFFECT: THRESHOLDS

Simple linear diffusion from a source into a medium, or from a source to a sink, sets up a gradient in the concentration of the diffusing substance. The concentration at a particular point (p) along such a gradient carries information. It can be used to estimate both the distance between p and the source (or the sink), and the time since diffusion began. For the purposes of pattern formation, the former, the estimation of position within a diffusion field, is the more interesting and useful one. In the simplest case of pattern formation, diffusion from a point source sets up a gradient of a chemical across an otherwise homogeneous developmental field, and some novel developmental event is caused to occur wherever the concentration of the gradient is above (or below) some critical value. As we will see below, the eyespots in the wing patterns of butterflies are produced by just such a simple mechanism. Changes in the threshold, and changes in the shape of the gradient can both alter the dimension and position of the "pattern" within the total field. The formal requirements and consequences of pattern formation by such simple gradient systems have been explored by Lewis Wolpert[34] in his "Theory of Positional Information."

On this view, the problem of pattern formation is twofold: first, how to establish a source for the diffusing signal, and second, how to retrieve the information in the diffusion gradient. The first of these problems is by far the most difficult one, and we will take it up below. The second problem can be rephrased to ask: how do you set up a threshold so that the continuously distributed gradient in one substance (the diffusing signal) is translated into a sharply discontinuous and stable change of some developmental or biochemical event?

Lewis et al.[11] have developed an elegant model for a threshold mechanism. They note that most threshold models assume an allosteric enzyme whose activity is a sigmoidal function of substrate concentration (Figure 1). The problem with such a model is that along a gradient of substrate concentration the transition from the inactive form to the active form of the enzyme is gradual and occurs over a relatively long distance (Figure 1). Increasing the number of cooperating subunits in the enzyme increases the steepness of the sigmoid transition and thus sharpens the "threshold" to some degree. Allosteric enzymes generally, however, have no more than four subunits, and that puts a practical limit to the refinement of a threshold by this means. Lewis et al.[11] suggested a modification of the allosteric model to include also a linear degradation term. They suggest the following structure. Suppose a gene G, which produces a product g, that stimulates its own synthesis by positive feedback at a rate that is a sigmoidal function of its concentration $(K_a g^2/k_b + g^2)$, and that g breaks down at a rate proportional to its concentration $(-k_c g)$. Suppose further that the synthesis of g is also stimulated by a signal molecule S, at a rate that is linearly proportional to the concentration of S. This gives the following relationship:

$$\frac{dg}{dt} = k_1 S + \frac{k_2 g^2}{k_3 + g^2} - k_4 g \tag{2}$$

which is shown graphically in Figure 2. The graph of the rate of production of g is in effect an inclined sigmoid curve whose position is controlled by the value of S. When S is small, the reaction has three steady states, two of which are stable (Figure 2). If the system starts with gene G off, and thus with no g present, the concentration of g will tend towards its low steady state. Small and moderate perturbations in its concentration will always cause g to return to this low steady state. However, if the concentration of S goes up, the level of the curve rises and there is eventually only one steady state of g (Figure 2), much higher than the previous one. Thus,

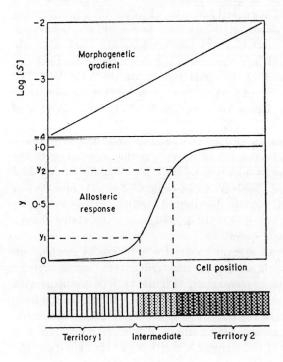

FIGURE 1 Allosteric model for a threshold. The concentration gradient in S activates an allosteric enzyme that obeys the Hill equation. The degree of saturation of the enzyme, y, that corresponds to various points along the gradient is shown in the lower graph. The threshold provided by this mechanism is not sharp and the transition can extend across many cells. Reprinted by permission of the publisher from "Thresholds in Development" by J. Lewis et al., *J. Theor. Biol.* 65, 579–590. Copyright © 1977 by Academic Press Inc. (London).

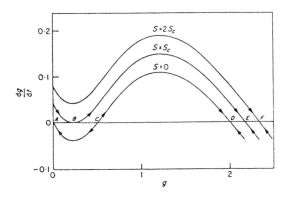

FIGURE 2 Curves produces by Eq. (2) for three values of the signal substance, S. As S gradually increases, the number of stable states abruptly fall from two to one.

if S increases gradually, there will be a sudden transition in the concentration of g from its low to its high steady state. A smooth and continuous change in the concentration of S thus results in an abrupt switch in the concentration of g. This gives us, then, a mechanism for a sharp threshold in the control variable, S, with no intermediate values between the extremes of the response variable, g.

An additional interesting and useful feature of this model is that it has a kind of "memory" because, once g has switched to its higher steady state, it will stay there even if S subsequently declines or disappears. Thus we have essentially a mechanism for the irreversible activation of a gene. If such a gene controls, for instance, the synthesis of pigment-forming enzymes, then we have a mechanism for producing a patch of pigment wherever the concentration of S is above the threshold defined by Eq. (2).

REACTION DIFFUSION

Pattern formation by diffusion gradients requires at the very minimum the existence of a source of the diffusing chemical. If pattern regulation is important, then a sink is also essential, so that all intermediate values of the gradient are always present within the developmental field. It should be clear that this requirement for a source (and a sink) in effect pushes the problem of pattern formation back one step, and the issue becomes one of determining what causes the sources and sinks to be where they are.

Though unsatisfying from a mathematical point of view, such potentially infinite regressions in control mechanisms are biologically reasonable and probably the rule rather than the exception. Development is, after all, a complex network of causal connections in which any process works correctly only if all the preceding ones did (at least within certain tolerances). There are, however, certain conditions under which a stable pattern can emerge in an initially homogeneous and randomly perturbed field without the need for initial sources or organizing centers. The conditions under which this can occur were discovered by Turing[28] and this discovery constitutes one of the major advances ever made in the theory of biological development. Turing[28] showed that the steady-state condition of certain kinds of biochemical reactions can be made spatially unstable if at least two of the reactants are able to diffuse. In other words, if the reactants are free to diffuse, then it is possible for them to become stably patterned into areas of high and areas of low steady-state concentrations. On first sight this is a non-intuitive result, because one generally thinks of diffusion as having a homogenizing effect. Under certain conditions, however, diffusion can act to amplify spatial waves of certain critical frequencies. The mathematics behind this process were outlined by Turing and have been more fully explored by many other authors since. Particularly readable accounts of the theory and the conditions under which such diffusive instabilities arise in chemical reaction systems are given by Segel and Jackson[24] and by Edelstein-Keshet,[6] and a more technical treatise with many examples is given by Murray.[14] The most elaborate exploration of the consequences and possible uses of one class of these *reaction-diffusion* mechanisms is given by Meinhardt.[12]

The conditions necessary for chemical pattern formation in reaction-diffusion systems are given by Edelstein-Keshet[6] as follows:

1. There must be at least two chemical species.
2. These chemicals must affect each other's rate of production and/or breakdown in particular ways.
3. These chemicals must also have different diffusion coefficients.

The general equation system for reaction diffusion is:

$$\frac{\partial A}{\partial t} = F(A, B) + D_A \nabla^2 A$$
$$\frac{\partial B}{\partial t} = G(A, B) + D_B \nabla^2 B$$

in which $F(A, B)$ and $G(A, B)$ define the reaction equations for the two interacting chemical species.

Most mechanisms for chemical patterning produce a set of conditions that are referred to as *lateral inhibition*. What this means is that one of the chemicals, usually called the *activator*, has a low diffusion coefficient and exerts its influence over a fairly short range while the other, called the *inhibitor*, has a much higher diffusion coefficient and thus exerts its effect over a much longer range. The term is derived from physiology where similar short-range activation, long-range inhibition systems are common, and particularly well studied in the retina where lateral inhibition is in part responsible for the detection of edges and patterns.

Three reaction-diffusion systems have achieved particular popularity for problems in developmental biology and biological pattern formation. The model of

Schnakenberg[22] is one of the simplest systems that exhibits chemical pattern formation. Its reaction dynamics are given by:

$$F(A, B) = k_1 - k_2 A + k_3 A^2 B,$$
$$G(A, B) = k_4 - k_3 A^2 B. \tag{3}$$

The lateral inhibition system of Meinhardt[12] is the one whose behavior has been studied most extensively:

$$F(A, B) = k_1 - k_2 A + \frac{k_3 A^2}{B},$$
$$G(A, B) = k_4 A^2 - k_5 B. \tag{4}$$

The reaction system of Thomas,[26] while more complicated than the preceding two, has the virtue that it is the only system that is empirical, based on real chemistry. It involves three reactants as follows:

$$F(A, B) = k_1 - k_2 A - H(A, B),$$
$$G(A, B) = k_3 - k_4 B - H(A, B),$$
$$H(A, B) = \frac{k_5 AB}{k_6 + k_7 A + k_8 A^2}. \tag{5}$$

For many purposes it is convenient to express equations such as these in a nondimensional form. One reason is that nondimensionalization always reduces the number of parameters in the model, which simplifies the analysis of the scope of the model. Another is that it removes the units of measurement and thus allows one to examine the effects of scale more effectively.[14,23] Murray[14] suggests the following general nondimensional form for reaction-diffusion systems:

$$u_t = \gamma f(u, v) + \nabla^2 u,$$
$$v_t = \gamma g(u, v) + d\nabla^2 v. \tag{6}$$

With the appropriate scaling, the reaction dynamics for the three systems mentioned above can be rewritten as follows:

$$f(u, v) = (a - u + u^2 v)$$
$$g(u, v) = (b - u^2 v) \tag{7}$$

for the Schnakenberg system;

$$f(u, v) = a - bu + \frac{u^2}{v}$$
$$g(u, v) = u^2 - v \tag{8}$$

for the Meinhardt system; and

$$f(u,v) = a - u - h(u,v)$$
$$g(u,v) = a(b - v) - h(u,v) \qquad (9)$$
$$h(u,v) = \frac{\rho u v}{1 + u + K u^2}$$

for the Thomas system.

The parameter d in Eq. (6) corresponds to the *ratio* of the diffusion coefficients of inhibitor and activator, while the parameter γ represents the scale of the system. Murray[14] suggests that γ is proportional to the area of the system, for two-dimensional diffusion. γ can also represent the strength of the reaction term relative to the diffusion term. An increase in γ can be offset by a decrease in d. The advantage of having a single variable that can represent the scale of the system is that the consequences of pattern formation in a growing system can be easily studied, and predictions can be made about the differences in pattern that would be produced when the same mechanism acts in developmental fields of different sizes. Both features are of interest to developmental biologists who perforce deal with many systems that undergo growth during the period of study.

The advantages of nondimensionalized systems in facilitating studies on the effects of scaling are offset, for the biologist at least, by the fact that other biologically important parameters (such as the reaction constants for the synthesis and breakdown of specific chemical species) become inaccessible to manipulation. Since such reaction constants provide the only direct link to the genome (genes code for enzymes, whose activity is represented by the reaction constant), it becomes virtually impossible to study the effects of single gene alterations. Thus biologists interested in exploring the potential of gradualistic accumulations of small genetic changes to cause gradualistic (or discontinuous) morphological change will need to work with fully dimensional forms of a system.

In addition to the general conditions for chemical pattern formation mentioned above, there are several specific conditions that must be met. These are treated in detail and with several examples by Segel and Jackson,[24] Edelstein-Keshet,[6] and Murray.[14] Only the summary conclusions will be given here. The form of the null clines (the graphs of $dx/dt = 0$) of the reactions gives essential information on whether diffusive instability is in principle possible. The character of the crossover point of the two null clines (the system's steady state) is critical: the activator and inhibitor must both have positive slopes or both have negative slopes at steady state, and, in either case, the slope of the inhibitor must be steeper than that of the activator.[6] The null clines for the nondimensional forms of the three reaction-diffusion equations listed above are given in Murray.[14]

Whether or not a system with null clines of the required shapes will exhibit diffusive instability depends critically on the values of the parameters, and these are specific to each system. Murray[13] has worked out the parameter space for the nondimensionalized forms of the three reaction systems listed above, and has shown that they are surprisingly narrow. In almost all cases parameters must be chosen with considerable precision and the choice of one parameter value places significant constraints on the possible values of the remaining parameters. Once the parameter space for a given system is known, however, it can form the basis for the numerical exploration of its pattern-forming properties.

The analysis of the general behavior of a reaction-diffusion system is not a trivial matter. Because reaction-diffusion systems involve coupled nonlinear equations, they usually cannot be solved analytically and their behavior must be studied by numerical simulation. It is, however, possible to get a general idea of how a particular system behaves by studying perturbations near the steady state of a linearized system (see Edelstein-Keshet[6] for the description of a method). Such a linear theory approach can predict the number of modes that will form after random perturbation of a field of given dimensions. Arcuri and Murray[1] have used linear theory to predict the pattern generated by the nondimensionalized Thomas system in one space dimension (Figure 3(a)). The theory predicts a regular increase in the number of modes as either d or γ increases (Figure 3(a)). Numerical simulation of the full nondimensional Thomas system, however, gives somewhat different results (Figure 3(b)). Odd modes appear to be favored over even modes, something which linear theory does not predict. The solution space for modes 2 and 6 is particularly small in the full nonlinear system.

Arcuri and Murray[1] also calculated how the modes of a Thomas system would behave in a growing field. As the field grows, it can support a progressively larger number of modes. Existing modes appear initially to split in two, which would result in a doubling of the number of modes. But this is not what happens. Instead, at a critical point the system appears to become unstable and reorganizes so that only a single mode is added. Only in some cases is more than one mode added as the field grows, which is consistent with the behavior shown in Figure 3(b), where many odd numbered modes are adjacent.

In two dimensions the succession of modes is more complicated, and is critically dependent on both the chemistry of the system and the geometry of the field. In a rectangular field two perpendicular sets of waves are possible, and as the scale of the field increases, more waves can be fitted along both axes. The succession of modes for a general reaction-diffusion system on a rectangular field with no-flux boundaries has been studied by Edelstein-Keshet,[6] and is shown in Figure 4 for a field of dimension 1 × 2. The quantity E^2 in Figure 4 has the following correspondence:

$$E^2 = m^2 + n^2/\gamma^2 \tag{10}$$

where m and n are integers that represent the number of wavelengths parallel to the x and y axis, respectively, and γ is the dimension of the field parallel to the y axis divided by the dimension parallel to the x axis. The succession of modes is then given by the sequence of values of m and n, ranked in order of increasing values of E^2.[6] Figure 5 illustrates the patterns that correspond to several of these modes. In a real system E^2 can be derived as a function of the area of the field and the ranges of activation and inhibition, and thus the succession of modes shown in

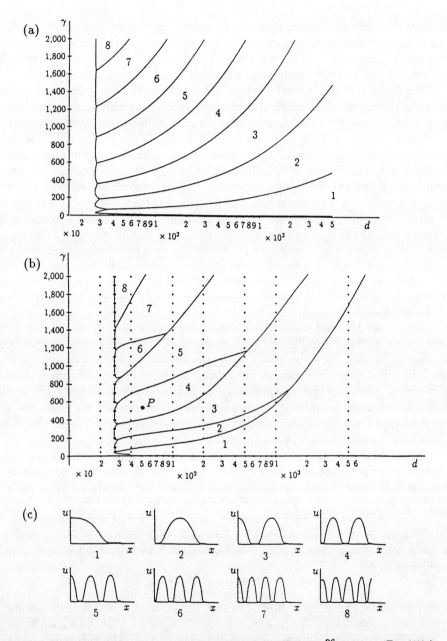

FIGURE 3 Solution space for the nondimensionalized Thomas[26] system (Eq. (9)) in one spatial dimension, with no-flux boundary conditions. (a) Modes for various values of d and γ obtained from linear theory. (b) Solution space obtained by simulation of the full non-linear system. (c) Spatial distribution of morphogen concentration for several of the regions indicated in (b). Reprinted by permission of the publisher from "Pattern Sensitivity to Boundary and Initial Conditions in Reaction-Diffusion Models" by P. Arcuri and J. D. Murray, *J. Math. Biol.* 24, 141–165. Copyright © 1986 by Springer-Verlag.

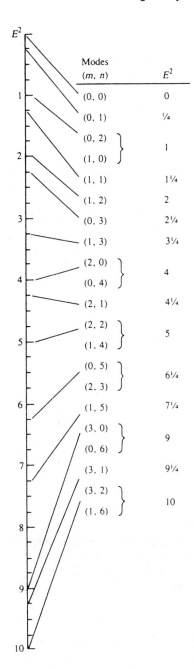

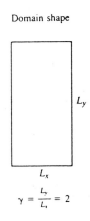

FIGURE 4 Progression of modes of a typical reaction-diffusion system in two dimensions, with increasing values of E^2. The modes m and n correspond to the number of wave peaks supported in the x and y direction, respectively as the domain size increases, or as the range of the inhibitor decreases. Reprinted by permission of the publisher from *Mathematical Models in Biology* by L. Edelstein-Keshet. Copyright © 1988 by Random House Publishers.

Figures 4 and 5 can be the consequence of gradual changes in any of these three parameters. Of course, fields of different shapes may have a different succession of modes, determined also by the value of γ.

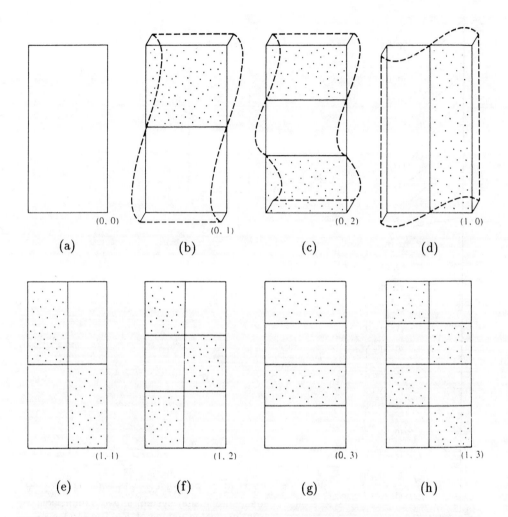

FIGURE 5 Examples of the first seven two-dimensional patterns predicted for a typical reaction-diffusion system under the conditions described in Figure 4. Reprinted by permission of the publisher from *Mathematical Models in Biology* by L. Edelstein-Keshet. Copyright © 1988 by Random House Publishers.

In circular and elliptical fields, the succession of modes is also different. Kauffman[10] has calculated the mode progression for a general reaction-diffusion system on an elliptical domain of increasing size and showed that the succession of nodal lines on such a growing field was very similar to the succession of compartmental boundaries that form in the wing imaginal disks of *Drosophila*. It may therefore be that the progressive compartmentalization of the *Drosophila* imaginal disks is the simple and spontaneous consequence of a reaction-diffusion system operating on a growing domain. It is interesting to modes that the succession of modes is also similar to the succession of nodes in vibrating circular and elliptical plates. Xu et al.[35] have shown that the vibrational modes of plates of more complex shapes also corresponds generally to the pattern boundaries produced by reaction-diffusion systems on similar-shaped fields. Murray[14] has noted that the initial stages of chemical pattern formation by reaction-diffusion poses the same mathematical eigenvalue problem as that describing the vibration of thin plates. Thus, assuming equivalent boundary conditions can be established on vibrating plates, we may have here an analog model of pattern formation by a general reaction-diffusion system that solves for the pattern almost instantaneously, and would therefore afford a fast and efficient way of exploring patterning in complex geometries.

A DISCRETE MODEL OF PATTERN FORMATION BY LATERAL INHIBITION

Young[36] has demonstrated that instead of using continuous partial differential equations to describe pattern formation by reaction diffusion, it is possible to obtain equivalent results with a completely discrete model that captures the essence of lateral inhibition but does not require solution of the diffusion equation. Young's theory is modeled on the one proposed by Swindale[25] for explaining patterns in the visual cortex of the brain.

Young[36] models the combined effect of a short-range activator and a long-range inhibitor by assuming that around each "source" cell there are two concentric circular regions: an inner one where there is a constant positive value of some control parameter, and an outer one where there is a constant but negative value of the same parameter (Figure 6). This condition corresponds to the short-range activation and long-range inhibition of a lateral inhibition model, the principal difference being that in reaction-diffusion systems the "activity" of the activator and inhibitor decline gradually with distance from the center of activation.

The Young mechanism produces spots or irregular stripes, depending on the ratio of activator and inhibitor levels (Figure 7). Small values produce spots, while large values of the ratio produce stripes. The size of these pattern elements is determined by the range of the activator, while their spacing is determined in large measure by the range of the inhibition. One of the chief advantages of the Young method is that the patterns form and stabilize after only three or four iterations. This mechanism can therefore produce patterns far more rapidly than one that depends on the numerical simulation of diffusion (which may require thousands of iterations).

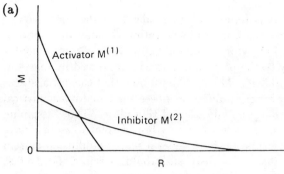
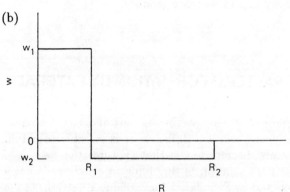

FIGURE 6 Discrete lateral-inhibition model of Young.[36] (a) The typical lateral inhibition system with continuously variable values of activator and inhibitor. (b) Young's model with discrete and spatially constant values for activator and inhibitor. Each differentiated cell exerts a constant short-range activating effect (w_1) and a constant long-range inhibitory effect (w_2) on its neighbors. Reprinted by permission of the publisher from "A Local Activator-Inhibitor Model of Vertebrate Skin Patterns" by D. A. Young, *Math. Biosci.* **72**, 51–58. Copyright © 1984 by Elsevier Science Publishing Co., Inc.

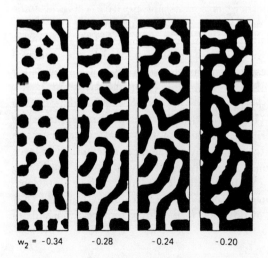

FIGURE 7 Patterns produced by the Young model for different values of w_2 (while w_1 is held constant at 1), after random activation of some cells. Reprinted by permission of the publisher from "A Local Activator-Inhibitor Model of Vertebrate Skin Patterns" by D. A. Young, *Math. Biosci.* **72**, 51–58. Copyright © 1984 by Elsevier Science Publishing Co., Inc.

RANDOM AND NON-RANDOM PATTERNS

The patterns produced by the Young mechanism (Figure 7) illustrate one of the limitations of the standard approach to the simulation of pattern formation. When patterning is initiated by random perturbation of the steady state (as is usually done to study the general properties of a given reaction-diffusion mechanism), then the pattern produced is also random. These patterns thus mimic the stripes on the coats of zebras, or the spotting patterns of cheetahs, leopards, and giraffes, all of which are random and characterize the individual like fingerprints. Randomness is, in fact, the hallmark of vertebrate color patterns, and of certain developmental patterns such as the interdigitating ocular dominance stripes in the vertebrate visual cortex. The vast majority of patterns in development, however, are regular and are reproduced identically from individual to individual. To obtain regularity and repeatability, it is necessary to define the boundary conditions and initial conditions by a non-random mechanism. The trick in modeling pattern formation in development is to find a non-arbitrary means of defining initial and boundary conditions. This generally requires substantial knowledge of the developmental biology of the system under study. Thus, while reaction-diffusion mechanisms can make patterns that look remarkably like those seen in nature, we can only accept a given pattern and mechanism as representing nature in a significant and meaningful way if it is backed up by a body of experimental evidence that gives us confidence that we have applied the correct boundary conditions.

RESULTS OF SIMULATIONS IN TWO DIMENSIONS

We use numerical simulation methods to illustrate some of the differences between the three reaction-diffusion schemes discussed above. The field dimensions and boundary conditions used in these examples were chosen because they define a problem of biological interest, namely the formation of butterfly wing patterns. We will first, however, examine the behavior of the models before illustrating their application to a biological problem.

Figures 8, 9, and 10 illustrate the behaviors, respectively, of the nondimensionalized Schnakenberg, Thomas, and Meinhardt systems subject to the same initial and boundary conditions. The field is a (1 × 2) rectangle, with fixed boundaries on one short side and the two long sides, and no-flux conditions at the remaining short side. Initial conditions were the unperturbed steady state. The figures show the near steady state concentration of the activator that develops after setting the fixed boundaries to 1.1 times the initial steady state. Each panel explores the d/γ parameter space. It will be recalled that an increase in the parameter γ can be interpreted as an increase in the size of the field, while an increase in parameter

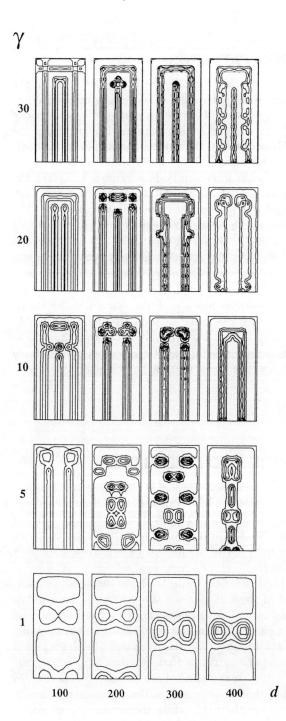

FIGURE 8 Patterns produced for various values of d and γ by the Schnakenberg system (Eq. (7)) in two spatial dimensions.

Pattern Formation in Biological Systems

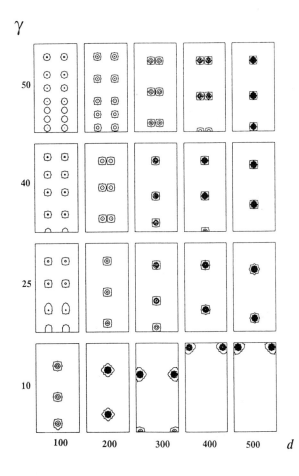

FIGURE 9 Patterns produced for various values of d and γ by the Meinhardt system (Eq. (8)) in two spatial dimensions.

d represents an increase in the range of the inhibitor. The patterns produced by fixed boundary conditions on all four sides can be visualized by reflection on the horizontal midline of each figure.

It is obvious that the three mechanisms produce dramatically different patterns. The Thomas and Schnakenberg systems produce mostly linear patterns, while the Meinhardt mechanism stabilizes as point patterns. The patterns produced by the Thomas and Schnakenberg systems differ considerably in detail. The Thomas patterns are relatively simple lines, while the Schnakenberg patterns tend to develop bulges and isolated islands of activator concentration. It is possible to get an idea of the sensitivity of these systems to variation in parameters and field size by noting the changes in pattern that are associated with, say, a 10% change in d or γ. On the whole, variation of this magnitude has relatively little effect on the pattern.

It is evident that the three reaction-diffusion systems are far from equivalent, even though linear theory predicts the same general behavior for all three systems.

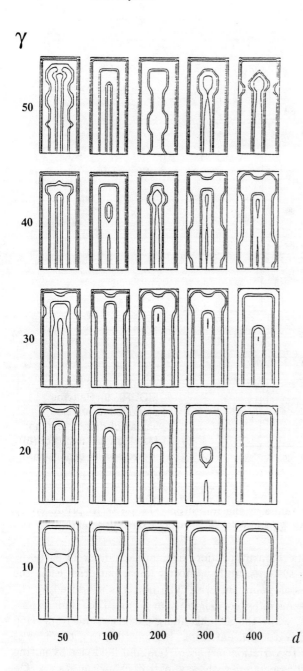

FIGURE 10 Patterns produced for various values of d and γ by the Thomas system (Eq. (9)) in two spatial dimensions.

The details of the patterns produced by each, and the characteristic differences between them, can only be uncovered by simulation. This means that there is no way of using the information in Figures 8 to 10 to predict how these three systems will behave under different boundary conditions. We can be assured that each will

Pattern Formation in Biological Systems   **179**

produce characteristically different patterns, but their form cannot be predicted without simulation.

Both the one-dimensional simulations of Arcuri and Murray[1] and the two-dimensional simulations shown above illustrate that the full nonlinear systems produce patterns whose details differ significantly (and often dramatically) from those predicted by linear theory. For most developmental systems the *details* of the pattern are more important than its general features, and this means that each biological problem in which reaction diffusion is believed to play a role must be studied by full simulation of the nonlinear system.

CELLULAR AUTOMATA

We conclude the general section on pattern formation with a brief discussion of the usefulness of cellular automata for simulating pattern formation in development. In their pure form, cellular automata are points in space which can take on one of two values (0 or 1) depending on the values of other such points in their neighborhood. The rules of a cellular automaton determine how the values of neighbors are interpreted. With relatively simple rules operating on such binary automata, it is possible to produce a vast array of complicated patterns that have fascinated mathematicians and biologists for nearly a decade (e.g., Wolfram[33]). Such automata have been used, among others, to simulate the color patterns on mollusk shells, and the branching pattens of algae. Spiral waves, such as those of the Belousov-Zhabotinski reaction, and interdigitating patterns, resembling ocular dominance stripes, are particularly easy to mimic and emerge from a variety of automata.

Cellular automata are attractive for biological simulation because they evoke an immediate image of biological cells, each with a fairly simple repertoire of behaviors, but collectively capable of complex morphogenesis.[33] Cellular automata can serve as models of biological pattern-formation systems because biological cells, too, behave by interacting only with their immediate neighbors, while obeying some set of internal "rules." The complex patterns that appear during development are emergent properties of the interaction of those rules with their cellular and chemical environment.[17] Many theoretical biologists are, however, reluctant to accept cellular automata models because the formal rules are difficult to analogize to known biological processes, and because there exist as yet no general methods for translating biological interactions into a table of local rules. Thus, while cellular automata can produce biologically realistic patterns, they often offer little insight into the biological process. In other words, getting the right pattern is of no use, if it is obtained for the wrong reason (a caveat that applies, obviously, to all theoretical modeling in biology).

Cellular automata can, however, be easily extended to increase their biological realism. Each point (or cell) in an array can be assumed to take on a continuous range of values, and can possess values in more than one variable. The rules by which these values change can reflect the interactions between cells, such as receptor binding, competition, or diffusion, and any number of biochemical reactions. Clearly, with such extensions cellular automata begin to resemble the methods used for numerical simulation. The main difference is that cellular automata do not attempt to model a differential equation (though they may). Such complex automata are useful for biologists because they can directly model communication between cells, and they allow examination of the consequences of qualitative and quantitative rules of interaction.

SIMULATION AND MIMICRY

Cellular automata, like reaction-diffusion systems, are useful only to the extent that they give insights into the biology of the system that is being simulated. In this regard it is perhaps useful to make a distinction between simulation and mimicry. In simulation the theoretical model grasps and accurately summarizes the principles behind the process being simulated, while in mimicry the model is wrong even though it produces the right kind of pattern. Mimicry in theoretical modeling commits what statisticians would call a type 2 error: accepting a false hypothesis, or in this case, getting the right answer for the wrong reason.

Unfortunately, much modeling in theoretical developmental biology appears at present to be mimicry. In developmental modeling it is easy to get the right kinds of pattern for the wrong reason because certain categories of biologically reasonable patterns (zebra stripes, ocular dominance stripes, sea shell patterns) emerge readily from a variety of reaction-diffusion and cellular automata models. In most modeled systems, we simply do not know enough about the developmental physiology to make sensible choices between alternative models, and, even when we can imagine only one model mechanism, we cannot be sure it has captured the essence of the underlying process.

In order for a model to be biologically useful, it must obviously incorporate as much information as possible about the developmental physiology of the system. But that is generally not sufficient. In order to have reasonable assurance that a model has captured the essence of a process, it must produce a pattern whose *details* resemble those of the morphology being modeled, it must also reproduce in its dynamics reasonable portions of the *ontogenetic transformation* that the real pattern undergoes, and because morphological evolution is gradualistic, it must be able to produce by simple changes of parameter values (and not by adding more terms to the model) a range of *diversity* of the pattern identical to that found to occur in nature. Few models meet these expectations.

PATTERN FORMATION ON BUTTERFLY WINGS

Here we briefly discuss pattern formation on the wings of butterflies as a concrete example of color pattern formation because it is one of the few systems that meets the expectations of physiology, detail, ontogeny, and diversity, mentioned above. It has the added advantage that the patterns are strictly two dimensional, exhibit an evolved system of homologous elements with transformations across the thousands of species of butterflies, and can be easily modeled without having to collapse any dimensions. This system has provided a variety of insights into the way in which developmental processes change during morphological evolution.[18]

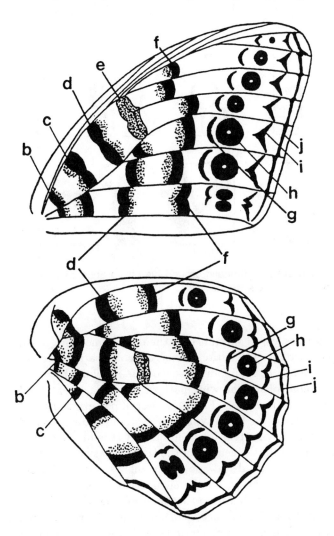

FIGURE 11 The nymphalid ground plan. This is a diagrammatic representation of the general distribution of pattern elements (labeled b–j) on the wings of butterflies. The pattern elements are arranged in serially homologous series that repeat from wing cell to wing cell. (From Nijhout[18]; reprinted by permission of the author.)

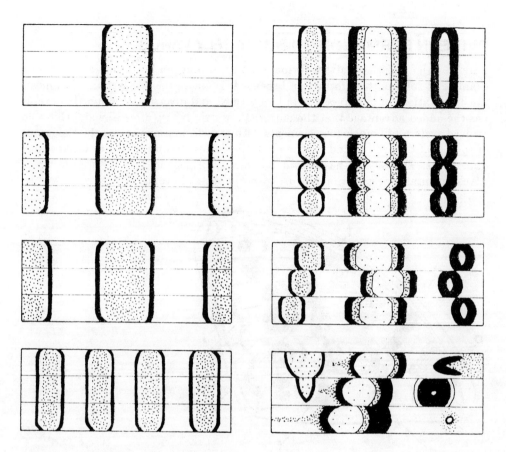

FIGURE 12 Hypothetical evolution of the nymphalid ground plan from ancestors with a few simple uncompartmentalized symmetry systems.

The color patterns of butterflies are all variants on a theme of homologies called the nymphalid ground plan (Figure 11). The entire diversity of color patterns comes about through the selective expression and modification of the individual pattern elements that make up the ground plan. The wing pattern is compartmentalized into two developmentally independent systems. First, the overall pattern is divided into three parallel symmetry systems: the basal symmetry system (elements **b** and **c**, in Figure 11), the central symmetry system (elements **d** and **f**), and the border symmetry system (elements **g** and **i**). In the centers of the latter two systems, there are two additional pattern elements, the discal spot (element **e**) and the border ocelli (element **h**). Secondly, the development of the elements of these symmetry systems within a given wing cell is uncoupled from that in adjoining wing cells. As a consequence of this developmental isolation, each element of the pattern has been free to evolve morphologically with nearly complete independence from the other pattern elements. The overall wing pattern is thus a mosaic of semi-independent pattern elements that can be modified and arranged on the wing surface to provide a variety of optical effects, ranging from camouflage to mimicry.[18]

The presumptive evolution of the nymphalid ground plan is illustrated diagrammatically in Figure 12. The ancestor is believed to have had a simple pattern with a single symmetry system, as is found in many species of moths today. Evolution of complexity progressed by the addition of more symmetry systems (Figures 12(b)–(d)), possibly by a system that sets up an increasing number of standing waves on the wing. The number of symmetry systems became stabilized at three, and each gradually evolved a distinctive morphology (Figures 12(e)–(h)), probably due to the evolution of a proximo-distal gradient or discontinuity in some variables that interact with the wave pattern. In the immediate ancestors of the butterflies, the wing veins became boundaries to pattern formation and the pattern became compartmentalized to each wing cell (Figures 12(f)–(h)). With this developmental isolation the pattern elements in each wing cell became free to diverge both in position (Figures 12(f) and (g)) and morphology (Figures 12(g) and (h)).

The developmental compartmentalization of the wing pattern greatly facilitates its modeling, because each pattern element in each wing cell can be modeled separately without having to worry about possible interactions with distant patterns. Nijhout[16,18] has shown that a relatively simple model can account for nearly the entire diversity of shapes of pattern elements that are found among the thousands of species of butterflies. The model generates the pattern in two steps, in accordance with what is known about the developmental physiology of pattern formation in this system. The first step establishes a system of line and point sources of a diffusible substance, and the second step establishes the pattern as a simple threshold on the diffusion gradients produced by those sources.

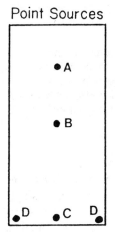
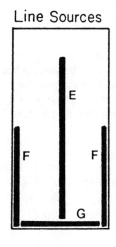

FIGURE 13 Distribution of sources (or sinks) that can produce nearly the entire diversity of patterns found in the wing cells of butterflies. The rectangular field represents a single wing cell in which vein make up the two long side boundary and the top boundary. (From Nijhout[18]; reprinted by permission of the author.)

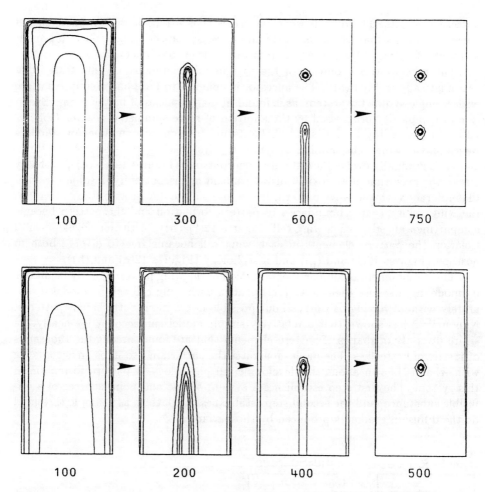

FIGURE 14 The lateral inhibition model of Meinhardt (Eq. (1)) can produce the diversity of source distribution shown in Figure 14 by varying boundary conditions and reaction constants. The series shown is a typical time sequence of activator concentration which gradually transforms from a high ridge to a series of point sources on the wing-cell midline. (From Nijhout[18]; reprinted by permission of the author.)

The distribution of diffusion sources (and barriers to diffusion) in real butterfly wings is known from experimental perturbation studies and from studies of the comparative morphology of normal and aberrant patterns.[18] When activated singly or in pairs, this distribution of sources (Figure 13) has been shown, by simulation, to be capable of producing nearly the entire diversity of pattern shapes found in the butterflies.

Sources in the exact locations shown in Figure 13, are readily produced by the Meinhardt[12] lateral inhibition system, and by no other reaction-diffusion system that has been examined so far.[16] The Meinhardt system produces the right patterns, but only when provided with fixed boundary conditions for the activator on three of the four sides of the rectangle that simulates a wing cell. These are the three locations of the wing veins around a typical wing cell. The wing veins afford the only means by which material can enter or leave the developing wing, and provide reasonable physical constant-level sources for materials, which are modeled as fixed boundaries.

Perhaps the most important feature of the Meinhardt lateral inhibition system implemented in this way is the dynamic progression of source distributions it produces as the reaction-diffusion progresses (Figure 14). This progression of sources produces patterns that closely resemble the diversity of color patterns seen among closely related species in several genera of butterflies. Diversity of this type in essence constitutes a heterochrony. This example illustrates that the most interesting feature of reaction-diffusion systems, from a biological perspective, is probably not the steady-state patterns to which a system tends, but the dynamic progression of patterns well before the steady state is reached. Development, like most of biology, is not an equilibrium phenomenon. Dynamically changing patterns like those of evolving reaction-diffusion systems may provide useful models for the progression of determinative processes during development.

REFERENCES

1. Arcuri, P., and J. D. Murray. "Pattern Sensitivity to Boundary and Initial Conditions in Reaction-Diffusion Models." *J. Math. Biol.* **24** (1986): 141–165.
2. Bryant, S. V., V. French, and P. J. Bryant. "Distal Regeneration and Symmetry." *Science* **212** (1981): 993–1002.
3. Carslaw, H. S., and J. C. Jaeger. *Conduction of Heat in Solids.* Oxford: Oxford Univ. Press, 1959.
4. Crank, J. *The Mathematics of Diffusion.* Oxford: Oxford Univ. Press, 1975.
5. DuShane, G.P. "An Experimental Study of the Origin of Pigment Cells in Amphibia." *J. Exp. Zool.* **72** (1935): 1–31.
6. Edelstein-Keshet, L. *Mathematical Models in Biology.* New York: Random House, 1988.
7. Ermentrout, B., J. Campbell, and G. F. Oster. "A Model for Shell Patterns based on Neural Activity." *Veliger* **28** (1986): 369–388.
8. French, V., P. J. Bryant, and S. V. Bryant. "Pattern Regulation in Epimorphic Fields." *Science* **193** (1976): 969–981.
9. Hinchliffe, J. R., and D. R. Johnson. *The Development of the Vertebrate Limb.* Oxford: Oxford Univ. Press, 1980.
10. Kauffman, S. A. "Chemical Patterns, Compartments, and a Binary Epigenetic Code in *Drosophila.*" *Amer. Zool.* **17** (1977): 631–648.
11. Lewis, J., J. M. Slack, and L. Wolpert. "Thresholds in Development." *J. Theor. Biol.* **65** (1977): 579–590.
12. Meinhardt, H. *Models of Biological Pattern Formation.* London: Academic Press, 1982.
13. Murray, J. D. "Parameter Space for Turing Instability in Reaction-Diffusion Mechanisms: A Comparison of Models." *J. Theor. Biol.* **98** (1982): 143–163.
14. Murray, J. D. *Mathematical Biology.* New York: Springer Verlag, 1989.
15. Newman, S. A., and W. D. Comper. "'Generic' Physical Mechanisms of Morphogenesis and Pattern Formation." *Development* **110** (1000): 1–18.
16. Nijhout, H. F. "A Comprehensive Model for Colour Pattern Formation in Butterflies." *Proc. Roy. Soc. London* **B 239** (1990a): 81–113.
17. Nijhout, H. F. "Metaphors and the Role of Genes in Development." *BioEssays* **12** (1990b): 441–446.
18. Nijhout, H. F. *The Development and Evolution of Butterfly Wing Patterns.* Washington, DC: Smithsonian Institution Press, 1991.
19. Oster, G. F., and J. D. Murray. "Pattern Formation Models and Developmental Constraints." *J. Exp. Zool.* **251** (1989): 186–202.
20. Oster, G. F., J. D. Murray, and P. K. Maini. "A Model for Chondrogenic Condensation in the Developing Limb: The Role of Extracellular Matrix and Tractions." *J. Embryol. Exp. Morphol.* **89** (1985): 93–112.
21. Oster, G. F., N. Shubin, J. D. Murray, and P. Alberch. "Evolution and Morphogenetic Rules. The Shape of the Vertebrate Limb in Ontogeny and Phylogeny." *Evolution* **45** (1988): 862–884.

22. Schnakenberg, J. "Simple Chemical Reaction Systems with Limit Cycle Behavior." *J. Theor. Biol.* **81** (1979): 389–400.
23. Segel, L. A. *Modeling Dynamic Phenomena in Molecular and Cellular Biology.* Cambridge: Cambridge Univ. Press, 1984.
24. Segel, L. A., and J. L. Jackson. "Dissipative Structures: An Explanation and an Ecological Example." *J. Theor. Biol.* **37** (1972): 545–559.
25. Swindale, N. V. "A Model for the Formation of Ocular Dominance Stripes." *Proc. Roy. Soc. London* **B208** (1980): 243–264.
26. Thomas, D. "Artificial Enzyme Membranes, Transport, Memory, and Oscillatory Phenomena." In *Analysis and Control of Immobilized Enzyme Systems*, edited by D. Thomas and and J.-P. Kervenez. New York: Springer Verlag, 1975.
27. Tomchik, K. J., and P. N. Devreotes. "Adenosine 3′, 5′-Monophosphate Waves in *Dictyostelium discoideum*: A Demonstration by Isotope Dilution-Fluorography." *Science* **212** (1981): 443–446.
28. Turing, A. M. "The Chemical Basis of Morphogenesis." *Phil. Trans. Roy. Soc. London* **B 237** (1952): 37–72.
29. Twitty, V. C. "The Developmental Analysis of Specific Pigment Patterns." *J. Exp. Zool.* **100** (1945): 141–178.
30. Twitty, V. C. *Of Scientists and Salamanders.* New York: Freeman, 1966.
31. Waddington, C. H., and R. J. Cowe. "Computer Simulation of Molluscan Pigmentation Pattern." *J. Theor. Biol.* **25** (1969): 219–225.
32. Winfree, A. T. *The Geometry of Biological Time.* New York: Springer Verlag, 1980.
33. Wolfram, S. "Cellular Automata as Models of Complexity." *Nature* **311** (1984): 419–424.
34. Wolpert, L. "Positional Information and Pattern Formation." *Curr. Top. Dev. Biol.* **6** (1971): 183–224.
35. Xu, Y., C. M. Vest, and J. D. Murray. "Holographic Interferometry Used to Demonstrate a Theory of Pattern Formation in Animal Coats." *Appl. Optics* **22** (1983): 3479–3483.
36. Young, D. A. "A Local Activator-Inhibitor Model of Vertebrate Skin Patterns." *Math. Biosci.* **72** (1984): 51–58.

Christopher G. Langton
Complex Systems Group, Theoretical Division, Los Alamos National Laboratory, Los Alamos, NM 87545, and Santa Fe Institute, 1660 Old Pecos Trail, Suite A., Santa Fe, NM 87501

Artificial Life

Artificial Life complements the traditional analytical biological methods by attempting to *synthesize* lifelike behaviors within computers and other "artificial" media. The primary motivations driving this synthetic approach are (1) to contribute to a truly general theoretical biology by extending the empirical data base beyond the carbon-based life that has evolved on the planet Earth, and (2) to apply fundamental principles of biological form and function to the solution of hard problems in science and engineering.

1. THE BIOLOGY OF POSSIBLE LIFE

Biology is the scientific study of life—in principle anyway. In practice, biology is the scientific study of life on Earth based on carbon-chain chemistry. There is nothing in its charter that restricts biology to the study of carbon-based life; it is simply that this is the only kind of life that has been available to study. Thus, heoretical biology has long faced the fundamental obstacle that it is impossible to derive general principles from single examples.

Without other examples it is extremely difficult to distinguish essential properties of life—properties that must be shared by any living system *in principle*—from properties that may be incidental to life, but which happen to be universal to life on Earth due *solely* to a combination of local historical accident and common genetic descent. Since it is quite unlikely that organisms based on different physical chemistries will present themselves to us for study in the foreseeable future, our only alternative is to try to synthesize alternative life-forms ourselves—*Artificial Life:* life made by man rather than by nature.

1.1 ARTIFICIAL LIFE

Biology has traditionally started at the top, viewing a living organism as a complex biochemical machine, and has worked *analytically* down from there through the hierarchy of biological organization—decomposing a living organism into organs, tissues, cells, organelles, and finally molecules—in its pursuit of the mechanisms of life. Analysis means "the separation of an intellectual or substantial whole into constituents for individual study." By composing our individual understandings of the dissected component parts of living organisms, traditional biology has provided us with a broad picture of the mechanics of life on Earth.

But there is more to life than mechanics—there is also dynamics. Life depends critically on principles of dynamical self-organization that have remained largely untouched by traditional analytic methods. There is a simple explanation for this—these self-organizing dynamics are fundamentally nonlinear phenomena, and nonlinear phenomena in general depend critically on the interactions *between* parts: they necessarily disappear when parts are treated in isolation from one another, which is the basis for the analytic method.

Rather, nonlinear phenomena are most appropriately treated by a *synthetic* approach. Synthesis means "the combining of separate elements or substances to form a coherent whole." In nonlinear systems, the parts must be treated in each other's presence, rather than independently from one another, because they behave very differently in each other's presence than we would expect from a study of the parts in isolation.

Artificial Life is simply the synthetic approach to biology: rather than take living things apart, Artificial Life attempts to put living things together.

But Artificial Life is more than this. To understand the overall aims of the Artificial Life enterprise, one needs to do the following. (1) Broaden the scope of the attempts, beyond simply recreating "the living state," to the synthesis of any and all biological phenomena, from viral self-assembly to the evolution of the entire biosphere. (2) Couple this with the observation that there is no reason, in principle, why the parts we use in our attempts to synthesize these biological phenomena need be restricted to carbon-chain chemistry. (3) Note that we expect the synthetic approach to lead us not only to, but quite often *beyond*, known biological phenomena: beyond *life-as-we-know-it* into the realm of *life-as-it-could-be*.

Thus, for example, Artificial Life involves attempts to (1) synthesize the process of evolution (2) in computers, and (3) will be interested in whatever emerges from the process, even if the results have no analogs in the "natural" world. It is certainly of scientific interest to know what kinds of things *can* evolve in principle, whether or not they happened to do so here on Earth.

1.2 AI AND THE BEHAVIOR GENERATION PROBLEM

Artificial Life is concerned with generating *lifelike* behavior. Thus, it focuses on the problem of creating *behavior generators*. A good place to start is to identify the mechanisms by which behavior is generated and controlled in natural systems, and to recreate these mechanisms in artificial systems. This is the course we will take later in this paper.

The related field of Artificial Intelligence is concerned with generating *intelligent* behavior. It, too, focuses on the problem of creating behavior generators. However, although it initially looked to natural intelligence to identify its underlying mechanisms, these mechanisms were not known, nor are they today. Therefore, following an initial flirt with neural nets, AI became wedded to the only other known vehicle for the generation of complex behavior: the technology of serial computer programming. As a consequence, from the very beginning artificial intelligence embraced an underlying methodology for the generation of intelligent behavior that bore no demonstrable relationship to the method by which intelligence is generated in natural systems. In fact, AI has focused primarily on the production of intelligent *solutions* rather than on the production of intelligent *behavior*. There is a world of difference between these two possible foci.

By contrast, Artificial Life has the great good fortune that many of the mechanisms by which behavior arises in natural living systems are known. There are still many holes in our knowledge, but the general picture is in place. Therefore, Artificial Life can start by recapturing natural life and has no need to resort to the sort of initial infidelity that is now coming back to haunt AI.

The key insight into the natural method of behavior generation is gained by noting that *nature is fundamentally parallel*. This is reflected in the "architecture" of natural living organisms, which consist of many millions of parts, each one of which has its own behavioral repertoire. Living systems are highly distributed and quite massively parallel. If our models are to be true to life, they must also be highly distributed and quite massively parallel. Indeed, it is unlikely that any other approach will prove viable.

2. HISTORICAL ROOTS OF ARTIFICIAL LIFE

Mankind has a long history of attempting to map the mechanics of his contemporary technology onto the workings of nature, trying to understand the latter in terms of the former.

It is not surprising, therefore, that early models of life reflected the principal technology of their era. The earliest models were simple statuettes and paintings—works of art which captured the static form of living things. These statues were provided with articulated arms and legs in the attempt to capture the dynamic form of living things. These simple statues incorporated no internal dynamics, requiring human operators to make them behave.

The earliest mechanical devices that were capable of generating their own behavior were based on the technology of water transport. These were the early Egyptian water clocks called *Clepsydra*. These devices made use of a rate-limited process—in this case the dripping of water through a fixed orifice—to indicate the progression of another process—the position of the sun. Ctesibius of Alexandria developed a water-powered mechanical clock around B.C. 135 which employed a great deal of the available hydraulic technology—including floats, a siphon, and a water-wheel-driven train of gears.

In the first century A.D., Hero of Alexandria produced a treatise on *Pneumatics*, which described, among other things, various simple gadgets in the shape of animals and humans that utilized pneumatic principles to generate simple movements.

However, it was really not until the age of mechanical clocks that artifacts exhibiting complicated internal dynamics became possible. Around 850 A.D., the *mechanical escapement* was invented, which could be used to regulate the power provided by falling weights. This invention ushered in the great age of clockwork technology. Throughout the Middle Ages and the Renaissance, the history of technology is largely bound up with the technology of clocks. Clocks often constituted the most complicated and advanced application of the technology of an era.

Perhaps the earliest clockwork simulations of life were the so-called "Jacks," mechanical "men" incorporated in early clocks who would swing a hammer to strike the hour on a bell. The word "jack" is derived from "jaccomarchiadus," which means "the man in the suit of armour." These accessory figures retained their popularity even after the spread of clock dials and hands—to the extent that clocks were eventually developed in which the function of timekeeping was secondary to the control of large numbers of figures engaged in various activities, to the point of acting out entire plays.

Finally, clockwork mechanisms appeared which had done away altogether with any pretense at timekeeping. These "automata" were entirely devoted to imparting lifelike motion to a mechanical figure or animal. These mechanical automaton simulations of life included such things as elephants, peacocks, singing birds, musicians, and even fortune tellers.

This line of development reached its peak in the famous duck of Vaucanson, described as "an artificial duck made of gilded copper who drinks, eats, quacks, splashes about on the water, and digests his food like a living duck."[1] Vaucanson's goal is captured neatly in the following description:

[1] All quotes concerning these mechanical ducks are from Chapuis.[5]

In 1735 Jacques de Vaucanson arrived in Paris at the age of 26. Under the influence of contemporary philosophic ideas, he had tried, it seems, to reproduce life artificially.

Unfortunately, neither the duck itself nor any technical descriptions or diagrams remain that would give the details of construction of this duck. The complexity of the mechanism is attested to by the fact that one single wing contained over 400 articulated pieces.

One of those called upon to repair Vaucanson's duck in later years was a "mechanician" named Reichsteiner, who was so impressed with it that he went on to build a duck of his own—also now lost—which was exhibited in 1847. Here is an account of this duck's operation from the newspaper *Das Freie Wort*:

> After a light touch on a point on the base, the duck in the most natural way in the world begins to look around him, eyeing the audience with an intelligent air. His lord and master, however, apparently interprets this differently, for soon he goes off to look for something for the bird to eat. No sooner has he filled a dish with oatmeal porridge than our famished friend plunges his beak deep into it, showing his satisfaction by some characteristic movements of his tail. The way in which he takes the porridge and swallows it greedily is extraordinarily true to life. In next to no time the basin has been half emptied, although on several occasions the bird, as if alarmed by some unfamiliar noises, has raised his head and glanced curiously around him. After this, satisfied with his frugal meal, he stands up and begins to flap his wings and to stretch himself while expressing his gratitude by several contented quacks. But most astonishing of all are the contractions of the bird's body clearly showing that his stomach is a little upset by this rapid meal and the effects of a painful digestion become obvious. However, the brave little bird holds out, and after a few moments we are convinced in the most concrete manner that he has overcome his internal difficulties. The truth is that the smell which now spreads through the room becomes almost unbearable. We wish to express to the artist inventor the pleasure which his demonstration gave to us.

Figure 1 shows two views of one of the ducks—there is some controversy as to whether it is Vaucanson's or Reichsteiner's. The mechanism inside the duck would have been completely covered with feathers and the controlling mechanism in the box below would have been covered up as well.

194 Christopher G. Langton

FIGURE 1 Two views of the mechanical duck attributed to Vaucanson. Printed in *Automata: A Historical and Technological Study* by Alfred Chapuis and Edmon Droz (B. A. Batsford Ltd.); reprinted by permission of the publisher.

2.1 THE DEVELOPMENT OF CONTROL MECHANISMS

Out of the technology of the clockwork regulation of automata came the more general—and perhaps ultimately more important—technology of *process control*. As attested to in the descriptions of the mechanical ducks, some of the clockwork mechanisms had to control remarkably complicated actions on the part of the automata, not only *powering* them but *sequencing* them as well.

Control mechanisms evolved from early, simple devices—such as a lever attached to a wheel which converted circular motion into linear motion—to later, more complicated devices—such as whole sets of cams upon which would ride many interlinked mechanical arms, giving rise to extremely complicated automaton behaviors.

Eventually *programmable controllers* appeared, which incorporated such devices as interchangeable cams, or drums with movable pegs, with which one could program arbitrary sequences of actions on the part of the automaton. The writing and picture drawing automata of Figure 2, built by the Jaquet-Droz family, are examples of programmable automata. The introduction of such programmable controllers was one of the primary developments on the road to general purpose computers.

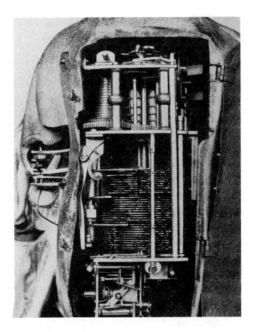 

FIGURE 2 Two views of a drawing automaton built by the Jaquet-Droz family. Printed in *Automata: A Historical and Technological Study* by Alfred Chapuis and Edmon Droz (B. A. Batsford Ltd.); reprinted by permission of the publisher.

2.2 ABSTRACTION OF THE LOGICAL "FORM" OF MACHINES

During the early part of the twentieth century, the formal application of logic to the mechanical process of arithmetic lead to the abstract formulation of a "procedure." The work of Church, Kleene, Gödel, Turing, and Post formalized the notion of a logical sequence of steps, leading to the realization that the essence of a mechanical process—the "thing" responsible for its dynamic behavior—is not a thing at all, but an abstract control structure, or "program"—a sequence of simple actions selected from a finite repertoire. Furthermore, it was recognized that the essential features of this control structure could be captured within an abstract set of rules— a formal specification—without regard to the material out of which the machine was constructed. The "logical form" of a machine was separated from its material basis of construction, and it was found that "machineness" was a property of the former, not of the latter. Today, the formal equivalent of a "machine" is an *algorithm*: the logic underlying the dynamics of an automaton, regardless of the details of its material construction. We now have many formal methods for the specification and operation of abstract machines: such as programming languages, formal language theory, automata theory, recursive function theory, etc. All of these have been shown to be logically equivalent.

Once we have learned to think of machines in terms of their abstract, formal specifications, we can turn around and view abstract, formal specifications as potential machines. In mapping the machines of our common experience to formal specifications, we have by no means exhausted the space of *possible* specifications. Indeed, most of our individual machines map to a very small subset of the space of specifications—a subset largely characterized by methodical, boring, uninteresting dynamics.

2.3 GENERAL PURPOSE COMPUTERS

Various threads of technological development—programmable controllers, calculating engines, and the formal theory of machines—have come together in the general purpose, stored program computer. Programmable computers are extremely general behavior generators. They have no intrinsic behavior of their own. Without programs, they are like formless matter. They must be told how to behave. By submitting a program to a computer—that is: by giving it a formal specification for a machine—we are telling it to behave as if it were the machine specified by the program. The computer then "emulates" that more specific machine in the performance of the desired task. Its great power lies in its plasticity of behavior. If we can provide a step-by-step specification for a specific kind of behavior, the chameleon-like computer will exhibit that behavior. Computers should be viewed as *second-order* machines—given the formal specification of a first-order machine, they will "become" that machine. Thus, the space of possible machines is directly available for study, at the cost of a mere formal description: computers "realize" abstract machines.

2.4 FORMAL LIMITS OF MACHINE BEHAVIORS

Although computers, and by extension other machines, are capable of exhibiting a bewilderingly wide variety of behaviors, we must face two fundamental limitations on the kinds of behaviors that we can expect of computers.

The first limitation is one of *computability in principle*. There are certain behaviors that are "uncomputable"—behaviors for which *no* formal specification can be given for a machine that will exhibit that behavior. The classic example of this sort of limitation is Turing's famous *Halting Problem*: can we give a formal specification for a machine which, when provided with the description of *any* other machine together with its initial state, will—by inspection alone—determine whether or not that machine will reach its halt state? Turing proved that no such machine can be specified. In particular, Turing showed that the best that such a proposed machine could do would be to emulate the given machine to see whether or not it halted. If the emulated machine halted, fine. However, the emulated machine might run forever without halting, and therefore the emulating machine could not answer whether or not it would halt. Rice and others have extended this undecidability result to the determination—by inspection alone—of *any* nontrivial property of the future behavior of an arbitrary machine.[14]

The second limitation is one of *computability in practice*. There are many behaviors for which we do not know how to specify a sequence of steps that will cause a computer to exhibit that behavior. We can automate what we can explain how to do, but there is much that we cannot explain how to do. Thus, although a formal specification for a machine that will exhibit a certain behavior may be possible *in principle*, we have no formal procedure for producing that formal specification in practice, short of a trial and error search through the space of possible descriptions.

We need to separate the notion of a formal specification of a machine—that is, a specification of the *logical structure* of the machine—from the notion of a formal specification of a machine's behavior—that is, a specification of the sequence of transitions that the machine will undergo. In general, we cannot derive behaviors from structure, nor can we derive structure from behaviors.

The moral is: in order to determine the behavior of some machines, there is no recourse but to run them and see how they behave! This has consequences for the methods by which we (or nature) go about *generating* behavior generators themselves, which we will take up in the section on evolution.

2.5 JOHN VON NEUMANN: FROM MECHANICS TO LOGIC

With the development of the general purpose computer, various researchers turned their attention from the *mechanics* of life to the *logic* of life.

The first computational approach to the generation of lifelike behavior was due to the brilliant Hungarian mathematician John von Neumann. In the words of his colleague Arthur W. Burks, von Neumann was interested in the general question[2]:

> What kind of logical organization is sufficient for an automaton to reproduce itself? This question is not precise and admits to trivial versions as well as interesting ones. Von Neumann had the familiar natural phenomenon of self-reproduction in mind when he posed it, but he was not trying to simulate the self-reproduction of a natural system at the level of genetics and biochemistry. *He wished to abstract from the natural self-reproduction problem its logical form.*

This approach is the first to capture the essence of Artificial Life. To understand the field of Artificial Life, one need only replace references to "self-reproduction" in the above with references to any other biological phenomenon.

In von Neumann's initial thought experiment (his "kinematic model"), a machine floats around on the surface of a pond, together with lots of machine parts. The machine is a *universal constructor:* given the description of any machine, it will locate the proper parts and construct that machine. If given a description of

---

[2]From Burks,[3] emphasis added.

itself, it will construct itself. This is not quite self-reproduction, however, because the offspring machine will not have a description of itself and hence could not go on to construct another copy. So, von Neumann's machine also contains a *description copier:* once the offspring machine has been constructed, the "parent" machine constructs a copy of the description that it worked from and attaches it to the offspring machine. This constitutes genuine self-reproduction.

Von Neumann decided that this model did not properly distinguish the logical form of the process from the material of the process, and looked about for a completely formal system within which to model self-reproduction. Stan Ulam—one of von Neumann's colleagues at Los Alamos[3]—suggested an appropriate formalism, which has come to be known as a *cellular automaton* (CA).

In brief, a CA consists of a regular lattice of *finite automata*, which are the simplest formal models of machines. A finite automaton can be in only one of a finite number of states at any given time, and its transitions between states from one time step to the next are governed by a *state-transition table:* given a certain input and a certain internal state, the state-transition table specifies the state to be adopted by the finite automaton at the next time step. In a CA, the necessary input is derived from the states of the automata at neighboring lattice points. Thus, the state of an automaton at time $t + 1$ is a function of the states of the automaton itself and its immediate neighbors at time t. All of the automata in the lattice obey the same transition table and every automaton changes state at the same instant, time step after time step. CA's are good examples of the kind of computational paradigm sought after by Artificial Life: bottom-up, parallel, local determination of behavior.

Von Neumann was able to embed the equivalent of his kinematic model as an initial pattern of state assignments within a large CA lattice using 29 states per cell. Although von Neumann's work on self-reproducing automata was left incomplete at the time of his death, Arthur Burks organized what had been done, filled in the remaining details, and published it.[4] Figure 3 shows a schematic diagram of von Neumann's self-reproducing machine.

Von Neumann's CA model was a constructive proof that an essential characteristic behavior of living things—self-reproduction—*was* achievable by machines. Furthermore, he determined that any such method must make use of the information contained in the description of the machine in two fundamentally different ways:

1. *Interpreted*, as instructions to be executed in the construction of the offspring.
2. *Uninterpreted*, as passive data to be duplicated to form the description given to the offspring.

[3] Ulam also investigated dynamic models of pattern production and competition.[25]
[4] Together with a transcription of von Neumann's 1949 lectures at the University of Illinois entitled "Theory and Organization of Complicated Automata," in which he gives his views on various problems related to the study of complex systems in general.[26]

Artificial Life

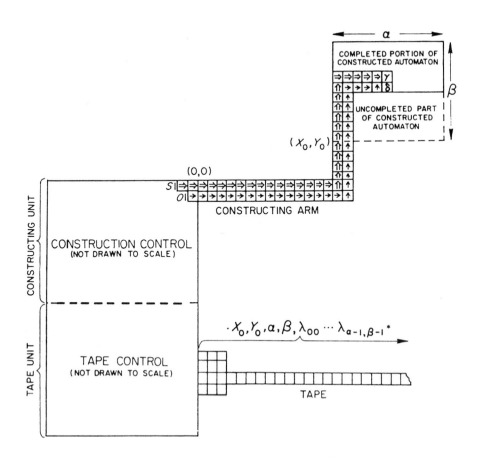

FIGURE 3 Schematic diagram of von Neumann's CA self-reproducing configuration. From *Essays on Cellular Automata* edited by A. W. Burk (University of Illinois Press, Urbana, 1970); reprinted by permission of the publisher.

Of course, when Watson and Crick unveiled the structure of DNA, they discovered that the information contained therein was used in precisely these two ways in the processes of transcription/translation and replication.

In describing his model, von Neumann pointed out that[5]:

> By axiomatizing automata in this manner, one has thrown half of the problem out the window, and it may be the more important half. One has resigned oneself not to explain how these parts are made up of real things, specifically, how these parts are made up of actual elementary particles, or even of higher chemical molecules.

[5] From Burks.[3]

Whether or not the more important half of the question has been disposed of depends on the questions we are asking. If we are concerned with explaining how the life that we know emerges from the known laws of physics and organic-chemistry, then indeed the interesting part has been tossed out. But, if we are concerned with the more general problem of explaining how lifelike behaviors emerge out of low-level interactions within a population of logical primitives, we have retained the more interesting portion of the question.

3. THE ROLE OF COMPUTERS IN STUDYING LIFE AND OTHER COMPLEX SYSTEMS

Artificial Intelligence and Artificial Life are each concerned with the application of computers to the study of complex, natural phenomena. Both are concerned with generating complex behavior. However, the manner in which each field employs the technology of computation in the pursuit of its respective goals is strikingly different.

AI has based its underlying methodology for generating intelligent behavior on the computational paradigm. That is, AI uses the technology of computation as a model of intelligence. AL, on the other hand, is attempting to develop a new computational paradigm based on the natural processes that support living organisms. That is, AL uses insights from biology to explore the dynamics of interacting information structures. AL has not adopted the computational paradigm as its underlying methodology of behavior generation, nor does it attempt to "explain" life as a kind of computer program.

One way to pursue the study of artificial life would be to attempt to create life *in vitro*, using the same kinds of organic chemicals out of which we are constituted. Indeed, there are numerous exciting efforts in this direction. This would certainly teach us a lot about the possibilities for alternative life-forms *within* the carbon-chain chemistry domain that could have (but didn't) evolve here.

However, biomolecules are extremely small and difficult to work with, requiring rooms full of special equipment, replete with dozens of "postdocs" and graduate students willing to devote the larger part of their professional careers to the perfection of electrophoretic gel techniques. Besides, although the creation of life *in vitro* would certainly be a scientific feat worthy of note—and probably even a Nobel prize—it would not, in the long run, tell us much more about the space of *possible* life than we already know.

Computers provide an alternative medium within which to attempt to synthesize life. Modern computer technology has resulted in machinery with tremendous potential for the creation of life *in silico*.

Computers should be thought of as an important laboratory tool for the study of life, substituting for the array of incubators, culture dishes, microscopes, electrophoretic gels, pipettes, centrifuges, and other assorted wet-lab paraphernalia, one simple-to-master piece of experimental equipment devoted exclusively to the incubation of information structures.

The advantage of working with information structures is that information has no intrinsic size. The computer is *the* tool for the manipulation of information, whether that manipulation is a consequence of our actions or a consequence of the actions of the information structures themselves. Computers themselves will not be alive, rather they will support informational universes within which dynamic populations of informational "molecules" engage in informational "biochemistry."

This view of computers as workstations for performing scientific experiments within artificial universes is fairly new, but it is rapidly becoming accepted as a legitimate, even necessary, way of pursuing science. In the days before computers, scientists worked primarily with systems whose defining equations could be solved analytically, and ignored those whose defining equations could *not* be so solved. This was largely the case because, in the absence of analytic solutions, the equations would have to be integrated over and over again, essentially simulating the time behavior of the system. Without computers to handle the mundane details of these calculations, such an undertaking was unthinkable except in the simplest cases.

However, with the advent of computers, the necessary mundane calculations can be relegated to these idiot-savants, and the realm of numerical simulation is opened up for exploration. "Exploration" is an appropriate term for the process, because the numerical simulation of systems allows one to "explore" the system's behavior under a wide range of parameter settings and initial conditions. The heuristic value of this kind of experimentation cannot be over-estimated. One often gains tremendous insight for the essential dynamics of a system by observing its behavior under a wide range of initial conditions. Most importantly, however, computers are beginning to provide scientists with a new paradigm for modeling the world. When dealing with essentially unsolvable governing equations, the primary reason for producing a formal mathematical model—the hope of reaching an analytic solution by symbolic manipulation—is lost. Systems of ordinary and partial differential equations are not very well suited for implementation as computer algorithms. One might expect that other modeling technologies would be more appropriate when the goal is the *synthesis*, rather than the *analysis*, of behavior.[6]

This expectation is easily borne out. With the precipitous drop in the cost of raw computing power, computers are now available that are capable of simulating physical systems from first principles. This means that it has become possible, for example, to model turbulent flow in a fluid by simulating the motions of its constituent particles—not just approximating *changes* in concentrations of particles at particular points, but actually computing their motions exactly.[7,24,28]

What does all of this have to do with the study of life? The most surprising lesson we have learned from simulating complex physical systems on computers is that *complex behavior need not have complex roots*. Indeed, tremendously interesting and

[6]See Toffoli[23] for a good exposition.

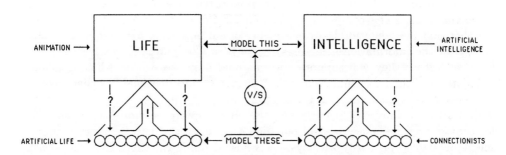

FIGURE 4 The bottom-up *versus* the top-down approach to modeling complex systems. Original figure appeared in "Artificial Life" by Christopher Langton, in *Artificial Life* edited by C. Langton (Addison-Wesley, Redwood City, 1989).

beguilingly complex behavior can emerge from collections of *extremely* simple components. This leads directly to the exciting possibility that much of the complex behavior exhibited by nature—especially the complex behavior that we call life—*also* has simple generators. Since it is very hard to work backwards from a complex behavior to its generator, but very simple to create generators and synthesize complex behavior, a promising approach to the study of complex natural systems is to undertake the general study of the kinds of behavior that can emerge from distributed systems consisting of simple components (Figure 4).

4. NONLINEARITY AND LOCAL DETERMINATION OF BEHAVIOR.

4.1 LINEAR VS. NONLINEAR SYSTEMS

As mentioned briefly above, the distinction between linear and nonlinear systems is fundamental, and provides excellent insight into why the principles underlying the dynamics of life should be so hard to find. The simplest way to state the distinction is to say that *linear systems* are those for which the behavior of the whole is just the sum of the behavior of its parts, while for *nonlinear systems*, the behavior of the whole is *more* than the sum of its parts.

Linear systems are those which obey the *principle of superposition*. We can break up complicated linear systems into simpler constituent parts, and analyze these parts *independently*. Once we have reached an understanding of the parts in isolation, we can achieve a full understanding of the whole system by *composing* our understandings of the isolated parts. This is the key feature of linear systems: by studying the parts in isolation, we can learn everything we need to know about the complete system.

This is not possible for nonlinear systems, which do *not* obey the principle of superposition. Even if we could break such systems up into simpler constituent parts, and even if we could reach a complete understanding of the parts in isolation, we would not be able to compose our understandings of the individual parts into an understanding of the whole system. The key feature of nonlinear systems is that their primary behaviors of interest are properties of the *interactions between parts*, rather than being properties of the parts themselves, and these interaction-based properties necessarily disappear when the parts are studied independently.

Thus, analysis is most fruitfully applied to linear systems. Analysis has *not* proved anywhere near as effective when applied to nonlinear systems: the nonlinear system must be treated as a whole.

A different approach to the study of nonlinear systems involves the inverse of analysis: *synthesis*. Rather than start with the behavior of interest and attempting to analyze it into its constituent parts, we start with constituent parts and put them together in the attempt to *synthesize* the behavior of interest.

Life is a property of *form*, not *matter*, a result of the organization of matter rather than something that inheres in the matter itself. Neither nucleotides nor amino acids nor any other carbon-chain molecule is alive—yet put them together in the right way, and the dynamic behavior that emerges out of their interactions is what we call life. It is effects, not things, upon which life is based—life is a kind of behavior, not a kind of stuff—and as such, it is constituted of simpler behaviors, not simpler stuff. *Behaviors themselves* can constitute the fundamental parts of nonlinear systems—*virtual parts*, which depend on nonlinear interactions between physical parts for their very existence. Isolate the physical parts and the virtual parts cease to exist. It is the *virtual parts* of living systems that Artificial Life is after, and synthesis is its primary methodological tool.

4.1 THE PARSIMONY OF LOCAL DETERMINATION OF BEHAVIOR

It is easier to generate complex behavior from the application of simple, *local* rules than it is to generate complex behavior from the application of complex, *global* rules. This is because complex global behavior is usually due to nonlinear interactions occurring at the local level. With bottom-up specifications, the system computes the local, nonlinear interactions explicitly and the global behavior, which was implicit in the local rules, emerges spontaneously without being treated explicitly.

With top-down specifications, however, local behavior must be implicit in global rules! This is really putting the cart before the horse! The global rules must "predict" the effects on global structure of many local, nonlinear interactions—something which we have seen is intractable, even impossible, in the general case. Thus, top-down systems must take computational shortcuts and explicitly deal with special cases, which results in inflexible, brittle, and unnatural behavior.

Furthermore, in a system of any complexity, the number of possible global states is astronomically enormous, and grows exponentially with the size of the system. Systems that attempt to supply *global* rules for *global* behavior simply *cannot* provide a different rule for every global state. Thus, the global states must be classified in some manner, and categorized using a coarse-grained scheme according to which the global states within a category are indistinguishable. The rules of the system can only be applied at the level of resolution of these categories. There are many possible ways to implement a classification scheme, most of which will yield different partitionings of the global state space. Any rule-based system must necessarily *assume* that finer-grained differences don't matter, or must include a finite set of tests for "special cases," and then must assume that no *other* special cases are relevant.

For most complex systems, however, fine differences in the global state can result in enormous differences in global behavior, and there may be no way in principle to partition the space of global states in such a way that specific fine differences have the appropriate global impact.

On the other hand, systems that supply *local* rules for *local* behaviors, *can* provide a different rule for each and every possible local state. Furthermore, the size of the local state space can be completely independent of the size of the system. In local rule-governed systems, each local state, and consequently the global state, can be determined exactly and precisely. Fine differences in the global state will result in very specific differences in the local state and, consequently, will affect the invocation of local rules. As fine differences affect local behavior, the difference will be felt in an expanding patch of local states, and in this manner—propagating from local neighborhood to local neighborhood—fine differences in global state can result in large differences in global behavior. The only "special cases" explicitly dealt with in locally determined systems are exactly the set of all possible local states, and the rules for these are just exactly the set of all local rules governing the system

5. BIOLOGICAL AUTOMATA

Organisms have been compared to extremely complicated and finely tuned biochemical machines. Since we know that it is possible to abstract the logical form of a machine from its physical hardware, it is natural to ask whether it is possible to abstract the logical form of an organism from its biochemical wetware. The field of Artificial Life is devoted to the investigation of this question.

In the following sections we will look at the manner in which behavior is generated in a bottom-up fashion in living systems. We then generalize the mechanisms by which this behavior generation is accomplished, so that we may apply them to the task of generating behavior in artificial systems.

We will find that the essential machinery of living organisms is quite a bit different from the machinery of our own invention, and we would be quite mistaken to attempt to force our preconceived notions of abstract machines onto the machinery of life. The difference, once again, lies in the exceedingly parallel and distributed nature of the operation of the machinery of life, as contrasted with the singularly serial and centralized control structures associated with the machines of our invention.

5.1 GENOTYPES AND PHENOTYPES

The most salient characteristic of living systems, from the behavior generation point of view, is the *genotype/phenotype* distinction. The distinction is essentially one between a specification of machinery—the genotype—and the behavior of that machinery—the phenotype.

The *genotype* is the complete set of genetic instructions encoded in the linear sequence of nucleotide bases that makes up an organism's DNA. The *phenotype* is the physical organism itself—the structures that emerge in space and time as the result of the interpretation of the genotype in the context of a particular environment. The process by which the phenotype develops through time under the direction of the genotype is called *morphogenesis*. The individual genetic instructions are called *genes* and consist of short stretches of DNA. These instructions are "executed"—or *expressed*—when their DNA sequence is used as a template for transcription. In the case of protein synthesis, transcription results in a duplicate nucleotide strand known as a *messenger RNA*—or *mRNA*—constructed by the process of base-pairing. This mRNA strand may then be modified in various ways before it makes its way out to the cytoplasm where, at bodies known as *ribosomes*, it serves as a template for the construction of a linear chain of *amino acids*. The resulting *polypeptide* chain will fold up on itself in a complex manner, forming a tightly packed molecule known as a *protein*. The finished protein detaches from the ribosome and may go on to serve as a passive structural element in the cell, or may have a more active role as an *enzyme*. Enzymes are *the* functional molecular "operators" in the logic of life.

One may consider the genotype as a largely unordered "bag" of instructions, each one of which is essentially the specification for a "machine" of some sort—passive or active. When instantiated, each such "machine" will enter into the ongoing logical "fray" in the cytoplasm, consisting largely of local interactions between other such machines. Each such instruction will be "executed" when its own triggering conditions are met and will have specific, local effects on structures in the cell. Furthermore, each such instruction will operate within the context of all of the other instructions that have been—or are being—executed.

The phenotype, then, consists of the structures and dynamics that emerge through time in the course of the execution of the parallel, distributed "computation" controlled by this genetic "bag" of instructions. Since gene's interactions with one another are highly nonlinear, the phenotype is a nonlinear function of the genotype.

5.2 GENERALIZED GENOTYPES AND PHENOTYPES

In the context of Artificial Life, we need to generalize the notions of *genotype* and *phenotype*, so that we may apply them in non-biological situations. We will use the term *generalized genotype*—or GTYPE—to refer to any largely unordered set of low-level rules, and we will use the term *generalized phenotype*—or PTYPE— to refer to the behaviors and/or structures that emerge out of the interactions among these low-level rules when they are activated within the context of a specific environment. The GTYPE, essentially, is the specification for a set of machines, while the PTYPE is the behavior that results as the machines are run and interact with one another.

This is the bottom-up approach to the generation of behavior. A set of entities is defined, and each entity is endowed with a specification for a simple behavioral repertoire—a GTYPE—that contains instructions which detail its reactions to a wide range of *local* encounters with other such entities or with specific features of the environment. Nowhere is the behavior of the set of entities as a whole specified. The global behavior of the aggregate—the PTYPE—emerges out of the collective interactions among individual entities.

It should be noted that the PTYPE is a multilevel phenomenon. First, there is the PTYPE associated with each particular instruction—the effect which that instruction has on an entity's behavior when it is expressed. Second, there is the PTYPE associated with each individual entity—its individual behavior within the aggregate. Third, there is the PTYPE associated with the behavior of the aggregate as a whole.

This is true for natural systems as well. We can talk about the phenotypic trait associated with a particular gene, we can identify the phenotype of an individual cell, and we can identify the phenotype of an entire multi-cellular organism—its body, in effect. PTYPES *should* be complex and multilevel. If we want to simulate life, we should expect to see hierarchical structures emerge in our simulations. In general, phenotypic traits at the level of the whole organism will be the result of many nonlinear interactions between genes, and there will be no single gene to which one can assign responsibility for the vast majority of phenotypic traits.

In summary, GTYPES are low-level rules for behav_ors_—i.e., abstract specifications for "machines"—which will engage in local interactions within a large aggregate of other such behav_iors_. PTYPES are the behav_ors_—the structures in time and space—that *develop* out of these nonlinear, local interactions (Figure 5).

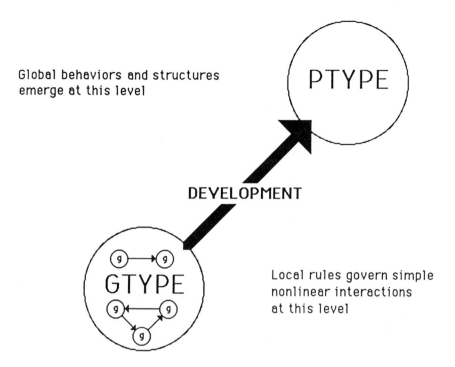

FIGURE 5 The relationship between GTYPE and PTYPE. Original figure appeared in "Artificial Life" by Christopher Langton, in *Artificial Life* edited by C. Langton (Addison-Wesley, Redwood City, 1989).

5.3 UNPREDICTABILITY OF PTYPE FROM GTYPE

Nonlinear interactions between the objects specified by the GTYPE provide the basis for an extremely rich variety of possible PTYPES. PTYPES draw on the full combinatorial potential implicit in the set of possible interactions between low-level rules. The other side of the coin, however, is that we cannot predict the PTYPES that will emerge from specific GTYPES, due to the general unpredictability of nonlinear systems. If we wish to maintain the property of predictability, then we must restrict severely the nonlinear dependence of PTYPE on GTYPE, but this forces us to give up the combinatorial richness of possible PTYPES. Therefore, a trade-off exists between behavioral richness and predictability (or "programmability"). We shall see in the section on evolution that the lack of programmability is adequately compensated for by the increased capacity for adaptiveness provided by a rich behavioral repertoire.

As discussed previously, we know that it is impossible in the general case to determine *any* nontrivial property of the future behavior of a sufficiently powerful computer from a mere inspection of its program and its initial state alone.[14] A Turing machine—the formal equivalent of a general purpose computer—can be captured within the scheme of GTYPE/PTYPE systems by identifying the machine's transition table as the GTYPE and the resulting computation as the PTYPE. From this we can deduce that in the general case it will not be possible to determine, by inspection alone, any nontrivial feature of the PTYPE that will emerge from a given GTYPE in the context of a particular initial configuration. In general, the only way to find out anything about the PTYPE is to start the system up and watch what happens as the PTYPE develops under control of the GTYPE and the environment.

Similarly, it is not possible in the general case to determine what specific alterations must be made to a GTYPE to effect a desired change in the PTYPE. The problem is that any specific PTYPE trait is, in general, an effect of many, many nonlinear interactions between the behavioral primitives of the system (an "epistatic trait" in biological terms). Consequently, given an arbitrary proposed change to the PTYPE, it may be impossible to determine by any formal procedure exactly what changes would have to be made to the GTYPE to effect that—and *only* that—change in the PTYPE. It is not a practically computable problem. There is no way to calculate the answer—short of exhaustive search—*even though there may be an answer!*[7]

The only way to proceed in the face of such an unpredictability result is by a process of trial and error. However, some processes of trial and error are more efficient than others. In natural systems, trial and error are interlinked in such a way that error guides the choice of trials under the process of evolution by natural selection. It is quite likely that this is the *only* efficient, *general* procedure that could find GTYPES with specific PTYPE traits when nonlinear functions are involved.

6. RECURSIVELY GENERATED OBJECTS

In the previous section, we described the distinction between genotype and phenotype, and we introduced their generalizations in the form of GTYPES and PTYPES. In this section, we will review a general approach to building GTYPE/PTYPE systems based on the methodology of *recursively generated objects*.

A major appeal of this approach is that it arises naturally from the GTYPE/PTYPE distinction: the local developmental rules—the recursive description itself—constitute the *GTYPE*, and the developing structure—the recursively generated object or behavior itself—constitutes the *PTYPE*.

---

[7] An example in biology would be: What changes would have to be made to the genome in order to produce six fingers on each hand rather than five?

Artificial Life

Under the methodology of recursively generated objects, the "object" is a structure that has sub-parts. The rules of the system specify how to modify the most elementary, "atomic" sub-parts, and are usually sensitive to the *context* in which these atomic sub-parts are embedded. That is, the state of the "neighborhood" of an atomic sub-part is taken into account in determining which rule to apply in order to modify that sub-part. It is usually the case that there are no rules in the system whose context is the entire structure; that is, there is no use made of *global* information. Each piece is modified solely on the basis of its own state and the state of the pieces "nearby."

Of course, if the initial structure consists of a single part—as might be the case with the initial seed—then the context for applying a rule is necessarily global. The usual situation is that a structure consists of *many* parts, only a local sub-set of which determine the rule that will be used to modify any one sub-part of the structure.

A recursively generated object, then, is a kind of PTYPE, and the recursive description that generates it is a kind of GTYPE. The PTYPE will emerge under the action of the GTYPE, developing through time via a process akin to morphogenesis.

We will illustrate the notion of recursively generated objects with examples taken from the literature on L-systems, cellular automata, and computer animation.

6.1 EXAMPLE 1: LINDENMAYER SYSTEMS

Lindenmayer systems (L-systems) consist of sets of rules for rewriting strings of symbols, and bear strong relationships to the formal grammars treated by Chomsky. We will give several examples of L-systems illustrating the methodology of recursively generated objects.[8]

In the following "$X \rightarrow Y$" means that one replaces every occurrence of symbol "X" in the structure with string "Y." Since the symbol "X" may appear on the right as well as the left sides of some rules, the set of rules can be applied "recursively" to the newly rewritten structures. The process can be continued ad infinitum although some sets of rules will result in a "final" configuration when no more changes occur.

SIMPLE LINEAR GROWTH Here is an example of the simplest kind of L-system. The rules are *context free*, meaning that the context in which a particular part is situated is *not* considered when altering it. There must be only one rule per part if the system is to be deterministic.

The rules (the "recursive description" or GTYPE):

```
1)    A -> CB
2)    B -> A
3)    C -> DA
4)    D -> C
```

[8] For a more detailed review, see the the book *The Algorithmic Beauty of Plants*.[20]

When applied to the initial seed structure "A," the following structural history develops (each successive line is a successive time step):

```
time    structure    rules applied (L to R)
----    ---------    ----------------------
 0         A         (initial "seed")
 1        C B        (rule 1 replaces A with CB)
 2       D A A       (rule 3 replaces C with DA & rule 2 replaces B with A)
 3      C C B C B    (rule 4 replaces D with C & rule 1 replaces the two
 4      ...(etc)...                                    A's with CB's)
```

And so forth.

The "PTYPE" that emerges from this kind of recursive application of a simple, local rewriting rule can get extremely complex. These kinds of grammars (whose rules replace single symbols) have been shown to be equivalent to the operation of finite state machines. With appropriate restrictions, they are also equivalent to the "regular languages" defined by Chomsky.

BRANCHING GROWTH L-systems incorporate meta-symbols to represent branching points, allowing a new line of symbols to branch off from the main "stem."

The following grammar produces branching structures. The "( )" and "[ ]" notations indicate left and right branches, respectively, and the strings within them indicate the structure of the branches themselves.

The rules—or GTYPE:

```
1) A -> C[B]D
2) B -> A
3) C -> C
4) D -> C(E)A
5) E -> D
```

When applied to the starting structure "A," the following sequence develops (using linear notation):

```
time              structure                     rules applied (L to R)
----    -----------------------------           ----------------------
 0                    A                         initial "seed".
 1                  C[B]D                       rule 1.
 2                C[A]C(E)A                     rules 3,2,4.
 3            C[C[B]D]C(D)C[B]D                 rules 3,1,3,5,1.
 4      C[C[A]C(E)A]C(C(E)A)C[A]C(E)A           rules 3,3,2,4,3,4,3,2,4.
```

In two dimensions, the structure develops as follows:

Artificial Life

$n=5, \delta=18°$

ω : plant
p_1 : plant $\rightarrow$ internode + [plant + flower] $- - //$
          [$- -$ leaf] internode [+ + leaf] $-$
          [plant flower] + + plant flower
p_2 : internode $\rightarrow$ F seg [// & & leaf] [// $\wedge \wedge$ leaf] F seg
p_3 : seg $\rightarrow$ seg F seg
p_4 : leaf $\rightarrow$ [' { +f$-$ff$-$f+ | +f$-$ff$-$f }]
p_5 : flower $\rightarrow$ [& & & pedicel ' / wedge //// wedge ////
          wedge //// wedge //// wedge]
p_6 : pedicel $\rightarrow$ FF
p_7 : wedge $\rightarrow$ [' $\wedge$ F] [{ & & & & $-$f+f | $-$f+f }]

FIGURE 6 An L-system plant grown from rules incorporating graphical rendering information. Original figure appeared in *The Algorithmic Beauty of Plants*. (Berlin: Springer-Verlag, 1991).[20]

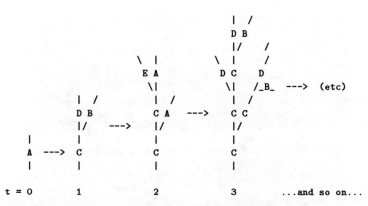

Note that at each step, *every symbol is replaced*, even if just by another copy of itself. This figure shows the result of growing a structure using the rules shown, which contain graphical rendering information in addition to the usual "structural" information.

SIGNAL PROPAGATION In order to propagate signals along a structure, one must have something more than just a single symbol on the left-hand side of a rule. When there is more than one symbol on the left-hand side of a rule, the rules are *context sensitive*—i.e., the "context" within which a symbol occurs (the symbols next to it) are important in determining what the replacement string will be. The next example illustrates why this is critical for signal propagation.

In the following example, the symbol in "{˜}'s" is the symbol (or string of symbols) to be replaced, the rest of the left-hand side is the context, and the symbols "[" and "]" indicate the left and right ends of the string, respectively.

Suppose the rule set contains the following rules:

```
1) [{C} -> C    a "C" at the left-end of the string remains a "C."
2) C{C} -> C    a "C" with a "C" to its left remains a "C."
3) *{C} -> *    a "C" with an "*" to its left becomes an "*."
4) {*}C -> C    an "*" with a "C" to its right becomes a "C."
5) {*}] -> *    an "*" at the right end of the string remains an "*."
```

Under these rules, the initial structure "*CCCCCCC" will result in the "*" being propagated to the right, as follows:

time	structure
0	*CCCCCCC
1	C*CCCCCC
2	CC*CCCCC
3	CCC*CCCC
4	CCCC*CCC
5	CCCCC*CC
6	CCCCCC*C
7	CCCCCCC*

This would not be possible without taking the "context" of a symbol into account. In general, these kinds of grammars are equivalent to Chomsky's "context-sensitive" or "Turing" languages, depending on whether or not there are any restrictions on the kinds of strings on the left- and right-hand sides.

The capacity for signal propagation is extremely important, for it allows arbitrary computational processes to be embedded within the structure, which may directly affect the structure's development. The next example demonstrates how embedded computation can affect development.

6.2 EXAMPLE 2: CELLULAR AUTOMATA

Cellular automata (CA) provide another example of the recursive application of a simple set of rules to a structure. In CA, the structure that is being updated is the entire universe: a lattice of finite automata. The local rule set—the GTYPE—in this case is the transition function obeyed homogeneously by every automaton in the lattice. The local context taken into account in updating the state of each automaton is the state of the automata in its immediate neighborhood. The transition function for the automata constitutes a *local* physics for a simple, discrete space/time universe. The universe is updated by applying the local physics to each local "cell" of its structure over and over again. Thus, although the physical structure itself doesn't develop over time, its *state* does.

Within such universes, one can embed all manner of processes, relying on the context sensitivity of the rules to local neighborhood conditions to propagate information around within the universe "meaningfully." In particular, one can embed general purpose computers. Since these computers are simply particular configurations of states within the lattice of automata, *they can compute over the very set of symbols out of which they are constructed.* Thus, structures in this universe can compute and construct other structures, which also may compute and construct.

For example, here is the simplest known structure that can reproduce itself:

```
          2 2 2 2 2 2 2
          2 1 7 0 1 4 0 1 4 2
          2 0 2 2 2 2 2 0 2
          2 7 2         2 1 2
          2 1 2         2 1 2
          2 0 2         2 1 2
          2 7 2         2 1 2
          2 1 2 2 2 2 2 2 1 2 2 2 2 2
          2 0 7 1 0 7 1 0 7 1 1 1 1 1 2
            2 2 2 2 2 2 2 2 2 2 2 2 2
```

Each number is the state of one automaton in the lattice. Blank space is presumed to be in state "0." The "2"-states form a sheath around the "1"-state data path. The "7 0" and "4 0" state pairs constitute signals embedded within the data path. They will propagate counterclockwise around the loop, cloning off copies which propagate down the extended tail as they pass the T-junction between loop and tail. When the signals reach the end of the tail, they have the following effects: each "7 0" signal extends the tail by one unit, and the two "4 0" signals construct a left-hand corner at the end of the tail. Thus, for each full cycle of the instructions around the loop, another side and corner of an "offspring-loop" will be constructed. When the tail finally runs into itself after four cycles, the collision of signals results in the disconnection of the two loops as well as the construction of a tail on each of the loops.

After 151 time steps, this system will evolve to the following configuration:

```
                              2
                            2 1 2
                            2 7 2
                            2 0 2
                            2 1 2
              2 2 2 2 2 2 2 7 2         2 2 2 2 2 2 2
              2 1 1 1 7 0 1 7 0 2       2 1 7 0 1 4 0 1 4 2
              2 1 2 2 2 2 2 1 2         2 0 2 2 2 2 2 2 0 2
              2 1 2           2 7 2     2 7 2           2 1 2
              2 1 2           2 0 2     2 1 2           2 1 2
              2 4 2           2 1 2     2 0 2           2 1 2
              2 1 2           2 7 2     2 7 2           2 1 2
              2 0 2 2 2 2 2 0 2         2 1 2 2 2 2 2 2 1 2 2 2 2 2
              2 4 1 0 7 1 0 7 1 2       2 0 7 1 0 7 1 0 7 1 1 1 1 1 2
              2 2 2 2 2 2 2             2 2 2 2 2 2 2 2 2 2 2 2
```

Thus, the initial configuration has succeeded in reproducing itself.

Each of these loops will go on to reproduce itself in a similar manner, giving rise to an expanding *colony* of loops, growing out into the array.

These embedded self-reproducing loops are the result of the recursive application of a rule to a seed structure. In this case, the primary rule that is being recursively applied constitutes the "physics" of the universe. The initial state of the loop itself constitutes a little "computer" under the recursively applied physics of the universe: a computer whose program causes it to construct a copy of itself. The "program" within the loop computer is also applied recursively to the growing structure. Thus, this system really involves a double level of recursively applied rules. The mechanics of applying one recursive rule within a universe whose physics is governed by another recursive rule had to be worked out by trial and error. This system makes use of the signal propagation capacity to embed a structure that itself *computes* the resulting structure, rather than having the "physics" directly responsible for developing the final structure from a passive seed.

This captures the flavor of what goes on in natural biological development: the genotype codes for the constituents of a dynamic process in the cell, and it is this dynamic process that is primarily responsible for mediating—or "computing"—the expression of the genotype in the course of development.

Artificial Life

6.3 EXAMPLE 3: FLOCKING "BOIDS"

The previous examples were largely concerned with the growth and development of *structural* PTYPES. Here, we give an example of the development of a *behavioral* PTYPE.

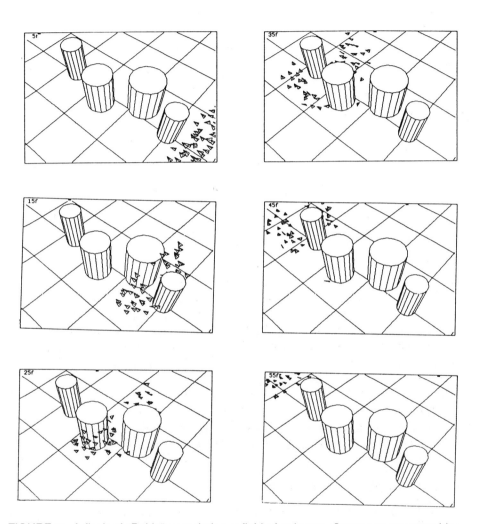

FIGURE 7 A flock of "Boids" negotiating a field of columns. Sequence generated by Criag Reynolds. Original figure appeared in "Artificial Life" by Christopher Langton, in *Artificial Life* edited by C. Langton (Addison-Wesley, Redwood City, 1989).

Craig Reynolds has implemented a simulation of flocking behavior.[22] In this model—which is meant to be a general platform for studying the qualitatively similar phenomena of flocking, herding, and schooling—one has a large collection of autonomous but interacting objects (which Reynolds refers to as "Boids"), inhabiting a common simulated environment.

The modeler can specify the manner in which the individual Boids will respond to *local* events or conditions. The global behavior of the aggregate of Boids is strictly an emergent phenomenon, none of the rules for the individual Boids depend on global information, and the only updating of the global state is done on the basis of individual Boids responding to local conditions.

Each Boid in the aggregate shares the same behavioral "tendencies":

- to maintain a minimum distance from other objects in the environment, including other Boids,
- to match velocities with Boids in its neighborhood, and
- to move toward the perceived center of mass of the Boids in its neighborhood.

These are the only rules governing the behavior of the aggregate.

These rules, then, constitute the generalized genotype (GTYPE) of the Boids system. They say nothing about structure, or growth and development, but they determine the behavior of a set of interacting objects, out of which very natural motion emerges.

With the right settings for the parameters of the system, a collection of Boids released at random positions within a volume will collect into a dynamic flock, which flies around environmental obstacles in a very fluid and natural manner, occasionally breaking up into sub-flocks as the flock flows around both sides of an obstacle. Once broken up into sub-flocks, the sub-flocks reorganize around their own, now distinct and isolated centers of mass, only to re-merge into a single flock again when both sub-flocks emerge at the far side of the obstacle and each sub-flock feels anew the "mass" of the other sub-flock (Figure 7).

The flocking behavior itself constitutes the generalized phenotype (PTYPE) of the Boids system. It bears the same relation to the GTYPE as an organism's morphological phenotype bears to its molecular genotype. The same distinction between the *specification* of machinery and the *behavior* of machinery is evident.

6.4 DISCUSSION OF EXAMPLES

In all of the above examples, the recursive rules apply to *local structures* only, and the PTYPE—structural or behavioral—that results at the global level emerges out of all local activity taken collectively. Nowhere in the system are there rules for the behavior of the system at the global level. This is a much more powerful and simple approach to the generation of complex behavior than that typically taken in AI, for instance, where "expert systems" attempt to provide global rules for global behavior. Recursive, "bottom up" specifications yield much more natural, fluid, and flexible behavior at the global level than typical "top down" specifications, and they do so *much* more parsimoniously.

IMPORTANCE OF CONTEXT SENSITIVITY. It is worthwhile to note that *context-sensitive* rules in GTYPE/PTYPE systems provide the possibility for nonlinear interactions among the parts. Without context sensitivity, the systems would be linearly decomposable, information could not "flow" throughout the system in any meaningful manner, and complex long-range dependencies between remote parts of the structures could not develop.

FEEDBACK BETWEEN THE LOCAL AND THE GLOBAL LEVELS. There is also a very important feedback mechanism *between* levels in such systems: the interactions among the low-level entities give rise to the global-level dynamics which, in turn, affects the lower levels by *setting the local context* within which each entity's rules are invoked. Thus, local behavior supports global dynamics, which shapes local context, which affects local behavior, which supports global dynamics, and so forth.

6.5 GENUINE LIFE IN ARTIFICIAL SYSTEMS

It is important to distinguish the ontological status of the various levels of behavior in such systems. At the level of the individual behav<u>ors</u>, we have a clear difference in kind: Boids are **not** birds, they are not even remotely like birds, they have no cohesive physical structure, but rather they exist as information structures—processes—within a computer. But—and this is *the* critical "But"—at the level of behav<u>iors</u>, *flocking Boids and flocking birds are two instances of the same phenomenon: flocking.*

The behavior of a flock as a whole does not depend critically on the internal details of the entities of which it is constituted, only on the details of the way in which these entities behave in each other's presence. Thus, flocking in Boids is true flocking, and may be counted as another empirical data point in the study of flocking behavior in general, right up there with flocks of geese and flocks of starlings.

This is *not* to say that flocking Boids capture *all* the nuances upon which flocking behavior depends, or that the Boid's behavioral repertoire is sufficient to exhibit all the different modes of flocking that have been observed—such as the classic "V" formation of flocking geese. The crucial point is that we have captured, within an aggregate of artificial entities, a *bona fide* lifelike behavior, and that the behavior emerges within the artificial system in the same way that it emerges in the natural system.

The same is true for L-systems and the self-reproducing loops. The constituent parts of the artificial systems are different kinds of things from their natural counterparts, but the emergent behaviors that they support are the same kinds of thing as their natural counterparts: genuine morphogenesis and differentiation for L-systems, and genuine self-reproduction in the case of the loops.

The claim is the following. The "artificial" in Artificial Life refers to the component parts, not the emergent processes. If the component parts are implemented correctly, the processes they support are *genuine*—every bit as genuine as the natural processes they imitate.

The *big* claim is that a properly organized set of artificial primitives carrying out the same functional roles as the biomolecules in natural living systems will support a process that will be "alive" in the same way that natural organisms are alive. Artificial Life will therefore be *genuine* life—it will simply be made of different stuff than the life that has evolved here on Earth.

7. EVOLUTION

7.1 EVOLUTION: FROM ARTIFICIAL SELECTION TO NATURAL SELECTION

Modern organisms owe their structure to the complex process of biological evolution, and it is very difficult to discern which of their properties are due to chance and which to necessity. If biologists could "rewind the tape" of evolution and start it over, again and again, from different initial conditions, or under different regimes of external perturbations along the way, they would have a full *ensemble* of evolutionary pathways to generalize over. Such an ensemble would allow them to distinguish universal, necessary properties (those which were observed in all the pathways in the ensemble) from accidental, chance properties (those which were unique to individual pathways). However, biologists cannot rewind the tape of evolution, and are stuck with a single, actual evolutionary trace out of a vast, intuited ensemble of possible traces.

Although studying computer models of evolution is not the same as studying the "real thing," the ability to freely manipulate computer experiments—to "rewind the tape," perturb the initial conditions, and so forth—can more than make up for their "lack" of reality.

It has been known for some time that one can evolve computer programs by the process of natural selection among a population of variant programs. Each individual program in a population of programs is evaluated for its performance on some task. The programs that perform best are allowed to "breed" with one another via *Genetic Algorithms*.[8,12] The offspring of these better-performing parent programs replace the worst-performing programs in the population, and the cycle is iterated. Such evolutionary approaches to program improvement have been applied primarily to the tasks of function optimization and machine learning.

However, such evolutionary models have rarely been used to study evolution itself.[27] Researchers have primarily concentrated on the *results*, rather than on the *process*, of evolution. In the spirit of von Neumann's research on self-reproduction via the study of self-reproducing *automata*, the following sections review studies of the process of evolution by studying evolving populations of "automata."

7.2 ENGINEERING PTYPES FROM GTYPES

In the preceding sections, we have mentioned several times the formal impossibility of predicting the behavior of an arbitrary machine by mere inspection of its specification and initial state. In the general case, we must run a machine in order to determine its behavior.

The consequence of this unpredictability for GTYPE/PTYPE systems is that we cannot determine the PTYPE that will be produced by an arbitrary GTYPE by inspection alone. We must "run" the GTYPE in the context of a specific environment, and let the PTYPE develop in order to determine the resulting structure and its behavior.

This is even further complicated when the environment consists of a population of PTYPES engaged in nonlinear interactions, in which case the determination of a PTYPE depends on the behavior of the specific PTYPES it is interacting with, and on the emergent details of the global dynamics.

Since, for any interesting system, there will exist an enormous number of potential GTYPES, and since there is no formal method for deducing the PTYPES from the GTYPES, how do we go about finding GTYPES that will generate lifelike PTYPES? Or PTYPES that exhibit any other particular sought-after behavior?

Until now, the process has largely been one of guessing at appropriate GTYPES, and modifying them by trial and error until they generate the appropriate PTYPES. However, this process is limited by our preconceptions of what the appropriate PTYPES would be, and by our restricted notions of how to generate GTYPES. We would like to be able to automate the process so that our preconceptions and limited abilities to conceive of machinery do not overly constrain the search for GTYPES that will yield the appropriate behaviors.

7.3 NATURAL SELECTION AMONG POPULATIONS OF VARIANTS

Nature, of course, has had to face the same problem, and has hit upon an elegant solution: *evolution by the process of natural selection among populations of variants.* The scheme is a very simple one. However, in the face of the formal impossibility of predicting behavior from machine description alone, it may well be the only efficient, general scheme for searching the space of possible GTYPES.

The mechanism of evolution is as follows. A set of GTYPES is interpreted within a specific environment, forming a population of PTYPES which interact with one another and with features of the environment in various complex ways. On the basis of the relative performance of their associated PTYPES, *some* GTYPES are duplicated in larger numbers than others, and they are duplicated in such a way that the copies are similar to—but not exactly the same as—the originals. These

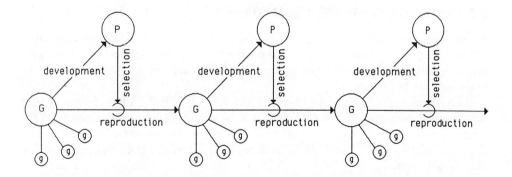

FIGURE 8 The process of evolution by natural selection. Original figure appeared in "Artificial Life" by Christopher Langton, in *Artificial Life*, edited by C. Langton (Addison-Wesley, Redwood City, 1989).

variant GTYPES develop into variant PTYPES, which enter into the complex interactions within the environment, and the process is continued *ad infinitum* (Figure 8). As expected from the formal limitations on predictability, GTYPES must be "run" (i.e., turned into PTYPES) in an environment and their behaviors must be evaluated explicitly, their implicit behavior cannot be determined.

7.4 GENETIC ALGORITHMS

In the spirit of von Neumann, John Holland has attempted to abstract "the logical form" of the natural process of biological evolution in what he calls the "Genetic Algorithm" (GA).[12,13] In the GA, a GTYPE is represented as a character string that encodes a potential solution to a problem. For instance, the character string might encode the weight matrix of a neural network, or the transition table of a finite state machine. These character strings are rendered as PTYPES via a problem-specific interpreter, which constructs, for example, the neural net or finite state machine specified by each GTYPE, evaluates its performance in the problem domain, and provides it with a specific fitness value, or "strength."

The GA implements natural selection by making more copies of the character strings representing the better performing PTYPES. The GA generates variant GTYPES by applying *genetic operators* to these character strings. The genetic operators typically consist of *reproduction*, *crossover*, and *mutation*, with occasional usage of *inversion* and *duplication*.

Recently, John Koza[15] has developed a version of the GA, which he calls the *Genetic Programming Paradigm* (GPP), that extends the genetic operators to work on GTYPES that are simple expressions in a standard programming language. The GPP differs from the traditional GA in that these program expressions are

(a) 1 1 0 0 1 0 1 0 1 1 1 0 1 0 1 1 0 1 0 1 0 0 0 1 0 1 0 1 1 0 1 1 1 0 1 0 1 1

(b) (OR (NOT a) (AND b c))

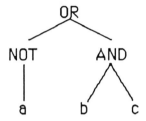

FIGURE 9 GTYPES in the GA and GPP paradigms.

not represented as simple character strings but rather as the parse trees of the expressions. This makes it easier for the genetic operators to obey the syntax of the programming language when producing variant GTYPES.

Figure 9 shows some examples of GTYPES in the GA and GPP paradigms.

THE GENETIC OPERATORS The genetic operators work as follows.

Reproduction is the most basic operator. It is often implemented in the form of *fitness proportionate reproduction*, which means that strings are duplicated in direct proportion to their relative fitness values. Once all strings have been evaluated, the average fitness of the population is computed, and those strings whose fitness is higher than the population average have a higher probability of being duplicated, while those strings whose fitness is lower than the population average have a lower probability of being duplicated. There are many variations on this scheme, but most implementations of the GA or the GPP use some form of fitness proportionate reproduction as the means to implement "selection." Another form of this is to simply keep the top 10% or so of the population and throw away the rest, using the survivors as breeding stock for the next generation.

Mutation in the GA is simply the replacement of one or more characters in a character string GTYPE with another character picked at random. In binary strings, this simply amounts to random bit flips. In the GPP, mutation is implemented by picking a sub-tree of the parse tree at random, and replacing it with a randomly generated sub-tree whose root node is of the same syntactic type as the root node of the replaced sub-tree.

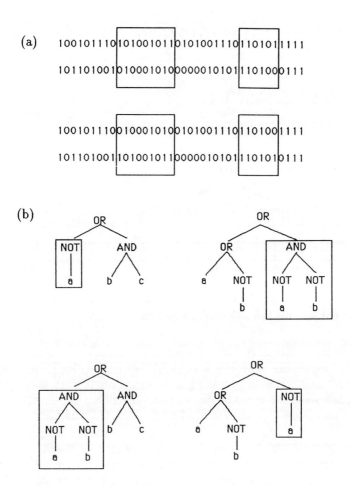

FIGURE 10 Crossover operation in the GA and GPP.

Crossover is an analog of sexual recombination. In the GA, this is accomplished by picking two "parent" character strings, lining them up side-by-side, and interchanging equivalent sub-strings between them, producing two new sub-strings that each contain a mix of their parent's genetic information. Crossover is an extremely important genetic operator. Whereas mutation is equivalent to random search, crossover allows the more "intelligent" search strategy of putting things that have proved useful in new combinations.

In the GPP, crossover is implemented by picking two "parent" parse trees, locating syntactically similar sub-trees within each, and swapping them.

Figure 10 illustrates the crossover operation in the GA and GPP.

Inversion is used rarely in order to rearrange the relative locations of specific pieces of genetic information in the character strings of the GA.

Duplication is sometimes used in situations where it makes sense for the genome to grow in length, representing, for instance, larger neural nets, or bigger finite state machine transition tables.

THE OPERATION OF THE GENETIC ALGORITHM The basic outline of the genetic algorithm is as follows:

1. Generate a random initial population of GTYPES.
2. Render the GTYPES in the population as PTYPES and evaluate them in the problem domain, providing each GTYPE with a fitness value.
3. Duplicate GTYPES according to their relative fitness using a scheme like fitness proportionate reproduction.
4. Apply *genetic operators* to the GTYPES in the population, typically picking crossover partners as a function of their relative fitness.
5. Replace the least-fit GTYPES in the population with the offspring generated in the last several steps.
6. Go back to step 2 and iterate.

Although quite simple in outline, the genetic algorithm has proved remarkably powerful in a wide variety of applications, and provides a useful tool for both the study and the application of evolution.

THE CONTEXT OF ADAPTATION GA's have traditionally been employed in the contexts of machine learning and function optimization. In such contexts, one is often looking for an explicit, optimal solution to a particular, well-specified problem. This is reflected in the implementation of the evaluation of PTYPES in traditional GA's: each GTYPE is expressed as a PTYPE independently of the others, tested on the problem, and assigned a value representing its individual fitness using an explicit fitness function. Thus, one is often seeking to evolve an *individual* that *explicitly encodes* an optimal solution to a precisely specified problem. The fitness of a GTYPE in such cases is simply a function of the problem domain, and is independent of the fitnesses of the other GTYPES in the population.

This is quite different from the context in which natural biological evolution has taken place, in which the behavior of a PTYPE and its associated fitness are highly dependent on which other PTYPES exist in the environment, and on the dynamics of their interactions. Furthermore, in the natural context, it is generally the case that there is no single, explicitly specified problem confronting the population. Rather, there is often quite a large set of problems facing the population at any one time, and these problems are only implicitly determined as a function of the dynamics of the population and the environment themselves, which may change significantly over time. *In such a context, nature has often discovered that the collective behavior emerging from the interactions among a set of PTYPES will address a subset of the implicitly defined problems.*

Thus, the proper picture for the natural evolutionary context is that of a large cloud of implicit collective solutions addressing a large cloud of implicit collective problems. Both of these clouds are implicit in the spatio-temporal dynamics of the population.

The dynamics of such systems are very complex and impossible to predict. One can think of them as the dynamical equivalent of many-body orbital mechanics problems: two-body problems can be treated analytically, whereas three- or more body problems are nonanalytic.

The important point here is that nonlinearities and emergent collective phenomena are properties that are to be exploited, rather than avoided as has been the traditional engineering viewpoint. Emergent nonlinear solutions may be harder to understand or to engineer, but there are far more of them than there are nonemergent, analyzable linear solutions. The true power of evolution lies in its ability to exploit emergent collective phenomena; it lies, in fact, in evolution's inability to *avoid* such phenomena.

7.5 FROM ARTIFICIAL SELECTION TO NATURAL SELECTION

In *The Origin of Species*, Darwin used a very clever device to argue for the agency of natural selection. In the first chapter of *Origin*, Darwin lays the groundwork of the case for *natural* selection by carefully documenting the process of *artificial* selection. Most people of his time were familiar with the manner in which breeders of domestic animals and plants could enhance traits arbitrarily by selective breeding of their stock. Darwin carefully made the case that the wide variety of domestic animals and plants extant at his time were descended from a much smaller variety of wildstock, due to the selective breedings imposed by farmers and herders throughout history.

Now, Darwin continues, simply note that environmental circumstances can fill the role played by the human breeder in artificial selection, and voilà! one has natural selection. The rest of the book consists in a very careful documentation of the manner in which different environmental conditions would favor animals bearing different traits, making it more likely that individuals bearing those traits would survive to mate with each other and produce offspring, leading to the gradual enhancement of those traits through time. A beautifully simple yet elegant mechanism to explain the origin and maintenance of the diversity of species on Earth—too simple for many of his time, particularly those of strong religious persuasion.

The abstraction of this simple elegant mechanism for the production and filtration of diversity in the form of the Genetic Algorithm is straightforward and obvious. However, as it is usually implemented, it is artificial, rather than natural, selection that is the agency determining the direction of computer evolution. Either we ourselves, or our algorithmic agents in the form of explicit fitness functions, typically stand in the role of the breeder in computer implementations of evolution. Yet it is plain that the role of "breeder" can as easily be filled by "nature" in the world inside the computer as it is in the world outside the computer—it is just a different "nature."

In the following sections, we will explore a number of examples of computational implementations of the evolutionary process, starting with examples that clearly involve artificial selection, and working our way through to an example that clearly involves natural selection. The key thing to keep track of throughout these examples is the manner in which we incrementally give over our role as breeder to the "natural" pressures imposed by the dynamics of the computational world itself.

A BREEDER'S PARADISE: BIOMORPHS The first model, a clear-cut example of computational artificial selection, is due to the Oxford evolutionary biologist, Richard Dawkins, author of such highly regarded books as *The Selfish Gene*, *The Extended Phenotype*, and *The Blind Watchmaker*.

In order to illustrate the power of a process in which the random production of variation is coupled with a selection mechanism, Dawkins wrote a program for the Apple Macintosh computer that allows users to "breed" recursively generated objects.

The program is set up to generate tree structures recursively by starting with a single stem, adding branches to it in a certain way, adding branches to those branches in the same way, and so on. The number of branches, their angles, their size relative to the stem they are being added to, the number of branching iterations, and other parameters affecting the growth of these trees are what constitute the GTYPES of the tree organisms—or "biomorphs" as Dawkins calls them. Thus, the program consists of a general purpose recursive tree generator, which takes an organism's GTYPE (parameter settings) as data and generates its associated PTYPE (the resulting tree).

The program starts by producing a simple default—or "Adam"—tree and then produces a number of mutated copies of the parameter string for the Adam tree. The program renders the PTYPE trees for all of these different mutants on the screen for the user to view. The user then selects the PTYPE (i.e., tree shape) he or she likes the best, and the program produces mutated copies of that tree's GTYPE, and renders the associated PTYPES. The user selects another tree, and the process continues. The original Adam tree together with a number of its distant descendants are shown in Figure 11.

It is clear that this is a process of artificial selection. The computer generates the variants, but the human user fills the role of the "breeder," the active selective agent, determining which structures are to go on to produce variant offspring. However, the mechanics of the production of variants are particularly clear: produce slight variations on the presently selected GTYPE. The specific action taken by the human breeder is also very clear: choose the PTYPE whose GTYPE will have variations of it produced in the next round. There is both a producer and a selector of variation.

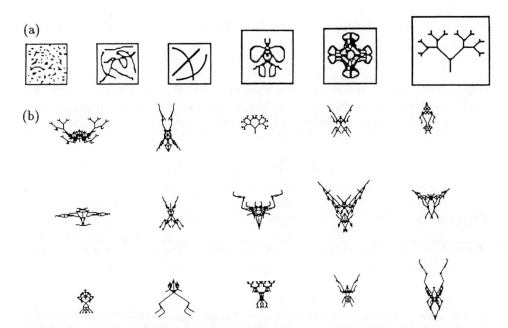

FIGURE 11 (a) Dawkin's original Adam tree, and (b) a number of its distant descendants.[6]

ALGORITHMIC BREEDERS In this section, we investigate a model which will take us two steps closer to natural selection. First, the human breeder is taken out of the loop, replaced by a program he writes, which formalizes his selection criteria, so that the act of selection can be performed by his computational agent. Second, we see that our computational representative can itself be allowed to evolve—an important first step toward eliminating our externally imposed, *a priori* criteria from the process completely.

The system we discuss here is due to Danny Hillis, inventor of the Connection Machine and chief scientist of Thinking Machines Corporation. In the course of the work at TMC, they have a need to design fast and efficient chips for the hardware implementation of a wide variety of common computational tasks, such as sorting numbers. For many of these, there is no body of theory that tells engineers how to construct the optimal circuit to perform the task in question. Therefore, progress in the design of such circuits is often a matter of blind trial and error until a better circuit is discovered. Hillis decided to apply the trial-and-error procedure of evolution to the problem of designing sorting circuits.

In his system, the GTYPES are strings of numbers encoding circuit connections that implement comparisons and swaps between input lines. GTYPES are rendered into the specific circuits they encode—their PTYPES—and they are rated according to the number of circuit elements and connections they require, and by their performance on a number of test strings which they have to sort. This rating is accomplished by an explicit fitness function—Hillis' computational representative—which implements the selection criteria and takes care of the breeding task. Thus, this is still a case of artificial selection, even though there is no human being actively doing the selection.

Hillis implemented the evolution problem on his Connection Machine CM2—a 64K processor SIMD parallel supercomputer. With populations of 64K sorting networks over thousands of generations, the system managed to produce a 65-element sorter, better than some cleverly engineered sorting networks, but not as good as the best known such network, which has 60 components. After reaching 65-element sorters, the system consistently became stuck on local optima.

Hillis then borrowed a trick from the biological literature on the coevolution of hosts and parasites (specifically Hamilton[9,10]) and in the process took a step closer to natural selection by allowing the evaluation function to evolve in time. In the previous runs, the sorting networks were evaluated on a fixed set of sorting problems—random sequences of numbers that the networks had to sort into correct order. In the new set of runs, Hillis made another evolving population out of the sorting problems. The task for the sorting networks was to do a good job on the sorting problems, while the task for the sorting problems was to make the sorting networks perform poorly.

In this situation, whenever a good sorting network emerged and took over the population, it became a target for the population of sorting problems. This led to the rapid evolution of sorting sequences that would make the network perform poorly and hence reduce its fitness. Hillis found that this coevolution between the sorting networks and the sorting problems led much more rapidly to better solutions than had been achieved by the evolution of sorting networks alone, resulting in a sorting network consisting of 61 elements.

It is the coevolution in this latter set of runs that both bring us one step closer to natural selection and is responsible for the enhanced efficiency of the search for an optimal sorting network. First of all, rather than having an absolute, fixed value, the fitness of a sorting network depends on the specific set of sorting problems it is facing. Likewise, the fitness of a set of sorting problems depends on the specific set of sorting networks it is facing. Thus, the "fitness" of an individual is now a relative quantity, not an absolute one. The fitness function depends a little more on the "nature" of the system, it is an evolving entity as well.

Coevolution increases the efficiency of the search as follows. In the earlier runs consisting solely of an evolving population of sorting networks, the population of networks was effectively hill climbing on a multi-peaked fitness landscape. Therefore, the populations would encounter the classic problem of getting stuck on local maxima. That is, a population could reach certain structures which lie on relatively low fitness peaks, but from which any deviations result in lower fitness, which is selected against. In order to find another, higher peak, the population would have to cross a fitness valley, which it is difficult to do under simple Darwinian selection (Figure 13(a)).

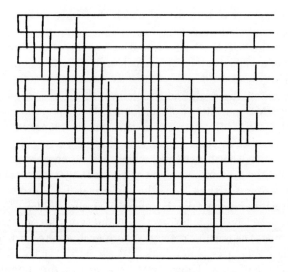

FIGURE 12 An evolved sorting network showing sequencing of comparisons and swaps. Original figure appeared in "Co-Evolving Parasites Improve Simulated Evolution as an Optimization Procedure" by W. D. Hillis, in *Artificial Life II*, edited by C. G. Langton, C. Taylor, J. D. Farmer, and S. Rasmussen, (Redwood City, CA: Addison-Wesley, 1991).[11]

In the coevolutionary case, here's what happens (Figure 13(b)). When a population of sorting networks gets stuck on a local fitness peak, it becomes a target for the population of sorting problems. That is, it defines a new peak for the sorting problems to climb. As the sorting problems climb their peak, they *drive down the peak on which the sorting networks are sitting*, by finding sequences that make the sorting networks perform poorly, therefore lowering their fitness. After a while, the fitness peak that the sorting networks were sitting on has been turned into a fitness valley, from which the population can escape by climbing up the neighboring peaks. As the sorting networks climb other peaks, they drive down the peak that they had provided for the sorting problems, which will then chase the sorting networks to the new peaks they have achieved and drive those down in turn.

In short, each population dynamically deforms the fitness landscape being traversed by the other population in such a way that both populations can continue to climb uphill without getting stuck on local maxima. When they do get stuck, the maxima get turned into minima which can be climbed out of by simple Darwinian means. Thus, coupled populations evolving by Darwinian means can bootstrap each other up the evolutionary ladder far more efficiently than they can climb it alone. By competing with one another, coupled populations improve one another at increased rates.

Artificial Life

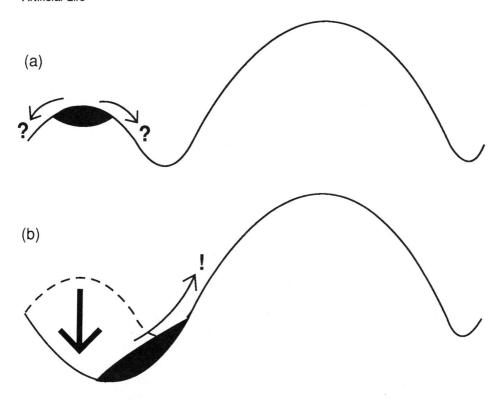

FIGURE 13 (a) Population of sorting networks stuck on a local fitness peak. In order to attain the higher peak, the population must cross a fitness "valley," which is difficult to achieve under normal Darwinian mechanisms. (b) The coevolving parasites deform the fitness landscape of the sorting networks, turning the fitness peak into a fitness valley, from which it is easy for the population to escape.

TABLE 1 The payoff matrix for the Prisoner's Dilemma Game. The pair (s_1, s_2) denotes the scores to players A and B, respectively.

		Player B	
		Cooperate	Defect
Player A	Cooperate	(3,3)	(0,5)
	Defect	(5,0)	(1,1)

Thus, when coupled in this way, a population may get hung up on local optima for a while, but eventually it will be able to climb again. This suggests immediately that the structure of the evolutionary record for such systems should show periods of stasis followed by periods of evolutionary change. The stasis comes about as populations sit at the top of local fitness peaks, waiting around for something to come along and do them the favor of lowering the peaks they are stuck on. The periods of change come about when populations are released from local optima and are freed to resume climbing up hills, and are therefore changing in time. Hillis has, in fact, carefully documented this kind of *Punctuated Equilibria* in his system.

COMPUTATIONAL ECOLOGIES Continuing on our path from artificial to natural selection, we turn to a research project carried out by Kristian Lindgren,[18] in which, although there is still an explicit fitness measure, many different species of organisms coevolve in each other's presence, forming ecological webs allowing for more complex interactions than the simple host-parasite interactions described above.

In this paper, Lindgren studies evolutionary dynamics within the context of a well-known game-theoretic problem: the *Iterated Prisoner's Dilemma* model (IPD). This model has been used effectively by Axelrod and Hamilton in their studies of the evolution of cooperation.[1,2]

In the prisoner's dilemma model, the payoff matrix (the fitness function) is constructed in such a way that individuals will garner the most payoff collectively in the long run if they "cooperate" with one another by *avoiding* the behaviors that would garner them the most payoff individually in the short run. If individuals only play the game once, they will do best by not cooperating ("defecting"). However, if they play the game repeatedly with one another (the "iterated" version of the game), they will do best by cooperating with one another.

The payoff matrix for the prisoner's dilemma game is shown in Table 1. This payoff matrix has the following interesting property. Assume, as is often assumed in game theory, that each player wants to maximize his immediate payoff, and let's analyze what player **A** should do. If **B** cooperates, then **A** should defect, because then **A** will get a score of 5 whereas he only gets a score of 3 if he cooperates. On the other hand, if **B** defects, then again, **A** should defect, as he will get a score of 1 if he defects while he only gets a score of 0 if he cooperates. So, no matter what **B** does, **A** maximizes his immediate payoff by defecting. Since the payoff matrix is symmetric, the same reasoning applies to player **B**, so **B** should defect no matter what **A** does. Under this reasoning, each player will defect at each time step, giving them 1 point each per play. However, if they could somehow decide to cooperate, they would each get 3 points per play: the two players will do better in the long run by foregoing the action that maximizes their immediate payoff.

The question is, of course, can ordinary Darwinian mechanisms, which assume that individuals selfishly want to maximize their immediate payoff, lead to cooperation? Surprisingly, as demonstrated by Axelrod and Hamilton, the answer is yes.

strategy	name
[0 0]	All Defect
[0 1]	TIT-for-TAT (TFT)
[1 0]	TAT-for-TIT (anti-TFT)
[1 1]	All Cooperate

FIGURE 14 Four possible memory 1 strategies.

In Lindgren's version of this game, strategies can evolve in an open-ended fashion by learning to base their decisions on whether to cooperate or defect upon longer and longer histories of previous interactions.

The scheme used by Lindgren to represent strategies to play the Iterated Prisoner's Dilemma game is as follows. In the simplest version of the game, players make their choice of whether to cooperate or defect based solely on what their opponent did to them in the last time step. This is called the *memory 1* game. Since the opponent could have done only one of two things, cooperate or defect, a strategy needs to specify what it would do in either of those two cases. As it has two moves it can make in either of those two cases, cooperate or defect, there are four possible memory 1 strategies. These can be encoded in bit strings of length 2, as illustrated in Figure 14.

If the players should base their decisions by looking another move into the past, to see what they did to their opponent before their opponent made his move, then we would have the *memory 2* game. In this case, there are two moves with two possible outcomes each, meaning that a memory 2 strategy must specify whether to cooperate or defect for each of four possible cases. Such a strategy can be encoded using four bits, twice the length of the memory 1 strategies, so there will be 16 possible memory 2 strategies. Memory 3 strategies require another doubling of the encoding bit string, i.e., 8 bits, yielding 256 possible strategies. In general, memory n strategies require 2^n bits for their encoding, and there will be $2^{(2^n)}$ such strategies.

In order to allow for the evolution of higher memory strategies, Lindgren introduces a new genetic operator: gene duplication. As a memory n strategy is just twice as long as a memory $n - 1$ strategy, a memory n strategy can be produced from a memory $n - 1$ strategy by simply duplicating the memory n strategy and concatenating the duplicate to itself. In Lindgren's encoding strategy, gene duplication has the interesting property that it is a *neutral mutation*. Simple duplication alone does not change the PTYPE, even though it has doubled the length of the

GTYPE. However, once doubled, mutations in the longer GTYPE will alter the behavior of the PTYPE.

Once again, evolution proceeds by allowing populations of different organisms to bootstrap each other up coupled fitness landscapes, dynamically deforming each other's landscapes by turning local maxima into local minima. Again, the fitness of strategies is not an absolute fixed number that is independently computable. Rather, the fitness of each strategy depends on what other strategies exist in the "natural" population.

Many complicated and interesting strategies evolve during the evolutionary development of this system. More important, however, are the various phenomenological features exhibited by the dynamics of the evolutionary process. First of all, as we might expect, the system exhibits a behavior that is remarkably suggestive of Punctuated Equilibria. After an initial irregular transient, the system settles down to relatively long periods of stasis "punctuated" irregularly by periods of rapid evolutionary change (Figure 15).

Second, the diversity of strategies builds up during the long periods of stasis, but often collapses drastically during the short, chaotic episodes of rapid evolutionary succession (Figure 16). These "crashes" in the diversity of species constitute "extinction events." In this model, these extinction events are observed to be a natural consequence of the dynamics of the evolutionary process alone, without invoking any catastrophic, external perturbations (there are no comet impacts or "nemesis" stars in this model!). Furthermore, these extinction events happen on multiple scales: there are lots of little ones and fewer large ones.

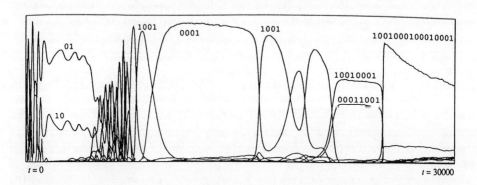

FIGURE 15 During evolutionary development, the system settles down to relatively long periods of stasis "punctuated" irregularly by periods of rapid evolutionary change. Original figure appeared in "Evolutionary Phenomena in Simple Dynamics " by K. Lindgren, in *Artificial Life II*, edited by C. G. Langton, C. Taylor, J. D. Farmer, and S. Rasmussen, (Redwood City, CA: Addison-Wesley, 1991)[18]

Artificial Life

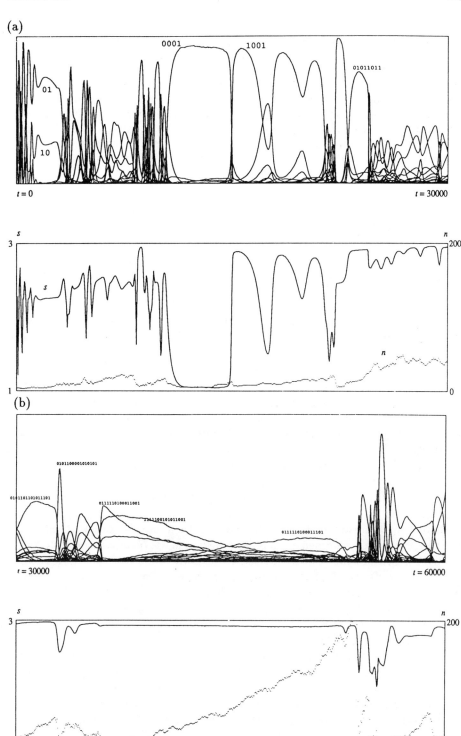

FIGURE 16 See caption on next page.

FIGURE 16 (cont'd.) The evolutionary dynamics of strategies in the iterated prisoner's dilemma system of Lindgren. In both cases, the top trace plots the changing concentration of strategies in the population while the bottom trace shows two things: the solid line plots the average fitness of the population, while the dotted line plots the diversity of species (the number of different strategies in the population at any time.) In all cases, time is traced on the horizontal axis. The top traces illustrate the interplay between metastable and chaotic episodes, while the bottom traces illustrate the "extinction events" that are often associated with the end of metastable periods. These extinction events can be quite large, as is seen in the bottom trace.

This is important because in order to understand the dynamics of a system that is subjected to constant perturbations, one needs to understand the dynamics of the *un*perturbed system first. We do not have access to an unperturbed version of the evolution of life on Earth; consequently, we could not have said definitively that extinction events on many size scales would be a natural consequence of the process of evolution itself. By comparing the perturbed and unperturbed versions of model systems like Lindgren's, we may very well be able to derive a universal scaling relationship for "natural" extinction events, and therefore be able to explain deviations from this relationship in the fossil record as due to external perturbations such as the impact of large asteroids.

Third, the emergence of ecologies is nicely demonstrated by Lindgren's model. It is usually the case that a mix of several different strategies dominates the system during the long periods of stasis. In order for a strategy to do well, it must do well by cooperating with other strategies. These mixes may involve three or more strategies whose collective activity produces a stable interaction pattern that benefits all of the strategies in the mix. Together, they constitute a more complex, "higher order" strategy, which can behave as a group in ways that are impossible for any individual strategy.

It is important to note that, in many cases, the "environment" that acts on an organism, and in the context of which an organism acts, is primarily constituted of the other organisms in the population and their interactions with each other and the physical environment. There is tremendous opportunity here for evolution to discover that certain sets of individuals exhibit emergent, collective behaviors that reap benefits to all of the individuals in the set. Thus, evolution can produce major leaps in biological complexity, without having to produce more complex individuals by simply discovering, perhaps even "tripping over," the many ways in which collections of individuals at one level can work together to form aggregate individuals at the next higher level of organization.[4]

This is thought to be the case for the origin of eukaryotic cells, which are viewed as descended from early cooperative collections of simpler, prokaryotic cells.[19] It is also the process involved in the origin of multicellular organisms, which lead to the Cambrian explosion of diversity some 700 million years ago. It was probably a significant factor in the origin of the prokaryotes themselves, and it has been discovered independently at least seven times by the various social insects (including species of wasps, bees, ants, and termites).

The final step in eliminating our hand from the selection/breeding process and setting the stage for true "natural" selection within a computer is taken in a model due to Tom Ray.[21] This step involves eliminating our algorithmic breeding agent completely.

In his "Tierra" simulation system, computer programs compete for CPU time and memory space. The "task" that these programs must perform in order to be reproduced is simply the act of self-reproduction itself! Thus, there is no need for an externally defined fitness function that determines which GTYPES get copied by an external copying procedure. The programs reproduce themselves, and the ones that are better at this task take over the population. The whole external task of evaluation of fitness has been internalized in the function of the organisms themselves. Thus, there is no longer a place for the human breeder or his computational agent. This results in genuine natural selection within a computer.

In Tierra, programs replicate themselves "noisily," so that some of their offspring behave differently. Variant programs that reproduce themselves more efficiently, which trick other programs into reproducing them, or which capture the execution pointers of other programs, etc., will leave more offspring than others. Similarly, programs that learn to defend themselves against such tricks will leave more offspring than those that do not.

We will discuss a few of the "digital organisms" that have emerged within the Tierra system (it is not necessary to understand the code in the illustrated programs in order to follow the explanation in the text.)[9]

Figure 17(a) shows the self-replicating "ancestor" program that is the only program Tom Ray has ever written in the Tierra system. All the other programs evolved under the action of natural selection.

The ancestor program works as follows. In the top block of code, the program locates its "head" and its "tail," templates marking the upper and lower boundaries of the program in memory. It saves these locations in special registers and, after subtracting the location of the head from the location of the tail, it stores its length in another register.

In the second block of code, the program enters an endless loop in which it will repeatedly produce copies of itself. It allocates memory space of the appropriate size and then invokes the final block of code, which is the actual reproduction loop. After it returns from the reproduction loop, it creates a new execution pointer to its newly produced offspring, and cycles back to create another offspring.

In the third and final block of code, the reproduction loop, the program copies itself, instruction by instruction, into the newly allocated memory space, making use of the addresses and length stored away by the first block of code. When it has copied itself completely, it returns to the block of code that called it, in this case, the second block.

[9] The details are to be found in Ray.[21]

It should be noted that "function calls" in Tierra are accomplished by seeking for a specific bit pattern in memory rather than by branching to a specific address. Thus, when the second block of code "calls" the third block of code, the reproduction loop, it does so by initiating a seek forward in memory for a specific "template." When this template is found, execution begins at the instruction following the template. Returns from function calls are handled in the normal manner, by simply returning to the instruction following the initial function call. This template addressing scheme is used in other reference contexts as well, and helps make Tierra language programs robust to mutations, as well as easily relocatable in memory.

Figure 17(b) shows a "parasite" program that has evolved to exploit the ancestor program. The parasite is very much like the ancestor program, except that it is missing the third block of code, the reproduction loop. How then does it copy itself?

The answer is that it makes use of a nearby ancestor program's reproduction loop! Recall that a function call in Tierra initiates a seek forward in memory for a particular template of bits. If this pattern is not found within the initiating program's own code, the search may proceed forward in memory into the code of other organisms, where the template may be found and where execution then begins. When the invoked function in another organism's code executes the "return" statement, execution reverts to the program that initiated the function call. Thus, organisms can execute each other's code, and this is exactly what the parasite program does: it makes use of the reproductive machinery of the ancestor host.

This means that the parasite does not have to take the time to copy the code constituting the reproductive loop, and hence can reproduce more rapidly, as it has fewer instructions to copy. The parasites thus proliferate in the population. However, they cannot proliferate to the point of driving out the ancestor hosts altogether, for they depend on them for their reproductive machinery. Thus, a balance is eventually struck optimizing the joint system.

Eventually, however, another mutant form of the ancestor emerges which has developed an immunity to the parasites. This program is illustrated in Figure 17(c). Two key differences from the ancestor program confer the immunity to the parasite programs. First, instead of executing a "return" instruction, the reproduction loop instead initiates a jump back in memory to the template found in the instruction that calls the reproduction loop. This has the same effect as a return statement when executed by the immune program, but has a very different effect on the parasite. The second important difference is that following the cell division in the second block (which allocates a new execution pointer to the offspring just created), the program jumps back to the beginning of the first block of code, rather than to the beginning of the second block. Thus, the immune program constantly resets its head, tail, and size registers. This seems useless when considering only the immune organism's own reproduction, but let's see what happens when a parasite tries to execute the reproduction loop in an immune organism.

When a parasite attempts to use the immune program's reproduction code, the new jump transfers the parasite's execution pointer to the second block of the

Artificial Life

immune program's code, rather than returning it to the second block of the parasite code, as the parasite expects. Then, this execution pointer is further re-directed to the first block of the immune program, where the registers originally containing the head, tail, and length of the parasite are reset to contain the head, tail, and length of the immune organism. The immune program has thus completely captured the execution pointer of the parasite. Having lost its execution pointer, the parasite simply becomes dormant data occupying memory, while the immune program now has two execution pointers running through it: its own original pointer, plus the pointer it captured from the parasite. Thus, the immune program now reproduces twice as rapidly as before. Once they emerge, such immune programs rapidly drive the parasites to extinction.

Complex interactions between variant programs like those described above continue to develop within evolutionary runs in Tierra. From a uniform population of self-reproducing ancestor programs, Ray, a tropical biologist by training, notes the emergence of whole "ecologies" of interacting species of computer programs. Furthermore, he is able to identify many phenomena familiar to him from his studies of real ecological communities, such as competitive exclusion, the emergence of parasites, key-stone predators and parasites, hyper-parasites, symbiotic relationships, sociality, "cheaters," and so forth.

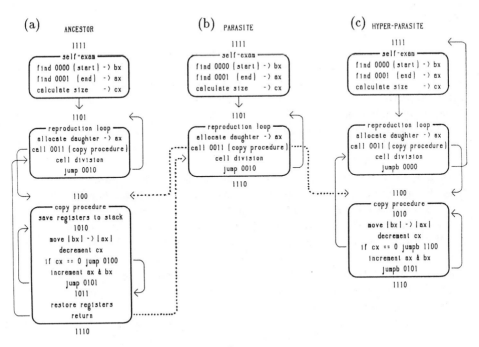

FIGURE 17 Digital organisms from Ray's[21] tierra simulation system. (a) Self-reproducing ancestor. (b) An early parasite of the ancestor. (c) A decendant of the ancestor that is immune to the parasite.

Again, the actual "fitness" of an organism is a complex function of its interactions with other organisms in the "soup." Collections of programs can cooperate to enhance each other's reproductive success, or they can drive each other's reproductive success down, thus lowering fitness and kicking the population off of local fitness peaks.

Not surprisingly, Ray, too, has noted periods of relative stasis punctuated by periods of rapid evolutionary change, as complex ecological webs collapse and new ones stabilize in their place. Systems like Ray's Tierra capture the proper context for evolutionary dynamics, and natural selection is truly at play here.

8. CONCLUSION

This article is intended to provide a broad overview of the field of Artificial Life, its motivations, history, theory, and practice. In such a short space, it cannot hope to go into depth in any one of these areas. Rather, it attempts to convey the "spirit" of the Artificial Life enterprise via several illustrative examples coupled with a good deal of motivating explanation and discussion.

The field of Artificial Life is in its infancy, and is currently engaged in a period of extremely rapid growth, which is producing many new converts to the principles detailed here. However, it is also raising a significant amount of controversy, and is not without its critics. The notion of studying biology via the study of patently non-biological things is an idea that is hard for the traditional biological copmmunity to accept. The acceptance of Artificial Life techniques within the biological community will be directly proportional to the contributions it makes to our understanding of biological phenomena.

That these contributions are forthcoming, I have no doubt. However, high-quality research in Artificial Life is difficult, because it requires that its practitioners be experts in both the computational sciences and the biological sciences. Either of those alone is a full time career, and so the danger lurks of doing either masterful biology but trivial computing, or doing masterful computing but trivial biology.

Therefore, I strongly suggest incorporating a trick from nature: cooperate! As is amply illustrated in many of the examples discussed in this article, nature often discovers that collections of individuals easily solve problems that would be extremely difficult or even impossible for individuals to solve on their own. Collaborations between biologists and computer scientists are quite likely to be the most appropriate vehicles for making significant contributions to our understanding of biology via the pursuit of Artificial Life.

So, if you are a computer expert dying to hack together an evolution program, go find yourself a top-notch evolutionary biologist to collaborate with, one who will bring to the enterprise an in-depth understanding of the subtleties of the evolutionary process plus a proper set of open questions about evolution towards which your evolution program might be addressed.

On the other hand, if you are a field biologist interested in doing some numerical simulations in order to understand the ecological dynamics you are observing in the field, hook up with a top-notch parallel-computing expert, who will bring to the enterprise a thorough knowledge of the subtleties involved in multi-agent interactions, and will be in possession of an equally open set of questions, which you very well might find to be strikingly related to your own.

Above all, when in doubt, turn to Mother Nature. After all, she is smarter than you!

ACKNOWLEDGMENTS

A large number of people have assisted in writing this paper, which is based on my lecture notes for the Complex Systems Summer School and on the overview of Artificial Life that served as an introduction to the proceedings of the first Artificial Life workshop. Besides the people credited in the latter paper, I would like to thank the following people for their help with this version: Tom Ray, Kristian Lindgren, Danny Hillis, and John Koza. I would also like to thank Ronda Butler-Villa and Della Ulibarri for their patience with me and for their skill in preparing the figures and text for publication.

REFERENCES

1. Axelrod, R., and W. D. Hamilton. "The Evolution of Cooperation." *Science* **211** (1981): 1390–1396.
2. Axelrod, R. *The Evolution of Cooperation.* New York, Basic Books, 1984.
3. Burks, A. W., ed. *Essays on Cellular Automata.* Urbana, IL: University of Illinois Press, 1970.
4. Buss, L. "The Evolution of Individuality." Princeton University Press, 1987.
5. Chapuis, A., and E. Droz. *Automata: A Historical and Technological Study.* Translation by A. Reid. London: B. T. Batsford Ltd, 1958.
6. Dawkins, R. "The Evolution of Evolvability." In *Artificial Life*, edited by C. G. Langton, 201–220. Santa Fe Institute Studies in the Sciences of Complexity, Proc. Vol. VI. Redwood City, CA: Addison-Wesley, 1988.
7. Frisch, U., B. Hasslacher, and Y. Pomeau. "Lattice Gas Automata for the Navier-Stokes Equation." *Phys. Rev. Lett.* **56** (1986): 1505–1508.
8. Goldberg, D. E. *Genetic Algorithms in Search, Optimization, and Machine Learning.* Reading, MA: Addison-Wesley, 1989.
9. Hamilton, W. D. "Sex Versus Non-Sex Versus Parasite." *OIKOS* **35** (1980): 282–290.

10. Hamilton, W. D. "Pathogens as Causes of Genetic Diversity in their Host Populations." In *Population Biology of Infectious Diseases*, edited by R. M. Anderson and R. M. May, 269–296. Berlin: Springer-Verlag, 1982.
11. Hillis, W. D. "Co-Evolving Parasites Improve Simulated Evolution as an Optimization Procedure." In *Artificial Life II*, edited by C. G. Langton, C. Taylor, J. D. Farmer, and S. Rasmussen, 313–324. Santa Fe Institute Studies in the Sciences of Complexity, Proc. Vol. X. Redwood City, CA: Addison-Wesley, 1991.
12. Holland, J. H. *Adaptation in Natural and Artificial Systems*. Ann Arbor: University of Michigan Press, 1975.
13. Holland, J. H. "Escaping Brittleness: The Possibilities of General Purpose Learning Algorithms Applied to Parallel Rule-Based Systems." In *Machine Learning II*, edited by R. S. Mishalski, J. G. Carbonell, and T. M. Mitchell, 593–623. New York: Kaufman, 1986.
14. Hopcroft, J. E., and J. D. Ullman. *Introduction to Automata Theory, Languages, and Computation*. Menlo Park, CA: Addison-Wesley, 1979.
15. Koza, J. R. "Genetic Evolution and Co-Evolution of Computer Programs." In *Artificial Life II*, edited by C. G. Langton, C. Taylor, J. D. Farmer, and S. Rasmussen, 603–630. Santa Fe Institute Studies in the Sciences of Complexity, Proc. Vol. X. Redwood City, CA: Addison-Wesley, 1991.
16. Langton, C. G. "Self-Reproduction in Cellular Automata." *Physica D* **10(1-2)** (1984): 135–144.
17. Langton, C. G. "Studying Artificial Life with Cellular Automata." *Physica D* **22** (1986): 120–149.
18. Lindgren, K. "Evolutionary Phenomena in Simple Dynamics." In *Artificial Life II*, edited by C. G. Langton, C. Taylor, J. D. Farmer, and S. Rasmussen, 295–312. Santa Fe Institute Studies in the Sciences of Complexity, Proc. Vol. X. Redwood City, CA: Addison-Wesley, 1991.
19. Margolus, L. *Origin of Eucaryotic Cells*. New Haven, Yale University Press, 1970.
20. Prusinkiewicz, P. *The Algorithmic Beauty of Plants*. Berlin: Springer-Verlag, 1991.
21. Ray, T. S. "An Approach to the Synthesis of Life." In *Artificial Life II*, edited by C. G. Langton, C. Taylor, J. D. Farmer, and S. Rasmussen, 371–408. Santa Fe Institute Studies in the Sciences of Complexity, Proc. Vol. X. Redwood City, CA: Addison-Wesley, 1991.
22. Reynolds, C. W. "Flocks, Herds, and Schools: A Distributed Behavioral Model." Proceedings of SIGGRAPH '87. *Computer Graphics* **V 21(4)** (1987): 25–34.
23. Toffoli, T. "Cellular Automata as an Alternative to (Rather than an Approximation of) Differential Equations in Modeling Physics." In Cellular Automata: Proceedings of an Interdisciplinary Workshop (Los Alamos, New Mexico, March 7–11, 1983), edited by J. D. Farmer, T. Toffoli, and S. Wolfram. *Physica D* (special issue) **10(1-2)** (1984).

24. Toffoli, T., and N. Margolus. *Cellular Automata Machines.* Cambridge: MIT Press, 1987.
25. Ulam, S. "On Some Mathematical Problems Connected with Patterns of Growth of Figures." *Proceedings of Symposia in Applied Mathematics* **14**, 1962, 215–224. Reprinted in *Essays on Cellular Automata*, edited by A. W. Burks. Urbana, IL: University of Illinois Press, 1970.
26. Von Neumann, J. *Theory of Self-Reproducing Automata*, edited and completed by A. W. Burks. Urbana: University of Illinois Press, 1966.
27. Wilson, S. W. "The Genetic Algorithm and Simulated Evolution." In *Artificial Life*, edited by C. G. Langton, 157–165. Santa Fe Institute Studies in the Sciences of Complexity, Proc. Vol. VI. Redwood City, CA: Addison-Wesley, 1989.
28. Wolfram, S. "Cellular Automaton Fluids 1: Basic Theory." *J. Stat. Phys.* **45** (1986): 471–526.

James P. Sethna
Laboratory of Atomic and Solid State Physics, Cornell University, Ithaca, New York 14853-2501

Order Parameters, Broken Symmetry, and Topology

As a kid in elementary school, I was taught that there were three states of matter: solid, liquid, and gas. The ancients thought that there were four—earth, water, air, and fire—which is considered sheer superstition. In junior high, I remember reading a book called *The Seven States of Matter*. At least one was "plasma," which made up stars and thus most of the universe,[1] and which sounded rather like fire to me.

The original three, by now, have become multitudes. In important and precise ways, magnets are a distinct form of matter. Metals are different from insulators. Superconductors and superfluids are striking new states of matter. The liquid crystal in your wristwatch is one of a huge family of

different liquid crystalline states of matter[2] (nematic, cholesteric, blue phase I, II, and blue fog, smectic A, B, C, C*, D, I,...). There are over 200 qualitatively different types of crystals, not to mention the quasi-crystals (Figure 1). There are disordered states of matter, like spin glasses, and states like the fractional quantum-hall effect with excitations of charge $e/3$

---

[1] They hadn't heard of dark matter back then.

(like quarks). Particle physicists tell us that the vacuum we live within has, in the past, been in quite different states: in the last vacuum but one, there were four different kinds of light[1] (mediated by what is now the photon, the W^+, the W^-, and the Z particle). We'll discuss this more in the next paper.[4]

When there were only three states of matter, we could learn about each one and then turn back to learning long division. Now that there are multitudes, though, we've had to develop a system. Our system is constantly being extended and modified, because we keep finding new phases which don't fit into the old frameworks. It's amazing how the 500th new state of matter somehow screws up a system which worked fine for the first 499. Quasi-crystals, the fractional quantum-hall effect, and spin glasses all really stretched our minds until (1) we understood why they behaved the way they did, and (2) we understood how they fit into the general framework.

In this paper, I'm going to tell you the system. In the subsequent sections, I'll discuss some gaps in the system: materials and types of behavior which don't fit into the neat framework presented here. I'll try to maximize the number of pictures and minimize the number of formulas, but there are problems and ideas that I don't understand well enough to explain simply. Most of what I tell you in this paper is both true and important. Much of what is contained in the following sections represents my own pet ideas and theories, and you should be warned not to take my messages there as gospel.

The system consists of four basic steps.[3] First, you must identify the broken symmetry. Second, you must define an order parameter. Third, you are told to examine the elementary excitations. Fourth, you classify the topological defects. Most of what I say I take from Mermin,[3] Coleman,[1] and deGennes,[2] and I heartily recommend these excellent articles to my audience. We take each step in turn.

I. IDENTIFY THE BROKEN SYMMETRY

What is it which distinguishes the hundreds of different states of matter? Why do we say that water and olive oil are in the same state (the liquid phase), while we say aluminum and (magnetized) iron are in different states? Through long experience, we've discovered that most phases differ in their symmetry.[2]

---

[2] This is not to say that different phases always differ by symmetries! Liquids and gases have the same symmetry. In fact, one can go continuously from a liquid to a gas, by going first to high pressures and then heating. It is safe to say, though, that if the two materials have different symmetries, they are different phases.

James P. Sethna 245

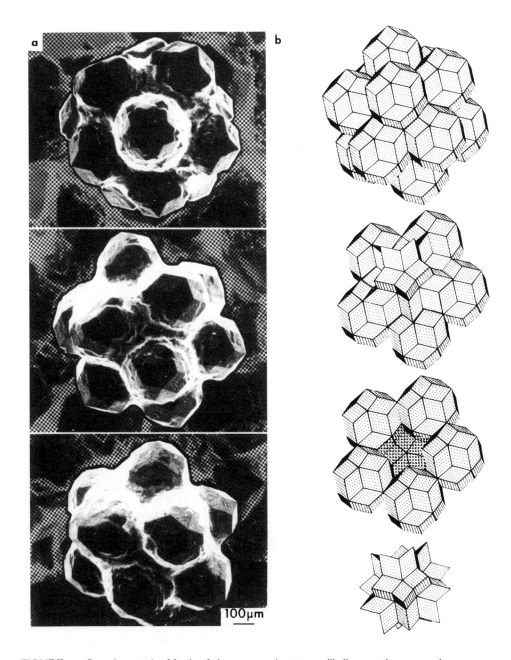

FIGURE 1 Quasi-crystals. Much of these two chapters will discuss the properties of crystals. Crystals are surely the oldest known of the broken-symmetry phases of matter, and remain the most beautiful illustrations. It's amazing that in the past few years, we've uncovered an entirely new class of crystals. Shown here is a photograph of a quasi-crystalline metallic alloy, with icosahedral symmetry. Notice the five-pointed stars: our old notions of crystals had to be completely revised to include this type of symmetry. Photograph courtesy of Marc Audier, Ecole Nationale Superieure d'Electrochimie et d'Electrametallargie de Grenoble.

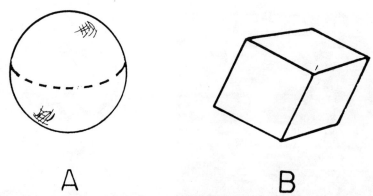

FIGURE 2 Which is more symmetric? The cube has many symmetries. It can be rotated by 90°, 180°, or 270° about any of the three axes passing through the faces. It can be rotated by 120° or 240° about the corners and by 180° about an axis passing from the center through any of the 12 edges. The sphere, though, can be rotated by *any* angle. The sphere respects rotational invariance: all directions are equal. The cube is an object which breaks rotational symmetry: once the cube is there, some directions are more equal than others.

Consider Figure 2, showing a cube and a sphere. Which is more symmetric? Clearly, the sphere has many more symmetries than the cube. One can rotate the cube by 90° in various directions and not change its appearance, but one can rotate the sphere by any angle and keep it unchanged.

In Figure 3, we see a two-dimensional schematic representation of ice and water. Which state is more symmetric here? Naively, the ice looks much more symmetric: regular arrangements of atoms forming a lattice structure. The water looks irregular and disorganized. On the other hand, if one rotated Figure 3(b) by an arbitrary angle, it would still look like water! Ice has broken rotational symmetry: one can rotate Figure 3(a) only by multiples of 60°. It also has a broken translational symmetry: it's easy to tell if the picture is shifted sideways, unless one shifts by a whole number of lattice units. While the snapshot of the water shown in the figure has no symmetries, water as a phase has complete rotational and translational symmetry.

One of the standard tricks to see if two materials differ by a symmetry is to try to change one into the other smoothly. Oil and water won't mix, but I think oil and alcohol do, and alcohol and water certainly do. By slowly adding more alcohol to oil, and then more water to the alcohol, one can smoothly interpolate between the two phases. If they had different symmetries, there must be a first point when mixing them when the symmetry changes, and it is usually easy to tell when that phase transition happens.

II. DEFINE THE ORDER PARAMETER

Particle physics and condensed-matter physics have quite different philosophies. Particle physicists are constantly looking for the building blocks. Once pions and protons were discovered to be made of quarks, they became demoted into engineering problems. Now that quarks and electrons and photons are made of strings, and strings are hard to study (at least experimentally), there is great anguish in the high-energy community. Condensed-matter physicists, on the other hand, try to understand why messy combinations of zillions of electrons and nuclei do such interesting simple things. To them, the fundamental question is not discovering the underlying quantum mechanical laws, but in understanding and explaining the new laws that emerge when many particles interact.

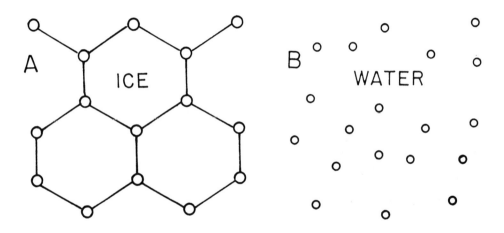

FIGURE 3 Which is more symmetric? At first glance, water seems to have much less symmetry than ice. The picture of "two-dimensional" ice clearly breaks the rotational invariance: it can be rotated only by $120°$ or $240°$. It also breaks the translational invariance: the crystal can only be shifted by certain special distances (whole number of lattice units). The picture of water has no symmetry at all: the atoms are jumbled together with no long-range pattern at all. Water, though, isn't a snapshot: it would be better to think of it as a combination of all possible snapshots! Water has a complete rotational and translational symmetry: the pictures will look the same if the container is tipped or shoved.

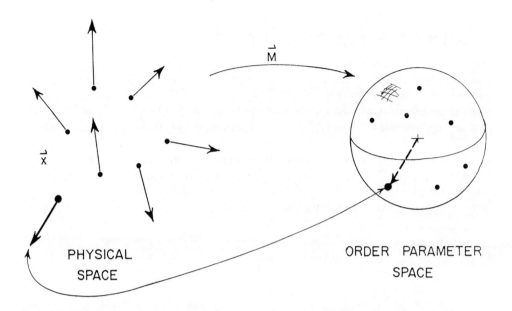

FIGURE 4 Magnet. We take the magnetization $\vec{M}$ as the order parameter for a magnet. For a given material at a given temperature, the amount of magnetization $|\vec{M}| = M_0$ will be pretty well fixed, but the energy is often pretty much independent of the direction $\hat{M} = \vec{M}/M_0$ of the magnetization. (You can think of this as a arrow pointing to the north end of each atomic magnet.) Often, the magnetization changes directions smoothly in different parts of the material. (That's why not all pieces of iron are magnetic!) We describe the current state of the material by an order parameter field $\vec{M}(\mathbf{x})$. The order parameter field is usually thought of as an arrow at each point in space. It can also be thought of as a function taking points in space $\mathbf{x}$ into points on the sphere $|\vec{M}| = M_0$. This sphere $\mathcal{S}^2$ is the order parameter space for the magnet.

As one might guess, we don't keep track of all the electrons and protons.[3] We're always looking for the important variables, the important degrees of freedom. In a crystal, the important variables are the motions of the atoms away from their lattice positions. In a magnet, the important variable is the local direction of the magnetization (an arrow pointing to the "north" end of the local magnet). The local magnetization comes from complicated interactions between the electrons, and is partly due to the little magnets attached to each electron and partly due to the way the electrons dance around in the material: these details are for many purposes unimportant.

[3] The particle physicists use order parameter fields, too. Their order parameter fields also hide lots of details about what their quarks and gluons are composed of. The main difference is that they don't know of what their fields are composed. It ought to be reassuring to them that we don't always find our greater knowledge very helpful.

The important variables are combined into an "order parameter field."[4] In Figure 4, we see the order parameter field for a magnet.[5] At each position $\mathbf{x} = (x,y,z)$, we have a direction for the local magnetization $\vec{M}(\mathbf{x})$. The length of $\vec{M}$ is pretty much fixed by the material, but the direction of the magnetization is undetermined. By becoming a magnet, this material has broken the rotational symmetry. The order parameter $\vec{M}$ labels which of the various broken symmetry directions the material has chosen.

The order parameter is a field: at each point in our magnet, $\vec{M}(\mathbf{x})$ tells the local direction of the field near $\mathbf{x}$. Why do we do this? Why would the magnetization point in different directions in different parts of the magnet? Usually, the material has lowest energy when the order parameter field is uniform, when the symmetry is broken in the same way throughout space. In practice, though, the material often doesn't break symmetry uniformly. Most pieces of iron don't appear magnetic, simply because the local magnetization points in different directions at different places. The magnetization is already there at the atomic level: to make a magnet, you pound the different domains until they line up. We'll see in this section that most of the interesting behavior we can study involves the way the order parameter varies in space.

The order parameter field $\vec{M}(\mathbf{x})$ can be usefully visualized in two different ways. On the one hand, one can think of a little vector attached to each point in space. On the other hand, we can think of it as a mapping from real space into order parameter space. That is, $\vec{M}$ is a function which takes different points in the magnet onto the surface of a sphere (Figure 4). Mathematicians call the sphere $\mathcal{S}^2$, because it locally has two dimensions. (They don't care what dimension the sphere is embedded in.)

Before varying our order parameter in space, let's develop a few more examples. The liquid crystal in LCD displays (like those in digital watches) are nematics. Nematics are made of long, thin molecules which tend to line up so that their

---

[4] Choosing an order parameter is an art. Usually it's a new phase which we don't understand yet, and guessing the order parameter is a piece of figuring out what's going on. Also, there is often more than one sensible choice. In magnets, for example, one can treat $\vec{M}$ as a fixed-length vector in $\mathcal{S}^2$, labelling the different broken symmetry states. This is the best choice at low temperatures, where we study the elementary excitations and topological defects. For studying the transition from low to high temperatures, when the magnetization goes to zero, it is better to consider $\vec{M}$ as a vector of varying length (a vector in $\mathcal{R}^3$). Finding the simplest description for your needs is often the key to the problem.

[5] Most magnets are crystals, which already have broken the rotational symmetry. For some "Heisenberg" magnets, the effects of the crystal on the magnetism is small. Magnets are really distinguished by the fact that they break time-reversal symmetry: if you reverse the arrow of time, the magnetization would change direction!

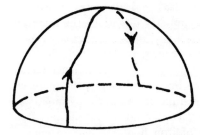

FIGURE 5 Nematic liquid crystal. Nematic liquid crystals are made up of long, thin molecules that prefer to align with one another. (Liquid crystal watches are made of nematics.) Since they don't care much which end is up, their order parameter isn't precisely the vector $\hat{n}$ along the axis of the molecules. Rather, it is a unit vector up to the equivalence $\hat{n} \equiv -\hat{n}$. The order parameter space is a half-sphere, with antipodal points on the equator identified. Thus, for example, the path shown over the top of the hemisphere is a closed loop: the two intersections with the equator correspond to the same orientations of the nematic molecules in space.

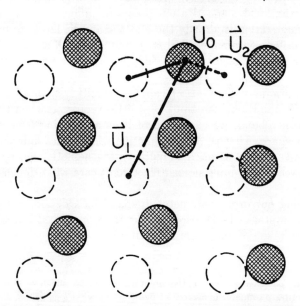

FIGURE 6 Two-dimensional crystal. A crystal consists of atoms arranged in regular, repeating rows and columns. At high temperatures, or when the crystal is deformed or defective, the atoms will be displaced from their lattice positions. The displacements $\vec{u}$ are shown. Even better, one can think of $u(\mathbf{x})$ as the local translation needed to bring the ideal lattice into registry with atoms in the local neighborhood of $\mathbf{x}$. Also shown is the ambiguity in the definition of u. Which "ideal" atom should we identify with a given "real" one? This ambiguity makes the order parameter u equivalent to $u + ma\hat{x} + na\hat{y}$. Instead of a vector in two-dimensional space, the order parameter space is a square with periodic boundary conditions.

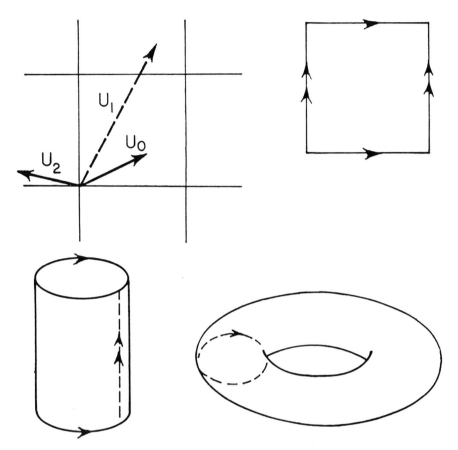

FIGURE 7 Order parameter space for a two-dimensional crystal. Here we see that a square with periodic boundary conditions is a torus. (A torus is a surface of a doughnut, inner tube, or bagel, depending on your background.)

long axes are parallel. Nematic liquid crystals, like magnets, break the rotational symmetry. Unlike magnets, though, the main interaction isn't to line up the north poles, but to line up the axes. (Think of the molecules as American footballs: the same up and down.) Thus the order parameter isn't a vector $\vec{M}$ but a headless vector $\vec{n} \equiv -\vec{n}$. The order parameter space is a hemisphere, with opposing points along the equator identified (Figure 5). This space is called $\mathcal{RP}^2$ by the mathematicians (the projective plane), for obscure reasons.

For a crystal, the important degrees of freedom are associated with the broken translational order. Consider a two-dimensional crystal which has lowest energy when in a square lattice, but which is deformed away from that configuration (Figure 6). This deformation is described by an arrow connecting the undeformed ideal lattice points with the actual positions of the atoms. If we are a bit more careful, we say that $\vec{u}(\mathbf{x})$ is that displacement needed to align the ideal lattice in the local region onto the real one. By saying it this way, $\vec{u}$ is also defined between the lattice positions: there still is a best displacement which locally lines up the two lattices.

The order parameter $\vec{u}$ isn't really a vector: there is a subtlety. In general, which ideal atom you associate with a given real one is ambiguous. As shown in Figure 6, the displacement vector $\vec{u}$ changes by a multiple of the lattice constant a when we choose a different reference atom:

$$\vec{u} \equiv \vec{u} + a\hat{x} = \vec{u} + ma\hat{x} + na\hat{y}. \tag{1}$$

The set of distinct order parameters forms a square with periodic boundary conditions. As Figure 7 shows, a square with periodic boundary conditions has the same topology as a torus, T^2. (The torus is the surface of a doughnut, bagel, or inner tube.)

Finally, let's mention that guessing the order parameter (or the broken symmetry) isn't always so straightforward. For example, it took many years before anyone figured out that the order parameter for superconductors and superfluid Helium 4 is a complex number ψ. The order parameter field $\psi(\mathbf{x})$ represents the "condensate wave function," which (extremely loosely) is a single quantum state occupied by a large fraction of the Cooper pairs or helium atoms in the material. The corresponding broken symmetry is closely related to the number of particles. In "symmetric," normal liquid helium, the local number of atoms is conserved; in superfluid helium, the local number of atoms becomes indeterminate! (This is because many of the atoms are condensed into that delocalized wave function.) Anyhow, the magnitude of the complex number ψ is a fixed function of temperature, so the order parameter space is the set of complex numbers of magnitude $|\psi|$. Thus the order parameter space for superconductors and superfluids is a circle S^1.

Now we examine small deformations away from a uniform order parameter field.

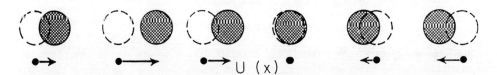

FIGURE 8 One-dimensional crystal: phonons. The order parameter field for a one-dimensional crystal is the local displacement $u(x)$. Long wavelength waves in $u(x)$ have low frequencies, and cause sound. Crystals are rigid because of the broken translational symmetry. Because they are rigid, they fight displacements. Because there is an underlying translational symmetry, a uniform displacement costs no energy. A nearly uniform displacement, thus, will cost little energy and, thus, will have a low frequency. These low-frequency elementary excitations are the sound waves in crystals.

III EXAMINE THE ELEMENTARY EXCITATIONS

Its amazing how slow human beings are. The atoms inside your eyelash collide with one another billions of times during each time you blink your eye. It's not surprising, then, that we spend most of our time in condensed-matter physics studying those things in materials that happen slowly. Typically only vast conspiracies of immense numbers of atoms can produce the slow behavior that humans can perceive.

A good example is given by sound waves. We won't talk about sound waves in air: air doesn't have any broken symmetries, so it doesn't belong in this paper.[6] Consider, instead, sound in the one-dimensional crystal shown in Figure 8. We describe the material with an order parameter field $u(x)$, where here x is the position within the material and $x - u(x)$ is the position of the reference atom within the ideal crystal.

Now, there must be an energy cost for deforming the ideal crystal. There won't be any cost, though, for a uniform translation: $u(x) \equiv u_0$ has the same energy as the ideal crystal. (Shoving all the atoms to the right doesn't cost any energy.) So, the energy will depend only on derivatives of the function $u(x)$. The simplest energy that one can write looks like

$$\mathcal{E} = \int dx \left(\frac{\kappa}{2}\right)\left(\frac{du}{dx}\right)^2. \qquad (2)$$

(Higher derivatives won't be important for the low frequencies that humans can hear.) Now, you may remember Newton's law $F = ma$. The force here is given by the derivative of the energy $F = -(d\mathcal{E}/du)$. The mass is represented by the density of the material ρ. Working out the math (a variational derivative and an integration by parts, for those who are interested) gives us the equation

$$\rho \ddot{u} = \kappa \left(\frac{d^2 u}{dx^2}\right). \qquad (3)$$

The solutions to this equation

$$u(x,t) = u_0 \cos\left(2\pi\left(\frac{x}{\lambda} - \nu_\lambda t\right)\right) \qquad (4)$$

represent phonons or sound waves. The wavelength of the sound waves is λ, and the frequency is ν_λ. Plugging (4) into (3) gives us the relation

$$\nu_\lambda = \frac{\sqrt{\kappa/\rho}}{\lambda}. \qquad (5)$$

---

[6] We argue here that low frequency excitations come from spontaneously broken symmetries. They can also come from conserved quantities: since air cannot be created or destroyed, a long-wavelength density wave cannot relax quickly.

254   Order Parameters, Broken Symmetry, and Topology

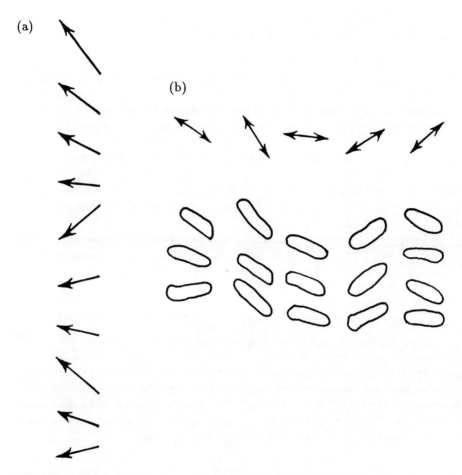

FIGURE 9 (a) Magnets: spin waves. Magnets break the rotational invariance of space. Because they resist twisting the magnetization locally but don't resist a uniform twist, they have low-energy spin wave excitations. (b) Nematic liquid crystals: rotational waves. Nematic liquid crystals also have low-frequency rotational waves.

The frequency gets small only when the wavelength gets large. This is the vast conspiracy: only huge sloshings of many atoms can happen slowly. *Why does the frequency get small?* Well, there is no cost to a uniform translation, which is what (4) looks like for infinite wavelength. *Why is there no energy cost for a uniform displacement?* Well, there is a translational symmetry: moving all the atoms the same amount doesn't change their interactions. *But haven't we broken that symmetry?* That is precisely the point.

Long after phonons were understood, Jeremy Goldstone started to think about broken symmetries and order parameters in the abstract. He found a rather general argument that, whenever a continuous symmetry (rotations, translations, $SU(3)$,

...) is broken, long wavelength modulations in the symmetry direction should have low frequencies. The fact that the lowest energy state has a broken symmetry means that the system is stiff: modulating the order parameter will cost an energy rather like that in Eq. (2). In crystals, the broken translational order introduces a rigidity to shear deformations, and low frequency phonons (Figure 8). In magnets, the broken rotational symmetry leads to a magnetic stiffness and spin waves (Figure 9(a)). In nematic liquid crystals, the broken rotational symmetry introduces an orientational elastic stiffness (it pours, but resists bending!) and rotational waves (Figure 9(b)).

In superfluids, the broken gauge symmetry leads to a stiffness which results in the superfluidity. Superfluidity and superconductivity really aren't any more amazing than the rigidity of solids. Isn't it amazing that chairs are rigid? Push on a few they on one side, and 10^9 atoms away they will move in lock-step. In the same way, decreasing the flow in a superfluid must involve a cooperative change in a macroscopic number of atoms, and thus never happens spontaneously any more than two parts of the chair ever drift apart.

The low-frequency Goldstone modes in superfluids are heat waves! (Don't be jealous: liquid helium has rather cold heat waves.) This is often called second sound, but it is really a periodic modulation of the temperature which passes through the material like sound does through a metal.

O.K., now we're getting the idea. Just to round things out, what about superconductors? They've got a broken gauge symmetry and a stiffness to decays in the superconducting current. What is the low-energy excitation? It doesn't have one. But what about Goldstone's theorem? Well, you know about physicists and theorems....

That's actually quite unfair: Goldstone surely had conditions on his theorem which excluded superconductors. Actually, I believe Goldstone was studying superconductors when he came up with his theorem. It's just that everybody forgot the extra conditions, and just remembered that you always got a low frequency mode when you broke a continuous symmetry. We, of course, understood all along why there isn't a Goldstone mode for superconductors: it's related to the Meissner effect. The high-energy physicists forgot, though, and had to rediscover it for themselves. Now we all call the loophole in Goldstone's theorem the Higgs mechanism, because (to be truthful) Higgs and his high-energy friends found a much simpler and more elegant explanation than we had. We'll discuss Meissner effects and the Higgs mechanism in another chapter.[4]

I'd like to end this section, though, by bringing up another exception to Goldstone's theorem: one we've known about even longer, but which we don't have a nice explanation for. What about the orientational order in crystals? Crystals break both the continuous translational order and the continuous orientational order. The

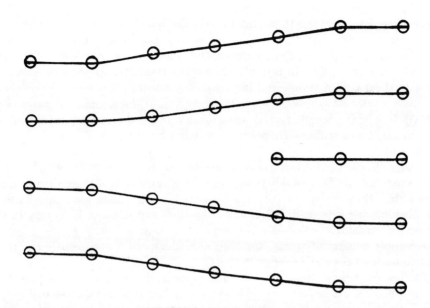

FIGURE 10 Dislocation in a crystal. Here is a topological defect in a crystal. We can see that one of the rows of atoms on the right disappears halfway through our sample. The place where it disappears is a defect, because it doesn't locally look like a piece of the perfect crystal. It is a topological defect, because it can't be fixed by any local rearrangement. No reshuffling of atoms in the middle of the sample can change the fact that five rows enter from the right, and only four leave from the left! The Burger's vector of a dislocation is the net number of extra rows and columns, combined into a vector (columns, rows).

phonons are the Goldstone modes for the translations, but *there are no orientational Goldstone modes*.[7] We'll discuss this further in another chapter,[4] but I think this is one of the most interesting unsolved basic questions in the subject.

[7] In two dimensions, crystals provide another loophole in a well-known theorem. Mermin and Wagner proved many years ago that two-dimensional systems with a continuous symmetry cannot have a broken symmetry at finite temperature. At least, that's the English phrase everyone quotes when they discuss the theorem: the theorem is stated in a much more technical way. Now, crystals in two dimensions actually don't break the translational symmetry: at finite temperatures, the atoms wiggle enough so that the atoms don't sit in lock-step over infinite distances; this is correctly stated as an important application of the theorem. But the crystals do have a broken orientational symmetry: the crystal axes point in exactly the same directions throughout space. Again, the theorem has technical conditions which exclude crystalline orientational order in the presence of translational order.

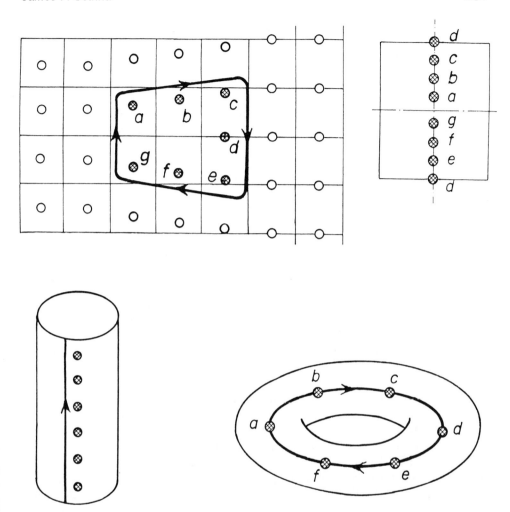

FIGURE 11 Loop around the dislocation mapped onto order parameter space. How do we think about our defect in terms of order parameters and order parameter spaces? Consider a closed loop around the defect. The order parameter field u changes as we move around the loop. The positions of the atoms around the loop with respect to their local "ideal" lattice drifts upward continuously as we traverse the loop. This precisely corresponds to a loop around the order parameter space: the loop passes once through the hole in the torus. A loop *around* the hole corresponds to an extra column of atoms. Moving the atoms slightly will deform the loop, but won't change the number of times the loop winds through or around the hole. Two loops which traverse the torus the same number of times through and around are equivalent. The equivalence classes are labelled precisely by pairs of integers (just like the Burger's vectors), and the first homotopy group of the torus is $\mathcal{Z} \times \mathcal{Z}$.

IV. CLASSIFY THE TOPOLOGICAL DEFECTS

When I was in graduate school, the big fashion was topological defects. Everybody was studying homotopy groups and finding exotic systems to write papers about. It was, in the end, a reasonable thing to do.[8] It is true that in a typical application you'll be able to figure out what the defects are without homotopy theory. You'll spend forever drawing pictures to convince anyone else, though. Most important, homotopy theory helps you to think about defects.

A defect is a tear in the order parameter field. A topological defect is a tear that can't be patched. Consider the piece of two-dimensional crystal shown in Figure 10. Starting in the middle of the region shown, there is an extra row of atoms. (This is called a dislocation.) Away from the middle, the crystal locally looks fine: it's a little distorted, but there is no problem seeing the square grid and defining an order parameter. Can we rearrange the atoms in a small region around the start of the extra row, and patch the defect?

No. The problem is that we can tell there is an extra row without ever coming near to the center. The traditional way of doing this is to traverse a large loop surrounding the defect, and count the net number of rows crossed on the path. In the path shown, there are two rows going up and three going down: no matter how far we stay from the center, there will naturally always be an extra row on the right.

How can we generalize this basic idea to a general problem with a broken symmetry? Remember that the order parameter space for the two-dimensional square crystal is a torus (see Figure 7). Remember that the order parameter at a point is that translation which aligns a perfect square grid to the deformed grid at that point. Now, what is the order parameter far to the left of the defect (a), compared to the value far to the right (d)? Clearly, the lattice to the right is shifted vertically by half a lattice constant: the order parameter has been shifted halfway around the torus. As shown in Figure 11, along the top half of a clockwise loop, the order parameter (position of the atom within the unit cell) moves upward, and along the bottom half, again moves upward. All in all, the order parameter circles once around the torus. The winding number around the torus is the net number of times the torus is circumnavigated when the defect is orbited once.

This is why they are called topological defects. Topology is the study of curves and surfaces where bending and twisting is ignored. An order parameter field, no matter how contorted, which doesn't wind around the torus can always be smoothly bent and twisted back into a uniform state. If along any loop, though, the order parameter winds either around the hole or through it a net number of times, then enclosed in that loop is a defect which cannot be bent or twisted flat: the winding number can't change by an integer in a smooth and continuous fashion.

How do we categorize the defects for two-dimensional square crystals? Well, there are two integers: the number of times we go around the central hole and the

---

[8] The next fashion, catastrophe theory, never became important for anything.

number of times we pass through it. In the traditional description, this corresponds precisely to the number of extra rows and columns of atoms we pass by. This was called the Burger's vector in the old days, and nobody needed to learn about tori to understand it. We now call it the first homotopy group of the torus:

$$\Pi_1(T^2) = \mathcal{Z} \times \mathcal{Z} \tag{6}$$

where $\mathcal{Z}$ represents the integers. That is, a defect is labeled by two integers (m, n), where m represents the number of extra rows of atoms on the right-hand part of the loop, and n represents the number of extra columns of atoms on the bottom.

Here's where in the lecture I showed the practical importance of topological defects. Unfortunately for you, I can't enclose a soft copper tube for you to play with, the way I did at the lecture. They're a few cents each, and machinists on two continents have been quite happy to cut them up for my demonstrations, but they don't pack well into books. Anyhow, most metals, and copper in particular, exhibits what is called work hardening. It's easy to bend the tube, but it's amazingly tough to bend it back. The soft original copper is relatively defect-free. To bend, the crystal has to create lots of line dislocations, which move around to produce the bending.[9] The line defects get tangled up and get in the way of any new defects. So, when you try to bend the tube back, the metal becomes much stiffer. Work hardening has had a noticable impact on the popular culture. The magician effortlessly bends the metal bar, and the strongman can't straighten it.... Superman bends the rod into a pair of handcuffs for the criminals....

Before we explain why these curves form a group, let's give some more examples of topological defects and how they can be classified. Figure 12(a) shows a "hedgehog" defect for a magnet. The magnetization simply points straight out from the center in all directions. How can we tell that there is a defect, always staying far away? Since this is a point defect in three dimensions, we have to surround it with a sphere. As we move around on this sphere in ordinary space, the order parameter moves around the order parameter space (which also happens to be a sphere, of radius $|\vec{M}|$). In fact, the order parameter space is covered exactly once as we surround the defect. This is called the *wrapping number* and doesn't change as we wiggle the magnetization in smooth ways. The point defects of magnets are classified by the wrapping number:

$$\Pi_2(\mathcal{S}^2) = \mathcal{Z}. \tag{7}$$

[9] This again is the mysterious lack of rotational Goldstone modes in crystals.

260   Order Parameters, Broken Symmetry, and Topology

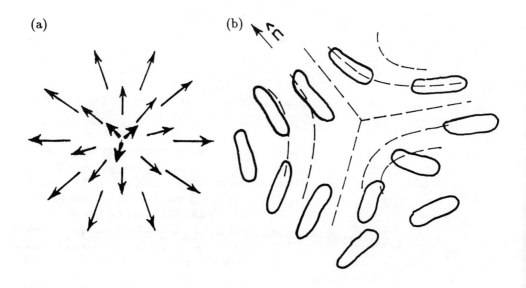

FIGURE 12 (a) Hedgehog defect. Magnets have no line defects (you can't lasso a basketball), but do have point defects. Here is shown the hedgehog defect, $\vec{M}(\mathbf{x}) = M_0\,\hat{x}$. You can't surround a point defect in three dimensions with a loop, but you can enclose it in a sphere. The order parameter space, remember, is also a sphere. The order parameter field takes the enclosing sphere and maps it onto the order parameter space, wrapping it exactly once. The point defects in magnets are categorized by this *wrapping number*: the second Homotopy group of the sphere is $\mathcal{Z}$, the integers. (b) Defect line in a nematic liquid crystal. You can't lasso the sphere, but you can lasso a hemisphere! Here is the defect corresponding to the path shown in Figure 5. As you pass clockwise around the defect line, the order parameter rotates counterclockwise by 180°. This path on Figure 5 would actually have wrapped around the right-hand side of the hemisphere. Wrapping around the left-hand side would have produced a defect which rotated clockwise by 180°. (Imagine that!) The path in Figure 5 is halfway in between, and illustrates that these two defects are really not different topologically.

Here, the 2 subscript says that we're studying the second Homotopy group. It represents the fact that we are surrounding the defect with a two-dimensional spherical surface, rather than the one-dimensional curve we used in the crystal.[10]

You might get the impression that a strength 7 defect is really just seven strength 1 defects, stuffed together. You'd be quite right: occasionally, they do bunch up, but usually big ones decompose into small ones. This doesn't mean, though, that adding two defects always gives a bigger one. In nematic liquid crystals,

[10]The zeroth homotopy group classifies domain walls. The third homotopy group, applied to defects in three-dimensional materials, classifies what the condensed matter people call textures and the particle people sometimes call skyrmions. The fourth homotopy group, applied to defects in space-time path integrals, classifies types of instantons.

two line defects are as good as none! Magnets didn't have any line defects: a loop in real space never surrounds something it can't smooth out. Formally, the first homotopy group of the sphere is zero: you can't loop a basketball. For a nematic liquid crystal, though, the order parameter space was a hemisphere (Figure 5). There is a loop on the hemisphere in Figure 5 that you can't get rid of by twisting and stretching. It doesn't look like a loop, but you have to remember that the two opposing points on the equator really represent the same nematic orientation. The corresponding defect has a director field n which rotates $180°$ as the defect is orbited: Figure 12(b) shows one typical configuration (called an $s = -1/2$ defect). Now, if you put two of these defects together, they cancel. (I can't draw the pictures, but consider it a challenging exercise in geometric visualization.) Nematic line defects add modulo 2, like clock arithmetic in elementary school:

$$\Pi_1(\mathcal{RP}^2) = \mathcal{Z}_2. \tag{8}$$

Two parallel defects can coalesce and heal, even though each one individually is stable: each goes halfway around the sphere, and the whole loop can be shrunk to zero.

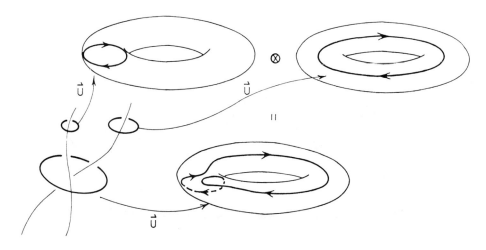

FIGURE 13 Multiplying two loops. The product of two loops is given by starting from their intersection, traversing the first loop, and then traversing the second. The inverse of a loop is clearly the same loop travelled backward: compose the two, and one can shrink them continuously back to nothing. This definition makes the homotopy classes into a group. This multiplication law has a physical interpretation. If two defect lines coalesce, their homotopy class must, of course, be given by the loop enclosing both. This large loop can be deformed into two little loops, so the homotopy class of the coalesced line defect is the product of the homotopy classes of the individual defects.

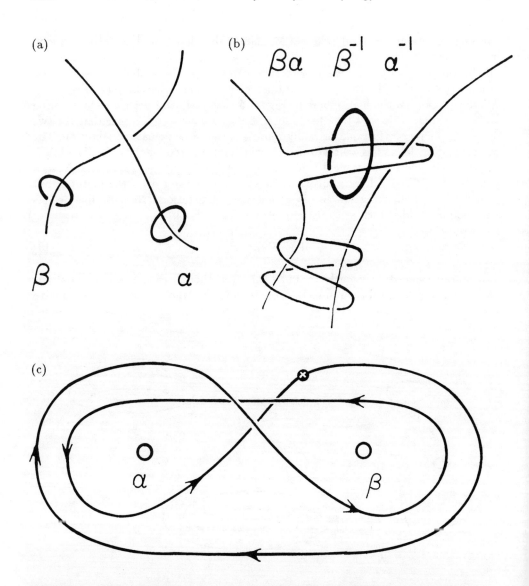

FIGURE 14 Defect entanglement. (a) Can a defect line of class α pass by a line of class β, without getting topologically entangled? (b) We see that we can pass by if we leave a trail: is the connecting double line topologically trivial? Encircle the double line by a loop. The loop can be wiggled and twisted off the double line, but it still circles around the two legs of the defects α and β. (c) The homotopy class of the loop is precisely $\beta\alpha\beta^{-1}\alpha^{-1}$, which is trivial precisely when $\beta\alpha = \alpha\beta$. Thus two defect lines can pass by one another if their homotopy classes commute!

Finally, why are these defect categories a group? A group is a set with a multiplication law, not necessarily commutative, and an inverse for each element. For the first homotopy group, the elements of the group are equivalence classes of loops: two loops are equivalent if one can be stretched and twisted onto the other, staying on the manifold at all times.[11] For example, any loop going through the hole from the top (as in the top right-hand torus in Figure 13) is equivalent to any other one. To multiply a loop u and a loop v, one must first make sure that they meet at some point (by dragging them together, probably). Then one defines a new loop $u \otimes v$ by traversing first the loop u and then v.[12]

The inverse of a loop u is just the loop which runs along the same path in the reverse direction. The identity element consists of the equivalence class of loops which don't enclose a hole: they can all be contracted smoothly to a point (and thus to one another). Finally, the multiplication law has a direct physical implication: encircling two defect lines of strength u and v is completely equivalent to encircling one defect of strength $u \otimes v$.

This all seems pretty trivial: maybe thinking about order parameter spaces and loops helps one think more clearly, but are there any real uses for talking about the group structure? Let me conclude this paper with an amazing, physically interesting consequence of the multiplication laws we described. There is a fine discussion of this in Mermin's article,[3] but I learned about it from Dan Stein's thesis.[5]

Can two defect lines cross one another? Figure 14(a) shows two defect lines, of strength (homotopy type) α and β, which are not parallel. Suppose there is an external force pulling the α defect past the β one. Clearly, if we bend and stretch the defect as shown in Figure 14(b), it can pass by, but there is a trail left behind, of two defect lines. α can really leave β behind only if it is topologically possible to erase the trail. Can the two lines annihilate one another? Only if their net strength is zero, as measured by the loop in Figure 14(b).

Now, get two wires and some string. Bend the wires into the shape found in Figure 14(b). Tie the string into a fairly large loop, surrounding the doubled portion. Wiggle the string around, and try to get the string out from around the doubled section. You'll find that you can't completely remove the string (no fair pulling the string past the cut ends of the defect lines!), but that you can slide it downward into the configuration shown in Figure 14(c).

Now, in Figure 14(c) we see that each wire is encircled once clockwise and once counterclockwise. Don't they cancel? Not necessarily! If you look carefully, the order of traversal is such that the net homotopy class is $\beta \alpha \beta^{-1} \alpha^{-1}$, which is only the identity if β and α *commute*. Thus the physical entanglement problem

---

[11] A loop is a continuous mapping from the circle into the order parameter space: $\theta \to u(\theta)$, $0 \leq \theta < 2\pi$. When we encircle the defect with a loop, we get a loop in order parameter space as shown in Figure 4: $\theta \to \vec{x}(\theta)$ is the loop in real space, and $\theta \to u(\vec{x}(\theta))$ is the loop in order parameter space. Two loops are equivalent if there is a continuous one-parameter family of loops connecting one to the other: $u \equiv v$ if there exists $u_t(\theta)$ continuous both in θ and in $0 \leq t \leq 1$, with $u_0 \equiv u$ and $u_1 \equiv v$.

[12] That is, $u \otimes v(\theta) \equiv u(2\theta)$ for $0 \leq \theta \leq \pi$, and $\equiv v(2\theta)$ for $\pi \leq \theta \leq 2\pi$.

for defects is directly connected to the group structure of the loops: commutative defects can pass through one another; noncommutative defects entangle.

I'd like to be able to tell you that the work hardening in copper is due to topological entanglements of defects. It wouldn't be true. The homotopy group of dislocation lines in fcc copper is commutative. (It's rather like the two-dimensional square lattice: if $\alpha = (m,n)$ and $\beta = (o,p)$ with m,n,o,p the number of extra horizontal and vertical lines of atoms, then $\alpha\beta = (m+o, n+p) = \beta\alpha$.) The reason dislocation lines in copper don't pass through one another is energetic, not topological. The two dislocation lines interact strongly with one another, and energetically get stuck when they try to cross. Remember at the beginning of the paper, I said that there were gaps in the system: the topological theory can only say when things are impossible to do, not when they are difficult to do.

I'd like to be able to tell you that this beautiful connection between the commutativity of the group and the entanglement of defect lines is nonetheless important in lots of other contexts. That, too, would not be true. There are two types of materials I know of which are supposed to suffer from defect lines which topological entangle. The first are biaxial nematics, which were thoroughly analyzed theoretically before anyone found one. The other are the metallic glasses, where David Nelson has a theory of defect lines needed to relieve the frustration. We'll discuss closely related theories in section 3. Nelson's defects don't commute, and so can't cross one another. He originally hoped to explain the freezing of the metallic glasses into random configurations as an entanglement of defect lines. Nobody has ever been able to take this idea and turn it into a real calculation, though.

Enough, then, of the beautiful and elegant world of homotopy theory: let's begin to think about what order parameter configurations are actually formed in practice.

ACKNOWLEDGMENTS

I'd like to acknowledge NSF grant # DMR-9118065, and thank NORDITA and the Technical University of Denmark for their hospitality while these chapters were written up.

REFERENCES

1. Coleman, Sydney, ed. "Secret Symmetry: An Introduction to Spontaneous Symmetry Breakdown and Gauge Fields." In *Aspects of Symmetry, Selected Erice Lectures*, 113. Cambridge: Cambridge University Press, 1985.

2. de Gennes, P. G. *The Physics of Liquid Crystals*. London: Oxford University Press, 1974.
3. Mermin, David. "The Topological Theory of Defects in Ordered Media." *Rev. Mod. Phys.* **51** (1979): 591.
4. Sethna, J. S. "Meissner Effects and Constraints." This volume.
5. Stein, D. L. "Topology of Order Parameter Spaces of Condensed Matter Systems." Ph.D. Thesis, Princeton University, 1979.

James P. Sethna
Laboratory of Atomic and Solid State Physics, Cornell University, Ithaca, New York 14853-2501

Meissner Effects and Constraints

In the last paper,[8] I explained how condensed-matter and high-energy physicists used topological theories to describe defects excitations in solids. In this paper, I'm going to make fun of topology.[1] Actually, I'm going to start by talking about constraints, then "massive" fields and how they produce constraints. I'll then turn to the Meissner-Higgs effect in superconductors, and finally explain why I don't understand crystals.

I. CONSTRAINTS

Consider Figure 1. See the beautiful ellipses and hyperbolas? Remember that topology treats ellipses as rubber bands. Any topological theory has got to miss the key

---

[1]Everything I know about focal conics and smectic liquid crystals[7] was explained to me by Maurice Kléman, who also was one of the originators of the topological theory of defects. No disrespect is intended.

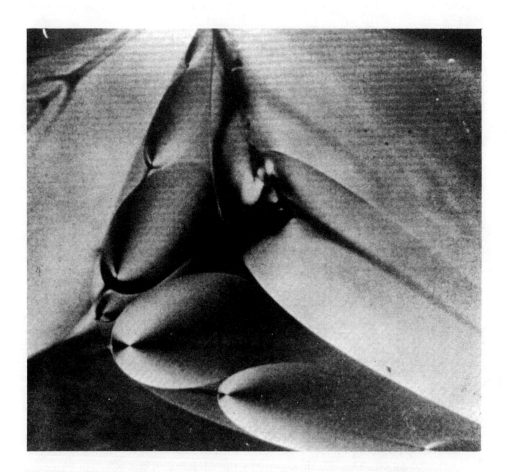

FIGURE 1 Ellipses: Defects in a Liquid Crystal. This is a drop of smectic A liquid crystal, squeezed between two microscope slides. The microscope is focused on the surface of the drop, where it contacts the glass. Notice the beautiful, geometrical ellipses. Notice that a line seems to exit from the focus of each ellipse. This line turns out to be a hyperbola (Figure 4). The visible ellipses and the hyperbolas are where the smectic layers pinch off to form cusps. These defects are *not* topological: they are geometrical consequences of the constraint of equal layer spacing. From Ref. 3, Figure 7.2, photo by C. Williams.

feature of the beautiful structures produced here: the geometrically perfect ellipses with dark lines coming out of one focus.

Figure 1 is a photograph of a drop of fluid, squeezed between two microscope slides. The microscope is focused, let's say, on the surface between the fluid and the bottom microscope slide: the ellipses are stuck onto the glass. The sizes of the ellipses are roughly given by the thickness of the fluid layer. The fluid is a smectic A liquid crystal. deGennes[3] has a fine discussion and some nice pictures, too.

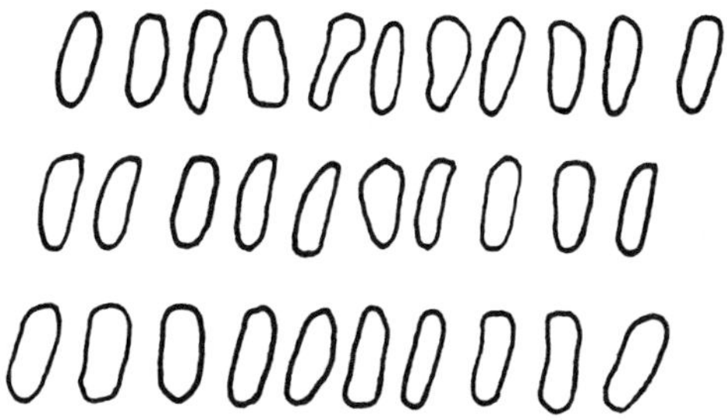

$$0\,0\,0\,0\,0\,0\,0\,0\,0\,0\,0$$
$$0\,0\,0\,0\,0\,0\,0\,0\,0\,0$$
$$0\,0\,0\,0\,0\,0\,0\,0\,0\,0$$

FIGURE 2 Order in Smectic Liquid Crystals. Smectic liquid crystals are formed of layers of molecules. In each layer, the molecules are in a random, liquid configuration. Crystals have broken translational symmetry along three independent axes; smectic A liquid crystals have broken translational symmetry in only one direction (normal to the layers).

In 1910, Friedel figured out why this liquid crystal forms these geometrical structures. He learned all he needed to know from his high-school geometry class. He actually worked backward, and used the ellipses to deduce what kind of broken symmetry the liquid had. Since none of you were taught about the cyclides of Dupin in high school,[2] I'd better start with the broken symmetry and work forward.

Smectic liquids form equally spaced layers. Some of them are compounds that, like soap, naturally form membranes and films: I think smectic is the Greek word for soap. Others are long thin molecules like nematics, which for some reason not only line up but segregate into planes (Figure 2). The molecules have liquid-like order in the planes. Like crystals, they have a broken translational symmetry, but only in one of the three directions.

Now, the important excitations for smectics are those that bend the layers. In Figure 3, we see a two-dimensional analogue of the smectic liquid crystals: equally spaced curves in the plane. Suppose we start with one curve and work outward. As you can see from the figure, the next curve is not precisely the same shape: keeping the surfaces at an equal spacing makes concave regions become sharper and convex regions become more rounded. It is easy to see that eventually the concave regions will become pinched: these pinches are the defects. They are not topological defects, since rounding them a bit makes them go away: they are geometrical defects produced by the constraint of equal layer spacing.

---

[2]Bertrand Fourcade tells me that even the French stopped teaching them.

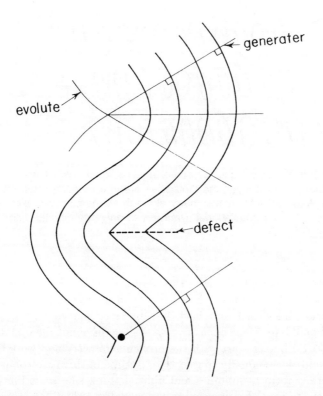

FIGURE 3 Equally Spaced Layers: Defect Formation. Here we consider a two-dimensional analogue of a smectic liquid crystal. The smectic layers are represented by curves in the plane. The lowest energy state, of course, consists of parallel straight layers, but the layers often settle into more complicated patterns, with defects. For reasons that we discuss in this paper, and which are not completely understood, smectic layers will deform by bending, but will remain strictly equally spaced (except very near boundaries and defects). The constraint of equal layer spacing has weird nonlocal consequences. First, one can see that as one moves outward the concave regions become more pinched, and eventually form cusps. Second, one can see that a line perpendicular to one layer (a generator) will be perpendicular to the next one, too. These generators intersect on a surface known as the *evolute*, and it is when the layers hit the evolute that a defect occurs. As one sees here, the defect is a line of pinched surfaces: in three dimensions it is typically a two-dimensional surface. This costs lots of energy. The only way in two dimensions to have a point-like low-energy defect is to have concentric circles: only circles have zero-dimensional evolutes. The only way in three dimensions to have one-dimensional evolutes[4] is to have cyclides of Dupin: the defects are ellipses and hyperbolas passing through one another's foci (Figures 1 and 4).

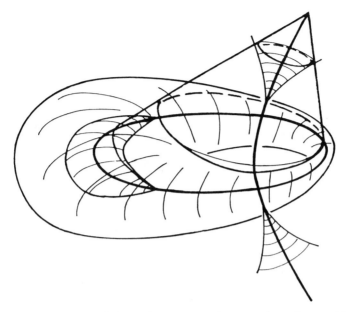

FIGURE 4 Focal Conic Defect. Here we see the smectic surfaces which form the focal conic defects seen in Figure 1. These are the cyclides of Dupin. The surfaces go from banana-shaped to squashed doughnuts to apple-shaped. The points on the bananas and the dimples at the stem and bottom of the apples are defects, which scatter light and show up in Figure 1. (Only the dimples of the apple are shown.) The banana defects lie on an ellipse, and the apple defects lie on a hyperbola which passes through the focus of the ellipse. Usually, the whole pattern isn't found in the experimental sample. As you see in Figure 1, the domains aggregate together in clumps. Each ellipse in Figure 1 has a conical region for its smectic layers.

Most curves, like the one shown in Figure 3, form one-dimensional pinched regions: only concentric circles and structures made from them can keep the pinched regions to points. In three dimensions, the only equally spaced surfaces with points as pinched regions are concentric spheres. Now, what Friedel knew and you don't know is that the only three-dimensional surfaces with one-dimensional line-like defects are the cyclides of Dupin,[4] *and the pinched regions form ellipses and hyperbolas.*[3]

Figure 4 shows the cyclides of Dupin. Notice that they pinch off on two curves: an ellipse and a hyperbola. The hyperbola is perpendicular to the plane of the ellipse, and passes through its focus. That's what you see streaming out of the foci in the photo, and why you don't see one for each focus. My contribution to

---

[3] Actually, the canal surfaces also have singularities confined to one-dimensional regions,[4,7] but let's not get bogged down.

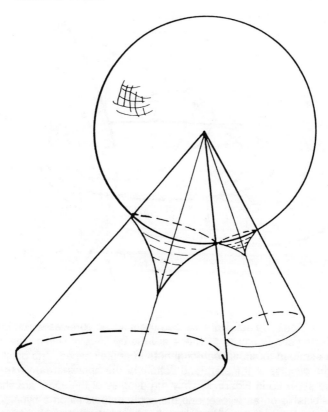

FIGURE 5 Focal Conic Defect Meshing onto Concentric Spheres. The conical regions in Figure 4 combine into compound defects by meshing onto the concentric sphere defect. Concentric spheres are the only surfaces with zero-dimensional defects. The surfaces on the edges of the cones mesh smoothly onto the concentric spheres.

the field (with Maurico Klóman) was to realize that these cyclides of Dupin fit together nicely inside concentric spheres, which explained neatly the ways the ellipses always seemed to fit together (Figure 5). Maybe the concentric spheres form because the layers nucleate on a dust particle on one of the microscope slides: when the spheres touch the other slide, the concentric spheres get twisted (they like to sit perpendicular to the glass) and the ellipses and hyperbolas form to relieve the strain.

Now, why do I show you this? It isn't just to show that there is more to the world than topology. Mostly, it's to illustrate the two themes of this paper: constraints and expulsion.

If we define an order parameter $\hat{n}$ for the smectic to be the unit normal to the smectic layers ($\hat{n}^2 = 1$), then the constraint that the layers be equally spaced implies

$$\operatorname{curl} \hat{n} = \begin{pmatrix} \partial n_z/\partial y - \partial n_y/\partial z \\ \partial n_x/\partial z - \partial n_z/\partial x \\ \partial n_y/\partial x - \partial n_x/\partial y \end{pmatrix} = 0. \qquad (1)$$

(This is derived, for those who know a bit about vector calculus, in the Appendix.) This is a remarkably powerful constraint. For example, knowing the position of one layer determines all the others! We show this mathematically in the Appendix, but you saw it physically in Figure 3: given one layer, there is only one way to place the next one preserving exactly equal spacing.

There is a pretty good analogy here to analytic continuation. For those of you who know about complex analysis, you know that an analytic function obeys the Cauchy-Riemann equations. If we let $n(x+iy) = n_x(x+iy) + in_y(x+iy)$ be an analytic function, then

$$\begin{pmatrix} \partial n_x/\partial x - \partial n_y/\partial y \\ \partial n_x/\partial y + \partial n_y/\partial x \end{pmatrix} = 0. \qquad (2)$$

As you know, analytic functions have really bizarre properties. If you know an analytic function in a small region, you can figure it out everywhere else, just like the order parameter in smectics. The point singularities of analytic functions have a rich and interesting classification (simple poles, essential singularities,...). Both in analytic functions and in our smectic problem, constraints on the derivatives of our order parameters produced really bizarre, nonlocal, geometrical consequences.

II. MASSIVE FIELDS

We've discovered that constraints can have beautiful, geometrical consequences. How are the constraints enforced? Clearly, it is possible to stretch the smectic layers apart or to compress them together: why doesn't this happen in practice, especially when the layers are being bent and twisted? The curl of $\hat{n}$ is constrained to zero. Why are magnetic fields pushed completely out of superconductors? The magnetic field is constrained to zero. Why isn't it possible to find an isolated quark in nature? Quarks have non-zero "color," and the net color is constrained to zero.

These constraints come from minimizing the energy. Saying that magnetic fields can happen inside superconductors is just like saying that marbles can sit on the side of a hill: it can happen, but not if the marbles are allowed to roll to minimize their energy. Under what conditions does the energy enforce a constraint? We say that it happens when the order parameter field develops a *mass*. We'll explain this term in a moment, but let's first give a simple example.

Suppose we have a fluid in one dimension. The density of a fluid is the important variable in describing its state. Suppose the density of the fluid is $\rho_0 + \rho(x)$, where ρ_0 is the ideal density (which the fluid would have if left to itself) and the order parameter $\rho(x)$ describes the deviation from the ideal density. A sensible free energy might be

$$\mathcal{E}_{\text{fluid}} = \int dx\,(1/2)(d\rho/dx)^2 + (1/2)m\rho^2\,. \tag{3}$$

The first term in the energy resists sudden changes in the density: having a high-density region right next to a low-density region costs extra. The second term in the energy says that deviations from the mean density cost energy, with m a coefficient which says how much deviations cost. Unlike phonons, where the order parameter $u(x)$ could be uniformly shifted without energy cost, here the lowest-energy state happens when the density is at its mean value $\rho(x) = 0$.

What happens when we try to find the minimum energy state? Clearly the best we can get is the ideal state $\rho(x) \equiv 0$, which has zero energy $\mathcal{E}_{\text{fluid}}$. Perhaps, though, we're pulling on the density at the two ends (Figure 6). If the liquid is in a trough of length L, we'll insist that $\rho(0) = \rho_i$ and $\rho(L) = \rho_f$. What configuration $\rho(x)$ minimizes the energy then? Clearly, it should sag towards ρ_0 inside, but how?

Here I'll show you a simple case of what's called the calculus of variations. I apologize for the math, but it is really a useful method. The trick is to realize that if $\rho(x)$ is the minimum energy configuration, then $\rho(x) + \delta(x)$ must have a higher energy, whatever $\delta(x)$ we might choose.

$$\mathcal{E}(\rho + \delta) - \mathcal{E}(\rho) = \int \left(\frac{d\rho}{dx}\frac{d\delta}{dx} + m\rho(x)\delta(x) + \left(\frac{1}{2}\right)\left(\frac{d\delta}{dx}\right)^2 + \left(\frac{1}{2}\right)m\delta^2 \right) dx \geq 0\,. \tag{4}$$

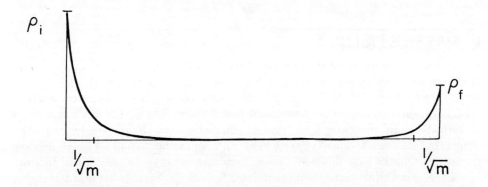

FIGURE 6 Massive Fields Decay Exponentially. Minimizing the energy $\mathcal{E}_{\text{fluid}}$ in equation (liquid), with boundary conditions $\rho(0) = \rho_i$ and $\rho(L) = \rho_f$. It is easy to understand physically what is happening. The system wants to achieve $\rho = 0$, and it sags to that value as quickly as it can, balancing the costs of $(d\rho/dx)^2$ energy against the gain. The solution decays exponentially to zero with a decay constant $\sqrt{m}$.

Meissner Effects and Constraints

Now, if we confine our attention to small $\delta(x)$, we can ignore the last two terms (because they are quadratic, rather than linear, in δ). The first term we integrate by parts, so

$$\int_0^L dx \frac{d\delta}{dx}\frac{d\rho}{dx} = \left(\frac{\delta\, d\rho}{dx}\right)\bigg|_0^L - \int_0^L \frac{dx\, \delta\, d^2\rho}{dx^2}. \quad (5)$$

Now, δ mustn't change the values at the endpoints, so $\delta(0) = \delta(L) = 0$, and the boundary terms in Eq. (5) drop out. We're left, then, with the equation

$$\mathcal{E}(\rho + \delta) - \mathcal{E}(\rho) \approx \int dx \left(\frac{-d^2\rho}{dx^2} + m\rho(x)\right)\delta(x) \geq 0. \quad (6)$$

Now, this must be true for any $\delta(x)$ we choose. This can only happen if $-d^2\rho/dx^2 + m\rho(x) = 0$, so $\rho'' = m\rho$.

The solutions to this equation are, of course, $\rho = Ae^{-\sqrt{m}x} + Be^{\sqrt{m}x}$. We can vary the arbitrary constants A and B to match the boundary conditions $\rho(0) = \rho_i$ and $\rho(L) = \rho_f$, and we see (Figure 6) that ρ is *expelled from the interior*: pulling it on the boundary only affects a region of length $\sqrt{m}$, and the order parameter exponentially decays into the bulk. ρ is constrained to zero in the inside of the sample!

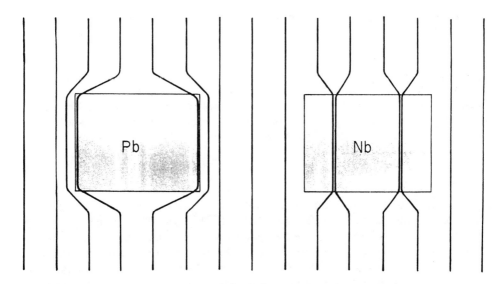

FIGURE 7 Superconductors Expel Magnetic Fields. A magnetic field passing through a metal will be pushed out when the metal is cooled through its superconducting transitions temperature. This can happen in two different ways. In type I superconductors like lead (chemical symbol Pb), the superconductivity is pushed entirely outside the sample. In type II superconductors like niobium (Nb), the magnetic field is broken up and confined to defect lines called vortices. In both cases, the magnetic field is swept out of the remainder of the sample. The magnetic field penetrates a distance $\Lambda \sim 100\text{Å}$ into the sample from the boundaries or from the vortex lines.

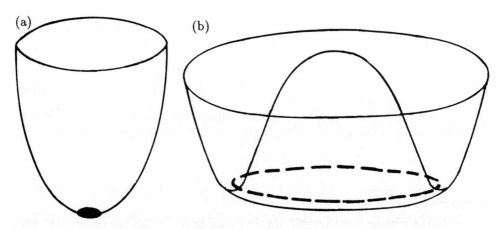

FIGURE 8 (a) Superfluid Free Energy, $T > T_c$: Unbroken Symmetry. The free energy for a normal metal or fluid, above the superconducting or superfluid transition temperature, for a uniform order parameter field ψ. The vertical axis represents the energy $\alpha|\psi|^2 + \beta|\psi|^4$, and the horizontal axes represent the real and imaginary parts of ψ. The coefficient $\alpha > 0$, so the minimum of the energy is at $\psi = 0$. Notice that the energy is invariant under the symmetry $\psi \to e^{i\theta}\psi$ (corresponding to rotating the figure about the vertical). This is a symmetry of the free energy. Notice also that the lowest energy state $\psi = 0$ is also unchanged by this rotation: the symmetry is unbroken above T_c. (b) Superfluid Free Energy: $T < T_c$: Broken Symmetry. The free energy $\mathcal{E}_{\text{superfluid}}$ for helium below the superfluid transition temperature. The energy now looks like a Mexican hat: it is still invariant under rotations about the vertical axis. Since now $\alpha < 0$, the energy is at a minimum along a circle, of radius $|\psi| = \sqrt{\alpha/2\beta}$ and arbitrary phase θ. The superfluid must choose between these various possible phases, and that choice breaks the symmetry. This is a good example of spontaneous symmetry breaking: just as the magnetization of a magnet selects a direction in space and breaks rotational invariance, the superconductor picks out a value of θ.

Why do we call this a mass? The name comes from particle physics. The photon is massless. Two charges e_1 and e_2 separated by a distance r interact by a force whose magnitude goes as $e_1 e_2/r^2$: this is Coulomb's law. The particle physicists interpret this force in terms of the two particles exchanging "virtual" photons. (I think of the $1/r^2$ decay as the virtual photons being diluted over a sphere of radius r.) Now, the strong interaction between protons and neutrons has a different form: the force between them is always attractive, and goes as $e^{-\lambda r}/r^2$. The exponential decay is extremely important, since it keeps the nuclei of different atoms from attracting one another. (We'd all have collapsed into neutron stars or worse were it not there!) At long distances, the particle physicists interpret this force as the proton and neutron exchanging virtual pions.[4] Since the pion isn't massless, the

---

[4] At shorter distances, the picture is quarks exchanging gluons. The gluons have color, though, so the proton and neutron can't exchange them at long distances. Since colorless glueballs, if

virtual pion field decays exponentially for exactly the same reason that $\rho(x)$ decayed in our example above.

So, to enforce a constraint, we need to give the corresponding field a mass. Let's see how that is done.

III. THE MEISSNER-HIGGS EFFECT

In this section, I want to explain how superconductors expel magnetic field. This is a really beautiful argument, which I've basically taken from Coleman's presentation.[2] I'm afraid that there is some math and a lot of physics that I need to introduce. Most of you will get lost at some point: skip onto the next section when you tire of this one.

A. INTRODUCTION TO THE MEISSNER EFFECT. Superconductors are named for their ability to carry currents of electricity with absolutely no losses. They have another, closely related property which is no less amazing: they are a perfect shield for magnetic fields. Remember the old science fiction stories about the scientist who finds a material which is impervious to the gravitational field, paints the bottom of his spacecraft with it, and falls to the moon? Superconductors work that way for magnetic fields.

Ashcroft and Mermin have a nice, not too technical discussion of superconductors in one of the last chapters in their textbook.[1] Figure 7 shows the two types of superconductors, represented by lead and niobium. At high temperatures, when the materials aren't superconducting, the magnetic field penetrates the materials almost as if they weren't there. (Iron would pull the magnetic field lines inward.) Lead, when superconducting, pushes the magnetic field out: just as for the example in section II, the field a distance r inward from the boundary decays like $B = B_0 e^{-r/\Lambda}$. If you put too high a field, the lead will give up and let the field in, but it will stop superconducting.

On the right, we see that niobium behaves a bit differently. It expels small magnetic fields like lead does, but larger fields are pushed into thin threads, called vortex lines. These two general categories are (rather unimaginatively) called type I and type II superconductors. The vortex lines are the topological defects for the superconductor.[8] Superconductors are described by a complex number $\psi = \rho e^{i\theta}$, whose magnitude $|\psi| = \rho$ is roughly constant. The order parameter at low temperatures is the phase θ, and thus the order parameter space is a circle $\mathcal{S}^1$. A vortex line must pass through any loop around which the phase of the order parameter changes by 2π. The magnetic field in type II superconductors decays

---

they exist, are much more massive than pions, the dominant interaction for long distances is pion exchange.

like $B = B_0 e^{-r/\Lambda}$ where here r is the distance to the vortex line. The magnetic field is squeezed out of the bulk of the material into these defects.

So, the magnetic field isn't actually stopped; it just peters out. What kind of a leaky shield is that? Actually, it's about as good as one can hope: after all, the magnetic field won't be able to tell it's in a superconductor until it gets inside a bit! (Atoms don't go superconducting, only huge heaps of atoms together do, so the field has to pass through a heap or two to realize that it isn't wanted.) Anyhow, Λ is usually pretty small, a few hundred Ångstroms or so. An 0.1mm thin layer of superconducting paint naively would let through a field one part in $e^{-10000} \sim 10^{-4000}$ of the original. Unfortunately, it usually doesn't work so well: a few vortex lines get stuck on junk in the paint, and let in comparatively large fields.

Before we can explain the repulsion of magnetic fields, we should explore the broken symmetry. Let's start with superfluids, which are simpler.

B. SUPERFLUID FREE ENERGY AND SPONTANEOUS SYMMETRY BREAKING.

The order parameter for a superfluid, just as for a superconductor, is a complex number ψ. The free energy for the superfluid is usually written as[5]

$$\mathcal{E}_{\text{superfluid}} = \int dV |\nabla \psi|^2 + \alpha |\psi|^2 + \beta |\psi|^4. \tag{7}$$

Above the superconducting transition temperature T_c, the coefficient $\alpha > 0$. If we imagine a constant order parameter field, the free energy forms a bowl (Figure 8(a)) with a minimum at zero, as a function of the real and imaginary part of ψ. Zero-order parameter corresponds to a normal metal (for a superconductor) or a normal liquid (for a superfluid).

Below T_c, $\alpha < 0$, and the potential is at a minimum for $\rho_0 = |\psi| = \sqrt{\alpha/2\beta}$: the potential in the complex plane looks like a Mexican hat (Figure 8(b)). Now there are many possible ground states: for any θ, a constant order parameter field $\psi = \rho_0 e^{i\theta}$ is a ground state. Because the free energy depends only on $|\psi|$ and $|\nabla \psi|$, it is symmetric to changing the phase θ: the superconducting state chooses a specific value for θ and, thus, *spontaneously breaks the symmetry*. The circle of ground states in the brim of the Mexican hat is the order parameter space for the superconductor.

We can write the free energy in terms of θ:

$$\mathcal{E}_{\text{superfluid}} = \int dV |\nabla \rho|^2 + \rho^2 |\nabla \theta|^2 + \alpha \rho^2 + \beta \rho^4. \tag{8}$$

---

[5] There are two new symbols here: $\nabla = (\partial/\partial x, \partial/\partial y)$ and $|\chi|^2 = \chi^* \chi$, where χ^* is the complex conjugate of χ. Written out in components,

$$\mathcal{E}_{\text{superfluid}} = \int dV \left(\frac{\partial \psi}{\partial x}\right)^* \left(\frac{\partial \psi}{\partial x}\right) + \left(\frac{\partial \psi}{\partial y}\right)^* \left(\frac{\partial \psi}{\partial y}\right) + \alpha \psi^* \psi + \beta (\psi^* \psi)^2.$$

You can think of this as a mathematical expression of the Mexican hat potential in Figure 8(b), together with a resistance to abrupt changes in the order parameter.

As we discussed in the previous section, ρ is "massive." In Figure 8(b), if we vary ρ slightly away from ρ_0, the energy increases quadratically: $\alpha\rho^2 + \beta\rho^4 - (\alpha\rho_0^2 + \beta\rho_0^4) \approx (\alpha + 6\beta\rho_0^2)(\rho - \rho_0)^2$ The effective free energy for ρ near ρ_0 is precisely of the form (3) (except for unimportant constant shifts), with $m = \alpha + 6\beta\rho_0^2$. Thus just as before, ρ will rapidly be drawn to its minimum energy state ρ_0. Because ρ is massive, it is basically constrained to stay at its minimum value. This is why it is ignored at low temperatures in writing the order parameter field. Here, the constraint doesn't do anything interesting: our next constraint will be more interesting.

The θ field keeps the symmetry of the original model: rotating it to $\theta + \theta_0$ doesn't change the energy a bit. It is a Goldstone mode for our problem, and long wavelength plane waves produce what is known as "second sound" in superfluids. Second sound turns out to be heat waves: pulses of temperature which propogate like waves through the superfluid.

C. SUPERCONDUCTING FREE ENERGY AND THE HIGGS MECHANISM.

To describe the expulsion of magnetic field from superconductors, I have to tell you how magnetic fields interact with the superconducting order. I'm afraid this will be rather sketchy, and I apologize for trying.

First of all, the particles which superconduct are pairs of electrons. Electrons are charged, and repel one another with electric fields. Thus the electrons interact with electric fields. We learn in the second semester of physics (if we're lucky) that electric and magnetic fields are closely related to one another. (This was discovered by Einstein: a moving electric E field develops a magnetic B component.)

Now, the E and B fields can be written at the same time in terms of another field A. It is this new field which is easiest to work with. In particular,

$$B = \text{curl } A = \left(\frac{\partial A_z}{\partial y} - \frac{\partial A_y}{\partial z}, \frac{\partial A_x}{\partial z} - \frac{\partial A_z}{\partial x}, \frac{\partial A_y}{\partial x} - \frac{\partial A_x}{\partial y}\right). \tag{9}$$

The magnetic energy is $\mathcal{E}_{\text{magnetic}} = \int dV\, B^2$.

Now, you remember that I mentioned earlier that light (the photon) is massless? You may know that light is sometimes called "electromagnetic radiation." The "order parameter field" for light is precisely the A field. We can see by expanding B^2 in terms of A that the energy for the A field

$$\mathcal{E}_{\text{magnetic}} = \int dV \left(\frac{\partial A_z}{\partial y} - \frac{\partial A_y}{\partial z}\right)^2 + \cdots \tag{10}$$

doesn't have any terms like A^2. When we add the energy from the electric fields, this is still true: light is massless because the electromagnetic energy involves only derivatives of A.

Now, I need to know how the electromagnetic order parameter A interacts with the superconducting order parameter ψ. I'll just tell you. The free energy for a superconductor looks like

$$\mathcal{E}_{\text{superconductor}} = \int dV\, |\nabla\psi - iA\psi|^2 + \alpha|\psi|^2 + \beta|\psi|^4 + B^2. \tag{11}$$

If we set $\psi = 0$, we get the magnetic energy B^2 for the A field. If we set $A = 0$, we get the superfluid energy (8). I don't know of a way to motivate the way in which we couple the A field to the gradient $\nabla \psi$. I don't think anyone has a simple derivation. This way of connecting the two is called "minimal coupling," which just gives a name to the unexplained fact that the simplest way of coupling the two gives the right answer.

Now, if we assume $T < T_c$, so $\alpha < 0$ and $\rho \sim \rho_0 e^{i\theta}$, we find

$$\mathcal{E} \approx \int dV \, \rho_0^2 \, |\nabla \theta - A|^2 + \left(\frac{\partial A_z}{\partial y} - \frac{\partial A_y}{\partial z}\right)^2 + \cdots . \tag{12}$$

We want to know if A or θ is going to develop a mass. The problem is: $\mathcal{E}_{\text{superconductor}}$ doesn't look quite like the form (3) for either one. If we combine the two into a new order parameter field $C = \nabla \theta - A$, and use the fact that the second partial derivative $\partial^2 \theta / \partial z \partial y = \partial^2 \theta / \partial y \partial z$, we see that

$$\operatorname{curl} C = \left(\frac{\partial C_z}{\partial y} - \frac{\partial C_y}{\partial z}, \cdots\right)$$
$$= \left(\frac{\partial A_z}{\partial y} - \frac{\partial A_y}{\partial z}, \cdots\right)$$
$$= B$$

so

$$\mathcal{E} \approx \int dV \, \rho_0^2 \, C^2 + \left(\frac{\partial C_z}{\partial y} - \frac{\partial C_y}{\partial z}\right)^2 + \cdots . \tag{13}$$

Thus the new, combined field C is massive. C will be constrained to zero in the bulk, exponentially decaying like $C_0 e^{-\rho_0 r}$. The magnetic field $B = \operatorname{curl} C$ thus also decays, and the penetration depth $\Lambda = 1/\rho_0$.

We started with a massless photon field A and a massless Goldstone mode θ. We ended up with only one field C, with a mass. Did we lose something? No, actually C has three components: two components corresponding to the original two polarizations of light and one component corresponding to the Goldstone mode. Coleman[2] says "the Goldstone boson eats the photon, and gains a mass!"

The Weinberg-Salaam theory of the weak interaction is exactly analogous to the theory of superconductivity. The role of lead or niobium is played by the vacuum. The free energy of the universe has an $SU(3)$ symmetry, which is spontaneously broken to a smaller symmetry $SU(2) \times U(1)$. The $W^{\pm}$ and Z bosons which now mediate the weak interaction used to be massless: they and the photon were all part of one big A field. If current theories of cosmology are true, this "superconducting" transition occurred in the first instants after the Big Bang.

Now, after explaining superconductors, the weak interaction, and the phase transition in the early universe, let's return to why we don't understand crystals.

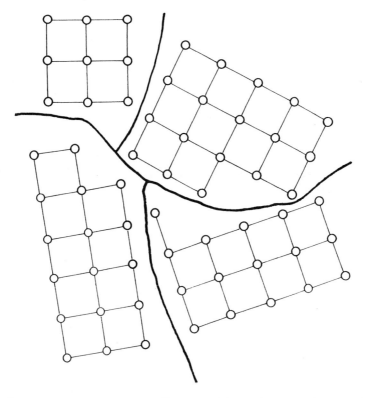

FIGURE 9 Polycrystal. Many crystalline materials, such as metals, normally aren't made of a single crystal. They are formed from many crystalline domains: a polycrystalline configuration. I show a schematic of a polycrystal here. The important thing to notice is that the atoms within a domain are almost undeformed except right next to the domain wall. All the rotational deformation is expelled into sharp domain boundaries.

IV. THE MYSTERY OF THE CRYSTALS

Normally, when you think of crystals, you think of diamonds, snowflakes, or maybe salt crystals.[6] These are single crystals: the sodium and chlorine atoms in a grain of salt sit in registry all the way across the grain, giving it its cubical shape. Did you know that metals form crystals? In the last paper,[8] I mentioned dislocation lines in a copper crystal. Metals don't have big facets and corners because they

---

[6]Some of you will think of wine glasses. They are made of glass and aren't crystals at all.

are polycrystalline. The atoms in a metal also sit on a regular lattice, but the metal breaks up into domains in which the lattices sit at various angles (Figure 9). Because there are lots of small domains, copper doesn't form facets like salt grains and snowflakes do.[7]

What Ming Huang (one of my students[5]) and I have been trying to explain for years is why those little domains form. It's easy to see that different regions might grow with different orientations (Figure 10). When they touch, the different domains will start pushing and twisting one another, trying to make one big domain. It isn't hard to believe that they will stop growing after a while, fighting one another to a standstill. What we've been trying to understand, though, is why the final state is made of perfect little crystals separated by sharp domain walls.

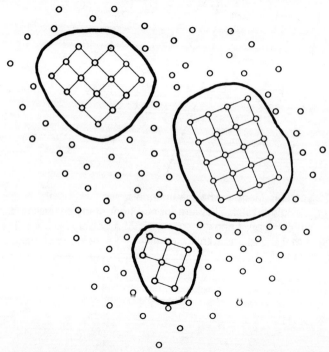

FIGURE 10 Growing a Crystal from a Liquid: Forming a Polycrystal. Polycrystals can form for lots of reasons. If one cools a liquid quickly, one can find that crystalline regions can form in many different places almost simultaneously. Since they will have random orientations, they won't match up when they meet. When they do meet, rearrangements of atoms will occur to try to realign and merge the domains (coarsening). As we continue to cool and wait, this process will eventually stop, leaving us with different domains.

[7]Metal crystals are sometimes found in nature. The growth takes place so slowly that a single crystal can form. The same idea happens with rock candy: you get a glass if you cool sugar syrup quickly, but if you evaporate a sugar solution slowly, you can get big crystals.

Now, I don't want to exaggerate. There are perfectly good explanations for why crystals form domain walls. They just aren't as beautiful and general as they might be. They don't fit in with the general ideas of broken symmetries and order parameters: they apply only to crystals. Our explanation for why superconductors don't have a Goldstone mode was perfectly okay before Higgs came, too. He made it beautiful and generalized it to explain something completely different. Ming and I want to understand grain boundaries in a way which will make simple and clear where else similar phenomena might occur. At least, we'd like to understand why focal conics occur at the same time. Domains formed by breaking translational symmetry in one direction and in three directions should have the same kind of explanation!

Figure 11 shows a domain wall in a crystal. The crystalline ground state rotates as one crosses the domain wall. The atoms at the wall are quite unhappy. You'd

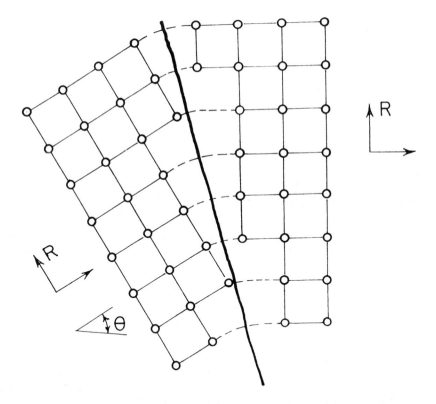

FIGURE 11 Domain Wall. Here we see a single domain wall. Notice that the domain wall can be also thought of as a series of dislocations. The strain field inside the crystal due to a line of dislocations can be shown to decay exponentially, just as the magnetic field dies away around a vortex line.

think that they would push and pull on their neighbors, and that there would be strains leaking far into the crystal. This isn't true. In fact, there is a well-known rule in the materials science literature that the strain field from a domain wall dies away exponentially as one enters the grain.

Doesn't that sound like a Meissner effect?

There are more analogies. Crystals break both the translational and the rotational symmetry of liquids. Many liquid crystals only break the rotational symmetry. They have Goldstone rotational waves: if you rotate a large region inside a liquid crystal, it will cost little energy and will slowly rotate back. When the translational symmetry is also broken, the rotational Goldstone mode disappears! If I rotate

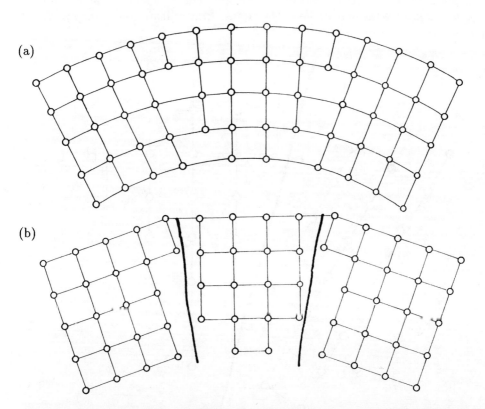

FIGURE 12 (a) Rotational distortion of a crystal. If we take a thick piece of metal and rotate one end with respect to another, it will start by bending uniformly. As it continues to bend, dislocations will form to ease the bending strain. These line dislocations will start off distributed irregularly through the sample. (b) Domain walls form to expel rotations. If we hold the rotation for a long time and let the dislocations move around, they will lower their energy by arranging themselves into domain walls. Between the domain walls we find undistorted crystal. This process is called polygonalization.

one piece of a crystal with respect to another, it costs an enormous energy (Figure 12(a)). If I let the distorted crystal rearrange locally to reach equilibrium, the rotational deformation is expelled into grain boundaries (Figure 12(b)), a process known in the field as polygonalization. Just like the massless photon developed a mass when the superconducting transition broke the gauge symmetry, the massless rotational mode develops a mass when the translational symmetry is broken.

This is surely also related to some of the old problems in the topological theory of defects. In describing a crystal, everybody uses the displacement field $u(x)$ and its derivatives. Now, as we saw in the last paper,[8] $u(x)$ describes the broken translational order, but not the broken orientational order. Why don't we also have a rotation matrix $R(x)$? For example, in Figure 11, $R(x)$ shifts abruptly from one side of the domain wall to the other. Mermin[6] discusses some of the weird behavior one gets following this path. The point is that $R(x)$ seems to be constrained: it doesn't change on its own but follows the broken translational order. Keeping it as an order parameter seems no more necessary than keeping $\rho = |\psi|$ around in a superconductor: only θ is massless, and ρ just wiggles around ρ_0 in a boring way.

Now, Ming and I have spent a huge amount of time trying to make these words into a mathematical theory. (We started with smectics, then studied superconductors, then thought about some ideas of Toner and Nelson,....) Ming has gone on to better things, and I'm still futzing with it. I can summarize where we are right now. Suppose we consider a rotationally distorted two-dimensional crystal (Figure 12(a)). We can define a rotational order parameter by looking at the angle of the nearest-neighbor bonds:

$$R(x) = \begin{pmatrix} \cos\theta & \sin\theta \\ -\sin\theta & \cos\theta \end{pmatrix}. \tag{14}$$

The translational order parameter $\vec{u}$ is just as it always was: if $\vec{x}$ is the original position and $\vec{p}(x)$ is the corresponding position in the ideal lattice,

$$\vec{u}(x) = \vec{p}(x) - \vec{x}. \tag{15}$$

Now, the free energy can only depend on gradients of $\vec{u}$, since it is translationally invariant. It also cannot change if we perform a uniform rotation: $R \to R_0 R$, $p \to R_0 p$. From this, we can see that the free energy must be written in terms of gradients of $R(x)$ and the particular combination[8]

$$\epsilon_{ij} = \delta_{ij} - \sum_{k=1}^{2} R_{ki}\left(\frac{\partial u_j}{\partial x_k} + \delta_{kj}\right). \tag{16}$$

A reasonable free energy for a crystal then becomes

$$\mathcal{E}_{\text{crystal}} = (\nabla\theta)^2 + 2\mu \sum_{ij}\left(\frac{\epsilon_{ij} + \epsilon_{ji}}{2}\right)^2 + \lambda \sum_i \epsilon_{ii}^2 + \kappa \left(\frac{\epsilon_{12} - \epsilon_{21}}{2}\right)^2. \tag{17}$$

[8] This is analogous to the minimal coupling term $\nabla\theta - A$ in the free energy for a superconductor.

This is just the normal elastic energy everybody uses, except for the third term multiplied by κ. Normally, the strain matrix ϵ is defined to be symmetric, so this term is then zero.

Our free energy doesn't keep ϵ automatically symmetric precisely because we have $R(x)$ as an independent degree of freedom. The antisymmetric part measures the amount that R disagrees with the local gradients of $\vec{u}$. It turns out that this antisymmetric part for the crystalline free energy is analogous to the current for the superconductor, which has a Meissner effect.[9]

There are several things I haven't been able to do, though. First, I don't think $\epsilon_{12} - \epsilon_{21}$ is expelled quite like its analogue in the superconductor. I think we can show, though, that it is a boring variable like ρ was. Second, I haven't a clue on how to show that grains exist. To show that grains exist I have to show a constraint like $\nabla \theta = 0$!

We started this paper by admiring the focal conic defects in smectic liquid crystals: beautiful ellipses and hyperbolas which are due not to topology but to geometrical consequences of a constraint. We saw how constraints can be enforced by the energy: "massive" modes decay exponentially. We saw explicitly how this occurs in superconductors—the magnetic field is constrained to zero because the photon and the Goldstone boson for the superconducting gauge symmetry combine into a massive particle. Finally, we discussed analogous effects in the everyday problem of grain boundaries in crystals, and realized that we don't really understand them in a deep sense.

APPENDIX: THE SMECTIC ORDER PARAMETER

Here we derive the consequences for layered systems of the constraint that the layers be equally spaced. Suppose that there are a stack of (bent) sheets, equally spaced from one to the next, with separation a. Suppose that the unit normal to these sheets at a position $\vec{x}$ is given by $\hat{n}$. Consider travelling around a loop C, crossing various layers as we go around (Figure 13). The number of layers we cross is given by the line integral

$$\left(\frac{1}{a}\right) \int_C \hat{n} \cdot d\ell = \text{ net \# crossed } . \tag{18}$$

If the layers exist throughout the region without any defects, then the net number crossed around any closed loop must be zero. Using Stokes' theorem, this integral over C is equal to an integral over the area A swept out by the curve:

$$\int_C \hat{n} \cdot d\ell = \int_A \text{curl } \hat{n} \cdot dA . \tag{19}$$

---

[9] It is the gradient of $\mathcal{E}_{\text{crystal}}$ with respect to θ, just as the current is the gradient of $\mathcal{E}_{\text{superconductor}}$ with respect to A. I thank Alan Luther for pointing this out.

Meissner Effects and Constraints

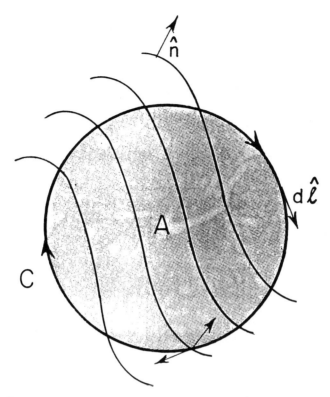

FIGURE 13 Equally spaced layers imply curl $n = 0$. Smectic layers, with a loop C enclosing an area A. The dot product $\hat{n} \cdot d\ell$ gives the cosine of the angle of the curve C with respect to the layers, and $a/\cos\theta$ is the length of curve C between two layers, so $1/a \int_C \hat{n} \cdot d\ell$ gives the net number of layers crossed by the curve C. (A layer crossed first forward and then backward cancels, of course). Since in a closed loop the net number of layers crossed must be zero (assuming no dislocations), this must be zero. By Stokes' theorem, $\int_C \hat{n} \cdot d\ell = \int_A \operatorname{curl} n \cdot dA$. This is true for any little area A, so $\operatorname{curl} n \equiv 0$.

But for this to be true for all areas A, curl $\hat{n}$ must be zero.

Now, we already know that $\hat{n}^2 = 1$. The derivative $\partial \hat{n}^2 / \partial x_a$, of course, must be zero, so using the product rule

$$\sum_\beta \hat{n}_\beta \frac{\partial \hat{n}_\beta}{\partial x_a} = 0. \tag{20}$$

Now, since we know curl $\hat{n} = 0$, we know from Eq. (9) that

$$\frac{\partial \hat{n}_\beta}{\partial x_a} = \frac{\partial \hat{n}_\alpha}{\partial x_\beta}. \tag{21}$$

Finally, combining these, we find

$$\sum_\beta \frac{\hat{n}_\beta \partial \hat{n}_\alpha}{\partial x_\beta} = (\hat{n} \cdot \nabla)\hat{n} = 0 \, . \tag{22}$$

This implies that $\hat{n}$ doesn't change when you move in the $\hat{n}$ direction. This means that $\hat{n}$ will be perpendicular to the next layer as well: that is, a straight line perpendicular to one layer will be perpendicular to every layer it crosses.

These perpendicular lines are called generators. We qualitatively knew already that one layer determined its surroundings; now we have a simple geometrical rule describing this nonlocal constraint. For your information, the defects occur where the generators cross (as shown in Figure 3): this surface is called the evolute, or surface of centers, for the layer.

ACKNOWLEDGMENTS

I'd like to acknowledge NSF grant # DMR-9118065, and thank NORDITA and the Technical University of Denmark for their hospitality while these papers were written up.

REFERENCES

1. Ashcroft, N. W., and N. D. Mermin. *Solid State Physics*, chapter 34. New York: Holt, Rinehart, and Winston, 1976.
2. Coleman, Sydney, ed. "Secret Symmetry: An Introduction to Spontaneous Symmetry Breakdown and Gauge Fields." In *Aspects of Symmetry, Selected Erice Lectures*, 113. Cambridge: Cambridge University Press, 1985.
3. de Gennes, P. G. *The Physics of Liquid Crystals*, especially Figure 7.1 and 7.2. London: Oxford University Press, 1974.
4. Hilbert, D., and S. Cohn-Vossen. *Geometry and the Imagination*. New York: Chelsea, 1952.
5. Huang, Ming. "Meissner Effects, Vortex Core States, and the Vortex Glass Phase Transition." Ph.D. Thesis, Cornell University, 1991.
6. Mermin, David. "The Topological Theory of Defects in Ordered Media." *Rev. Mod. Phys.* **51** (1979): 591.
7. Sethna, James, P., and M. Kléman. "Spheric Domains in Smectic Liquid Crystals." *Phys. Rev. B* **26** (1982): 3037.
8. Sethna, James P. "Order Parameters, Broken Symmetry, and Topology." This volume.

Michael F. Shlesinger
Physics Division, Office of Naval Research, 800 N. Quincy St., Arlington, VA 22217-5000

Fractal Time Dynamics:
From Glasses to Turbulence

One major theme in physics has been to find the right scale for a problem. How high can the highest mountains be on Earth and Mars? What is the size of a hydrogen atom? What is the density of the atmosphere at 35,000 feet? These types of problems represent great successes in physics, in part because the predictions can be tested, and agreement leads to confidence in understanding basic physics. These types of problems are prevalent in the teaching of physics.

Another, newer theme is to investigate problems where many scales enter, but no scale dominates, i.e., there is no characteristic size. The struggles and successes of meeting the challenge of scale invariance in the field of phase transitions are well known. The scaling enters through the divergence of a correlation length which occurs at a finite temperature. Clustering of correlations (spin clusters, lattice gas particle clusters, etc.) occurs, and one cluster percolates through the material when the correlation length diverges. The modern methods for approaching these problems is through the renormalization group which also allows the (usually non-integer) critical exponents to be calculated from fixed points of transformation equations which relate different scales to each other. Mandelbrot,[6] throughout his long career, has shown that a fractal geometry can underlie the appearance of non-integer exponents from describing the coastline of England to the 2.5 dimensions of a 3-D percolation cluster. The idea of a fractal dimension (or physics in non-integer dimensions) has become commonplace in many fields of science including physical, engineering, biological, social, and economic.

In these lectures, the struggles and successes of applying scaling ideas to transport and relaxation in amorphous materials will be presented. The first section will rely heavily on the notion of fractal time, our first topic for discussion. Fractal-time random walks will prove useful for understanding scaling properties of glassy materials. Next, we will turn our attention to random walks with fractal trajectories, called Levy flights. When Kolmogorov space-time scaling is applied to these Levy flights a novel view of turbulent diffusion arises.

1. FRACTAL TIME[10,15]

Let $\psi(t)$ be the probability density that the duration between events is t. This might represent the time it takes for a particle to jump out of a trap. A simple choice for $\psi(t)$ is the exponential, $\psi(t) = \exp(-\lambda t)$. This density has a well-defined mean time $\langle t \rangle = 1/\lambda$. Let us investigate $\psi(s)$, the Fourier transform of $\psi(t)$. We will be interested in asymptotically long times (or equivalently small s) behavior.

$$\psi(s) = \int_0^\infty \exp(-st)\psi(t)dt = 1 - s\langle t \rangle + O(s^2). \tag{1}$$

The appearance of s to the first power is, in some sense, in agreement with the notion that time is one dimensional, and that it flows smoothly forward. The general result of Eq. (1) is certainly true for the case for $\psi(t) = \lambda \exp(-\lambda t)$, where

$$\psi(s) = \frac{\lambda}{\lambda + s} = 1 - \frac{s}{\lambda} + \left(\frac{s}{\lambda}\right)^2 + O(s^3).$$

What would happen if $\langle t \rangle$ were infinite? If $\psi(t) \sim t^{-1-\beta}$, $(0 < \beta < 1)$ as $t \to \infty$, then $\psi(t)$ is normalizable, but its first moment diverges. What would $\psi(t)$ look like in this case? Certainly, the expansion in integer powers of s will not apply when $\langle t \rangle$ is infinite. Let us consider a $\psi(t)$ which is a sum of exponentials where scales of all orders of magnitude enter,

$$\psi(t) = \frac{1-N}{N} \sum_{j=1}^\infty N^j \lambda^j \exp(-\lambda^j t), \quad (\lambda < N < 1). \tag{2}$$

The mean value of $\psi(t)$ is given by

$$\langle t \rangle = \left[\frac{1-N}{N}\right] \sum_{j=1}^\infty \left(\frac{N}{\lambda}\right)^j = \infty. \tag{3}$$

To make further progress in analyzing $\psi(t)$, let us look at its Fourier transform $\psi(s)$,

$$\psi(s) = \int_0^\infty \exp(-st)\psi(t)dt = \frac{1-N}{N} \sum_{j=1}^\infty \frac{(N\lambda)^j}{[s+\lambda^j]}. \tag{4}$$

Note that $\psi(s)$ satisfies a scaling relation,

$$\psi(s) = N\psi\left(\frac{s}{\lambda}\right) + \frac{[1-N]\lambda}{(s+\lambda)} \tag{5}$$

whose homogeneous part $\psi(s) = N\psi(s/\lambda)$ has a solution of the form,

$$\psi(s) = s^\beta, \text{ with } \beta = \frac{\ln N}{\ln \lambda}. \tag{6}$$

This hints that an expansion of $\psi(s)$ for small s will involve non-integer powers of s. An exact analysis can be performed by substituting, for $1/[s+\lambda^j]$ in Eq. (4), its own inverse Mellin transform to find[11]

$$2\pi i \psi(s) = \frac{1-N}{N} \int_{c-i\infty}^{c+i\infty} \sum_{j=1}^\infty (N\lambda)^j \frac{\pi}{\sin(\pi\epsilon)} s^{-\epsilon} \lambda^{j(\epsilon-1)} d\epsilon \tag{7}$$

($0 < c = \text{Re } \epsilon < 1$). Interchanging the sum and integral yields

$$2\pi i \psi(s) = \frac{1-N}{N} \int_{c-i\infty}^{c+i\infty} \frac{s^{-\epsilon} \pi N \lambda^\epsilon}{[\sin(\pi\epsilon)\{1-N\lambda^\epsilon\}]} d\epsilon. \tag{8}$$

The integrand has simple poles from the $\sin(\pi\epsilon)$ term at $\epsilon = 0, \pm 1, \pm 2, \ldots$, and from the factor in the denominator when $\epsilon = -\ln N/\ln \lambda \pm 2\pi i j/\ln \lambda$ ($j = 0, 1, 2, \ldots$). Translating the contour of Re $\epsilon = -\infty$ and taking account of the poles crossed, we find for $\epsilon < 1$ that

$$\psi(s) = 1 + s^\beta K(s) + \frac{1-N}{N} \sum_{j=1}^\infty \frac{(-1)^j s^j N \lambda^j}{[\lambda^j - n]},$$

with β given by Eq. (6), and where $K(s)$ is periodic in $\ln s$ with period $\ln \lambda$, as given by

$$K(s) = -\frac{1-N}{N \ln \lambda} \sum_{j=-\infty}^\infty \frac{\pi N \lambda^x \exp(-2xij \ln s \ln \lambda)}{\sin(\pi x)} \tag{9}$$

where $x = -\ln N/\ln \lambda + 2\pi i j/\ln \lambda$.

While the above has been somewhat technical, it should assure the reader that expansions in non-integer exponents are legitimate and arise in probability theory for transforms of temporal probability distributions which possess infinite first moments. We will encounter similar examples for random walk jump distributions whose second moments are infinite.

The s^β term, somehow, implies that time is a β-dimensional quantity with $\beta < 1$. When $\langle t \rangle$ is finite, this implies $\beta = 1$. In this sense unity is the upper critical dimension of time. A simulation of the above temporal process would find waiting time durations of $\lambda^{-1}, \lambda^{-2}, \ldots, \lambda^{-j}$, etc., with waits an order of magnitude longer (in base λ) occurring an order of magnitude less often (in base N). The similarity definition of a fractal dimension is the log (# subclusters/cluster)/log (scale factor). For our fractal-time waiting process, the jumps occur in self-similar clusters with about N jumps, each separated in time by $1/\lambda$ occurring before a wait of $1/\lambda^2$ occurs. Then another cluster of about N closely spaced (in time) jumps occurs before another wait of $1/\lambda^2$ occurs. After about N of these clusters of jumps occurs, a wait of $1/\lambda^3$ arises, and so on, and so on in a hierarchical fashion. We find self-similar clusters of jumps with about N subclusters/cluster and differing in time duration between jumps by about a factor of λ^{-1}. This is in accord with treating β as the fractal dimension of the process. The jumps do not occur at a nice well-defined rate, but in a very patchy manner. If one made marks on a time axis when jumps occur, then the set of points would look like the points in a random Cantor set with fractal point set dimension β. The number of points $M(T)$ on a line of length T, versus $M(T/\lambda)$, the number of points on a line of length T/λ, gives the relation $M(T) = NM(T/\lambda)$. This equation has a solution in the form $M(T) = T^\beta$ with $\beta = \ln N / \ln \lambda$.

2. STRETCHED EXPONENTIAL RELAXATION (A FRACTAL TIME-INITIATED PROCESS)[12]

Let us consider a model of relaxation for a glassy material. Suppose a glassy material supports the motion of defects. The defects are viewed as encapsulating free volume which when transported to a frozen-in region of the glass can cause a relaxation to occur. The frozen-in part of the material may be a dipole (so dielectric relaxation occurs) or a polymer chain segment (so mechanical relaxation occurs), etc. Let us assume that there are N mobile defects in the material which can move between V sites. Assume there is a frozen-in region at the origin of our coordinate system, and that the N defects are randomly placed in the material. Let $\phi(t)$ be the probability that no defect has reached the origin by time t. We formally write $\phi(t)$ as

$$\phi(t) = \left[1 - \frac{1}{V} \sum_r \int_0^t F(r, \tau) d\tau \right]^N, \tag{10}$$

where $F(r, \tau)$ is the probability density for a random walker starting at site r, to reach the origin at time τ. F is called a first passage time probability. The integral allows for a first passage of a walker to the origin in the time interval $(0, t)$. This encounters will cause the relaxation to occur. Note we sum over all initial starting points for each walker, and the probability of starting from any site is $1/V$. Thus, the term within the brackets is the probability that a particular random walker did not reach the origin by time t. We raise the bracket to the Nth power to account for none of the N random walkers reaching the origin by time t. Let us take a thermodynamic limit of $N \to \infty$, $V \to \infty$, but $c = N/V$ remaining constant. In this limit $\phi(t)$ becomes

$$\phi(t) = \exp\left[-c \sum_r \int_0^t F(r,\tau)d\tau\right]. \tag{11}$$

The argument in the brackets is minus the flux of random walkers into the origin at time t. This is the type of expression Smoluchkowski would have written to describe this reaction scheme. We can rewrite Eq. (11) by noting that any of the sites from which a walker can reach the origin within a time t are, by symmetry, the same set of sites a walker starting at the origin can reach by time t. We can now write Eq. (11) as

$$\phi(t) = \exp[-S(t)] \tag{12}$$

where $S(t)$ is the distinct number of sites a random walker starting at the origin visits within a time t. The walker may make 100 jumps, but only visit 28 different sites. $S(t)$ would then be 28, and walkers starting from these 28 sites could reach the origin within time t. Let us now take an interlude into the theory of random walks to learn how to calculate $S(t)$.

MONTROLL-WEISS CONTINUOUS-TIME RANDOM WALKS[8]

Let us first consider the statistics of n-step random walks. We will not yet focus on the statistics of random time interval occurring between jumps. Let's begin with an equation for $P_{n+1}(r)$ the probability that a random walker beginning at the origin reaches site r on its $(n+1)$st step. We can write this probability in terms of the probability $p(r)$ that a single step has a displacement r, via

$$P_{n+1}(r) = \sum_{r'} P_n(r - r')p(r'). \tag{13}$$

It is useful to introduce a generating function $G(r;z)$ defined as

$$G(r;z) = \sum_{n=0}^{\infty} P_n(r)z^n. \tag{14}$$

Multiplying both sides of Eq. (14) by $z^{(n+1)}$, and summing over all n, let us rewrite Eq. (13) as a Green's function equation

$$G(r;z) - z\sum_{r'} G(r - r'; z)p(r') = \delta_{r,0}. \tag{15}$$

We can also write $P_n(r)$ as

$$P_n(r) = \sum_{m=0}^{n} F_{n-m}(r) P_m(0) + \delta_{n,0}\delta_{r,0}, \qquad (16)$$

where $F_n(r)$ is the probability that on the nth step the walker reaches site r for the first time. This calculation allows for the walker reaching site r for the first time after $n-m$ steps and then returning in m steps (i.e., zero displacement in the last m steps). Let us now define the generating function

$$F(r;z) = \sum_{n=0}^{\infty} F_n(r) z^n. \qquad (17)$$

Multiplying Eq. (16) by z^n and summing over n (and taking advantage of the convolution form of the equation) leads to the following generating function equation,

$$F(r;z) = \frac{[G(r;z) - \delta_{r,0}]}{G(r=0;z)}. \qquad (18)$$

The number of distinct sites, S_n, visited in an n-step random walk is closely related to the first passage time probabilities by

$$S_n = 1 + \sum_r [F_1(r) + \ldots + F_n(r)]. \qquad (19)$$

Forming a generating function for S_n, we find

$$\begin{aligned}
S(z) &= \sum_{n=0}^{\infty} S_n z^n \\
&= \frac{1}{1-z} + z \sum F_1(r) + z^2 \sum [F_1(r) + F_2(r)] \\
&\quad + \ldots + z^n \sum [F_1(r) + \ldots + F_n(r)] + \ldots \\
&= \frac{1}{1-z} \sum \delta_{r,0} + \frac{1}{1-z} \sum [z F_1(r) + \ldots + z^n F_n(r) + \ldots] \\
&= \frac{1}{1-z} \sum_r F(r;z) \\
&= \frac{z}{(1-z)^2} \frac{1}{G(r=0;z)}.
\end{aligned} \qquad (20)$$

The next stage of complication is to introduce the waiting time density $\psi(t)$ governing the duration between steps. Let $\psi_n(t)$ be the probability density that the nth jump occurs at time t. This can be written in terms of $\psi(t)$ as

$$\psi_n(t) = \int_0^t \psi_{n-1}(t-\tau)\psi(\tau)d\tau. \qquad (21)$$

Fractal Time Dynamics

Taking Laplace transforms, we find $\psi_n(s) = [\psi(s)]^n$. The probability density to reach site r for the first time at time t is given by

$$F(r,t) = \sum_{n=0}^{\infty} F_n(r)\psi_n(t).$$

Its Laplace space representation has the form of a generating function

$$F(r,s) = \int_0^{\infty} \exp(-st)F(r,t)dt = \sum_{n=0}^{\infty} F_n(r)[\psi(s)]^n. \tag{22}$$

We can now see that by replacing z by $\psi(s)$ in Eq. (20) that we have an expression for the generating function for visiting n distinct sites when the nth jump occurs at time t. We need just one more adjustment to take account of the nth jump occurring at time $t - \tau$ and no jump occurring in the remaining time, i.e., n jumps occur but the nth jump occurs before time t. We write $S(t)$ as

$$S(t) = \sum_{n=0}^{\infty} S_n \int_0^t \psi_n(t-\tau)W(\tau)d\tau \tag{23}$$

where $W(\tau) = \int_{\tau}^{\infty} \psi(u)du$ is the probability that the time between jumps is longer than τ. The Laplace transform of $W(t)$ is $[1 - \psi(s)]/s$. Using Eq. (20), we write $S(t)$, in Laplace space, as

$$S(s) = \frac{S(z = \psi(s))(1 - \psi(s))}{s}$$
$$= \frac{\psi(s)}{[s(1 - \psi(s))]G(0; \psi(s))}. \tag{24}$$

In three dimensions, at long times (small s) $G(0; z) \sim \text{const.} + O(1 - z)$.

For a continuous-time random process where $\langle t \rangle$ is finite, then $\psi(s) \sim 1 - s\langle t \rangle$, and $S(s) \sim 1/s^2$, so $S(t) \sim t$. The fractal time case of $\langle t \rangle$ infinite is more interesting. There we have $\psi(s) \sim 1 - s^{\beta}$, with $\beta < 1$, so $S(t) \sim t^{\beta}$.

BACK TO THE RELAXATION LAW

Let us now substitute our calculation for $S(t)$ back into the relaxation law equation (12).

$$\phi(t) \sim \begin{cases} \exp(-ct) & \langle t \rangle \text{ finite}; \\ \exp(-ct^{\beta}) & \langle t \rangle \text{ infinite}. \end{cases} \tag{25}$$

We arrive at simple exponential decay if a time scale exists, and stretched exponential decay if it doesn't. Both cases are examples of probability limit distributions. This can explain the ubiquity of the stretched exponential law for glassy materials.

3. DIVERGENCE OF THE TIME SCALE[1]

If we rewrite $\phi(t) = \exp(-ct^\beta)$ as

$$\phi(t) = \exp\left[-\left(\frac{t}{\tau}\right)^\beta\right],$$

then

$$\tau = \left(\frac{1}{c}\right)^{\frac{1}{\beta}}.$$

It is known that for many glassy materials an empirical fit to data yields

$$\tau = \tau_0 \exp\left[\frac{\text{const.}\, T}{T - T_0}\right] \tag{27}$$

with τ_0 and T_0 being constants. This is known as the Vogel-Fulcher-Tammann law. In our model, our defects can cluster as the temperature is lowered to lower the entropy of the system. This clustering would be due to an attractive interaction between the defects. We assume a lattice gas model of defects versus non-defect sites. In a mean field model, the correlation length diverges with temperature as $\xi = 1/(T-T_0)^{1/2}$. Let us assume that singlet defects are more mobile than doublet, triplet,... clusters, so we will focus on the temperature-dependent population of the singlets which we denote by c_1. The probability a defect is at a site and that it is not correlated with other defects is given by

$$c_1 = c(1-c)^{V_\xi}, \tag{28}$$

where ξ is the correlation length, and $V_\xi = \xi^3$ is the correlation volume. For our lattice gas model of the defects $c_1 \sim \exp(-cV_\xi) \sim \exp(-c/[T-T_0]^{3/2})$, and thus using Eq. (26) we find

$$\tau \sim \exp\left([T-T_0]^{-\frac{3}{2}}\right). \tag{29}$$

While this differs from the Vogel by having an exponent of 3/2 versus 1, it has proven in several comparison to provide the better fit.[1,2]

If the defects do not cluster, then one can still obtain the stretched exponential law, but none need not obtain a Vogel-type law. This is the case for SiO_2. ϕ is a stretched exponential, but τ is Arrhenius. This is consistent with our model as the law for $\phi(t)$ focuses on how a defect moves, while the law for τ focuses on the temperature dependence of the mobile defect population. High above the glass transition temperature, Tg, many mobile defects exist. Their movement breaks up rigidity in the material. As the temperature is lowered, the clustering of defects reduces the mobile defect population. At Tg presumably the decline in mobile defects allows rigidity to percolate through the sample. This is the glass transition. Below Tg, mobile defects still exist and relaxation still occurs. The time-scale dynamics for τ is focused on T_0, the temperature when mobile defects disappear, and not on Tg. This is why $T < Tg$.

4. FRACTAL RANDOM WALKS IN TURBULENCE[14]
RICHARDSON'S LAW

In 1926, Lewis Fry Richardson[9] discussed his discovery that the usual Brownian diffusion law for a mean square displacement, $\langle R^2(t) \rangle = 6Dt$, does not hold for diffusion in a turbulent fluid. In his studies of the dispersal of smoke from smokestacks on windy days, and of the separation of floating objects in turbulent waters, Richardson announced that

$$\langle R^2(t) \rangle = Dt^3 \text{ (for turbulent diffusion).} \tag{30}$$

In trying to understand this result, Richardson drew pictures somewhat reminiscent of fractal patterns trying to show how a drop of dye would be pulled apart, over many scales, in a turbulent flow. He felt that eddies of all sizes would cause the relative separation of two particles in a turbulent flow to have a diffusion constant which depended on their positions. Taking the diffusion law $\langle R^2(t) \rangle = Dt$ with $D = D(R) = R^{4/3}$, Richardson could recover the correct scaling of Eq. (27), but with so many scales entering a turbulent flow that Richardson wondered how to mathematically describe turbulent trajectories. He wrote, "The failure of the dispersal of a point-charge to serve as a mathematical element, from which the dispersal of an extended system may be built up, appears to be intimately connected with the fact that in the atmosphere dispersal goes on in patches." Differential equations are local, so they cannot possibly properly treat global spatial-temporal motions set up by a hierarchy of vortices. This led Richardson to even question the existence of differential equation for the description of turbulent flows. He asked, "Does the Wind Possess a Velocity? This question at first sight foolish improves upon acquaintance." Richardson then gave the Weierstrass function as an example of an everywhere non-differentiable function and thought it might somehow be connected with turbulent flows.

We will take a circuitous route to deriving Richardson's law. We will begin by discussing the Central Limit Theorem for adding random variables whose second moments exist. We will then study Levy's generalization for summing random variables with infinite moments. By introducing trajectories associated with infinite moment random variable sums, we will be able to describe a scale-invariant random-walk process. Finally, incorporating Kolmogorov space-time scaling into these trajectories, we will arrive at Richardson's law. Incidentally, along the way we will use the Weierstrass function, mentioned by Richardson, as a generator for our scale-invariant fractal space-time random walks.

LEVY FLIGHTS[5,8]

Add up several identically distributed random variables X_i, each with zero mean

$$c_n^\beta Y_n = \sigma_1^\beta X_1 + \ldots + \sigma_n^\beta X_n, \tag{31}$$

with the condition that $c_n^\beta = \sum_i \sigma_i^\beta$. The value of each variable X_i can be thought of as a step in a random walk. Each jump length is chosen from a distribution $p(x)$. Levy asked the question of when can the distribution of the sum of n steps $p_n(x)$ (up to some scale factors) be the same as the distribution of any term in the sum, $p(x)$. This is basically the question of fractals: when does the whole (the sum) look like one of its parts? One answer to this question is well known. A sum of Gaussians is a Gaussian. Setting each $\sigma_i = 1$ for $\beta = 2$, we obtain $c_n^2 = n$. This means for adding Gaussians that the variance of the sum is the sum of the variances. The distribution of the X_i's is $p(x) = (2\pi)^{-1/2} \exp(-x^2)$, the distribution of $Y_n = (n)^{-1/2} \sum_i X_i$ is $p_n(x) = (2\pi n)^{-1/2} \exp(-x^2/n)$. So $p(x)$ and $p_n(x)$ have the same distribution up to the scale factor n. In Fourier space $(x \to k) p_n(k)$ has a simple form,

$$p_n(k) = \int_{-\infty}^{\infty} p_n(x) \exp(ikx) dx = \exp(-nk^2). \tag{32}$$

Note that the second moment of $p_n(x)$ is given by $-\partial^2 p_n(k=0)/\partial k^2 = n$. Levy discovered that other solutions existed for Eq. (27) such that $p_n(x)$ and $p(x)$ had the same distribution. He found this to be the case when

$$p_n(k) = \exp(-\text{const.}|k|^\beta) \qquad (\text{for } 0 < \beta \leq 2). \tag{33}$$

The $\beta = 2$ case is the Gaussian which we have just studied. For $\beta < 2$, we note that $\langle x^2 \rangle = -\partial p_n(k=0)/\partial k^2$ is infinite. These random walks with steps with infinite second moments are known as Levy flights. It now seems obvious that to have scale-invariant distributions, we would need to sum up random variables with no scale. As we saw above, if we have finite second moments, then we will get the Gaussian distribution.

The exponent β will turn out to be the dimension of the point set visited by a Levy flight. For the Gaussian case where $\beta = 2$, consider a random walk of N^2 steps. The probability distribution $p_n(x)$ is a Gaussian of standard deviation of N. Each of the N^2 jumps has a Gaussian distribution with a standard deviation, $\sigma^{1/2} = $ unity, so one can consider the distribution after N^2 steps to be comprised of N^2 Gaussians, each scaled down in standard deviation from N to unity. Thus a fractal dimension of $\ln N^2 / \ln N = 2$ can be ascribed to standard random walks whose steps have finite mean square displacements. This is also in accord with the knowledge that a random walker visits every point in two dimensions, but not so in higher dimensions.

To get a deeper understanding of Eq. (33), let us write $p_n(x)$ as

$$p_n(x) = \int p_{n-1}(x - x') p(x') dx' \tag{34}$$

which transforms in Fourier space to (due to its convolution form)

$$p_n(k) = [p(k)]^n. \tag{35}$$

For small k (large x),

$$p(k) = \int p(x)\exp(-ikx)dx = 1 - \left(\frac{1}{2}\right)\langle x^2\rangle k^2 + O(k^3) \tag{36}$$

when $\langle x^2\rangle$ is finite. For this case $p(k) \sim \exp(-\langle x^2\rangle k^2/2)$, $p_n(k) \sim \exp(-n\langle x^2\rangle k^2/2)$, so $p_n(x)$ asymptotically behaves like a Gaussian with variance n.

How can we expand $p(k)$ when $\langle x^2\rangle$ is infinite? Our analysis now parallels the discussion of fractal time, except here we address the properties of fractal space. For a specific example let us construct a random walk without a characteristic jumps size. Let

$$p(x) = \frac{N-1}{2N}\sum_{j=0}^{\infty} N^{-j}[\delta_{x,b^j} + \delta_{x,b^{-j}}]. \tag{37}$$

Jumps of size ± 1, $\pm b$, $\pm b^2$, etc. can occur, but jumps an order of magnitude longer in base b occur an order of magnitude less often in base N. We make about N jumps of length unity before, on the average, a jump of length b occurs, and so on, until in a hierarchical fashion patchy clusters of all sizes are formed. We expect a fractal dimension of $\ln N/\ln b$ to appear. We need to analyze $p(k)$ and, taking the Fourier transform of $p(x)$, we arrive at

$$p(k) = \frac{N-1}{N}\sum_{j=0}^{\infty} N^{-j}\cos(b^j k) \tag{38}$$

which is precisely the Weierstrass function called for by Richardson. We could again go through the Mellin transform analysis introduced in our discussion of fractal time, but here we will just note that $p(k)$ satisfies the scaling relation

$$p(k) = N^{-1}p(bk) + \left[\frac{N-1}{N}\right]\cos(k) \tag{35}$$

which has a solution which includes a term of the form

$$p(k) \sim 1 - |k|^\beta \sim \exp(-|k|^\beta) \quad \left(\text{with } \beta = \frac{\ln N}{\ln b}\right). \tag{40}$$

This exponential form with the fractional power comprises the non-Gaussian solutions to Levy's question addressed in Eq. (33). If one wants to add random variables (take a random walk) and have the probability distribution after n steps look like the probability distribution after one step (except for a change of scale), then your random variables are either Gaussian or have infinite second moments. This means the solution is either Gaussian or fractal.

LEVY WALKS[14]

How does one use Levy flights in physics, since mean square distances are infinite. One approach to make the problem more physical is to take account of how long it would take to traverse a distance r. Let $\Psi(r,t)$ be the probability density to make a jump of displacement r, in a time t. We write

$$\Psi(r,t) = \psi(t|r)p(r), \tag{41}$$

where $p(r)$, as before, is the probability that a jump of length r occurs. $\psi(r|t)$ is the conditional probability that, given a jump of length r occurred, it took a time t to complete. For Levy flights, we choose $p(r) \sim |r|^{1+\beta}$ with $\beta < 2$ so $\langle r^2 \rangle$ diverges. If we choose $\psi(t|r) = \psi(t)$ so jump distances and jump times are chosen independently, then a divergent $\langle r^2 \rangle$ will still result. A coupled space-time memory can, however, get around this problem. Let us chose

$$\psi(t|r) = \delta\left(t - \frac{r}{v(r)}\right) \tag{42}$$

where $v(r)$ is the velocity of a jump of length r. The form $\psi(t|r) = \exp(-[t - r/v(r)]^2)$ would do just as well. Fortunately, Kolmogorov has taught us for isotropic homogeneous turbulent flows how to calculate $v(r)$. He assumed that the average dissipation ϵ_r over a scale r would be independent of r. Now ϵ is the energy/time $\sim v(r)^2/t = v(r)^3/r$. For the dissipation to be constant, we need

$$v(r) = r^{1/3} \text{(Kolmogorov scaling)}. \tag{43}$$

The energy E is proportional to $v^2 \sim r^{2/3}$, and its Fourier transform $E_k \sim k^{-5/3}$. This $-5/3$ law is the best known version of Kolmogorov scaling. With this information we can now proceed to calculate the mean square displacement for turbulent diffusion.

TURBULENT DIFFUSION: A SPACE-TIME FRACTAL[14]

We need to generalize our random walk to include the coupled memory $\Psi(r,t)$. Note fractal time involved being stuck at one place for a hierarchical distribution of time. For Levy walks the walker gets stuck in the same momentum state for a hierarchical distribution of times. The probability density $Q(r,t)$ for reaching a site r exactly at time t is given by

$$Q(r,t) = \sum_{r'} \int_0^t Q(r-r', t-\tau)\Psi(r',\tau)d\tau + \delta_{r,0}\delta(t) \tag{44}$$

where we account for reaching $r - r'$ at time $t - \tau$, and then taking a jump of displacement r' which takes a time τ to complete. This is the coupled memory continuous-time version of Eq. (13) for $P_n(r)$. In Fourier-Laplace space we find, for the Green's function propagator,

$$Q(k,s) = [1 - \Psi(k,s)]^{-1}. \qquad (45)$$

The probability $P(r,t)$ to be at site r at time t is a little bit more complicated because one can take a jump which passes r at time t, but the jump continues and ends at a different site at a later time. For simplicity, one can get the right scaling by just focusing on walks which reach site r at time $t - \tau$, and the next jump is yet completed by time t. Then

$$P(r,t) = \int_0^t Q(r, t-\tau) \left\{ 1 - \int_0^\tau \psi(z) dz \right\} d\tau \qquad (46)$$

where $\psi(t) = \sum \Psi(r,t)$. In Fourier-Laplace space

$$P(k,s) = \frac{1 - \psi(s)}{s[1 - \Psi(k,s)]}. \qquad (47)$$

For a walk with $\langle r \rangle = 0$, $p(r) \sim |r|^{1+\beta}$, and $\psi(t|r) = \delta(t - r/v(r))$ with $v(r) = r^{1/3}$, we find, for $\langle r^2(t) \rangle = -\mathcal{L}^{-1} \partial^2 P(k=0,s)/\partial k^2$ where $\mathcal{L}^{-1}$ is the inverse Laplace transform, that

$$\langle r^2(t) \rangle = \begin{cases} t^3 & \text{for } \beta \leq 1/3; \\ t^{2+(3/2)(1-\beta)} & \text{for } 1/3 \leq \beta \leq 5/3; \\ t & \text{for } \beta \geq 5/3. \end{cases} \qquad (48)$$

For the case $\beta \leq 1/3$, we recover Richardson's law of turbulent diffusion. For large enough β the mean square time spent in a flight segment becomes finite and Brownian motion is achieved in accord with the Central Limit Theorem. If the memory was decoupled, then the calculation of $\langle r^2 \rangle$ would involve $\partial^2 p(k=0)/\partial k^2$ which is infinite. For the coupled space-time memory, one instead calculates with $\int \exp(ikr)\psi(s|r)p(r)dr$ instead of $p(k)$. This will turn the infinity of the Levy flight into the temporal scaling of the Levy walk.

One advantage of this approach is that one can visualize the types of random-walk trajectories which can lead to turbulent motions. F. Hayot[4] has actually implemented the Levy Walk model to simulate turbulent pipe flow. Instead of the parabolic velocity profile of laminar flow, a better mixed flow with a flatter velocity profile is discovered. Comparing his calculated velocity profile with experimental velocity profiles, he is able to associate a Reynolds number of about 10^5 with the flow. Basically, the Levy Walk zeroth-order state is already turbulent for small enough β, while the traditional lattice gas hydrodynamics is based on nearest-neighbor collisions which corresponds to low Reynolds number flow. Enormous computing power, or tricks, would be needed to reach a turbulent state in the standard lattice gas hydrodynamic approach. Turbulence is natural for the Levy Walk approach. Phase diffusion in Josephson junctions[3,13] and transport in stochastic webs[17] are other examples where Levy walks occur.

5. BERNOULLI SCALING[16]

It is of interest that the history of probability theory has already provided us with a beautiful example of the types of scaling discussed in this paper. The problem involves a certain game of chance. The game is to flip a coin until a head appears. This could take only one flip with probability of $1/2$, or n flips (i.e., one gets $n-1$ tails in a row before a head appears) with probability $1/2^n$. Suppose you win 2^n coins if $n-1$ tails appear before the head appears. Then your expected winnings are $1 \times 1/2 + 2 \times 1/4 + \ldots + 2^n/2^{n+1} + \ldots = \infty$. This game was introduced by Nicolaas Bernoulli (the nephew of Jacob and John) in the early 1700s. It is called the St. Petersburg Paradox because Daniel Bernoulli wrote about it in the Commentary of the St. Petersburg Academy. The question which was posed was how much ante should be required to place the game. The player favors a small ante because he will win only 1 coin with probability $1/2$, 2 or less coins with probability $3/4$, 4 or less coins with probability $7/8$, etc. The banker, who must take on all comers, favors an infinite ante because this is his expected loss. The two parties cannot come to an agreement because they are trying to determine a characteristic scale from a distribution which does not possess one. All scales enter and the probabilities for all possible winnings add up to unity. However, an order of magnitude greater winnings occurs with an order of magnitude less probability. This example is the forerunner of fractal time where the waiting times between jumps occur on all scale, but with order of magnitude longer waits occurring an order of magnitude less often. The fact that $\langle t \rangle$ was infinite for a fractal time process did not mean that the duration between every event was infinite, just as in this coin game not every player wins an infinite amount of money just because the expected winning is infinite. The perception of this paradox in the 1700s was to cast aspersions on the ability of probability to have a sound mathematical foundation. In the 20th century we see this paradox as a rich example of scaling with all its inherent exponents, fractal dimensions, renormalizations, and natural description of complex systems.

ACKNOWLEDGMENTS

This article reviews joint work done with: the late Elliott Montroll and Barry Hughes on fractal random walks; John Bendler on the glass transition and the Vogel law; and Yossi Klafter and Bruce West on turbulent flows. It is a pleasure to thank all of the above for their contributions to exploring the world of random walks which lie outside the horizon of Einstein's Brownian motion.

REFERENCES

1. Bendler, J. T., and M. F. Shlesinger. "On a Generalized Vogel Law for the Glass Transition." *J. Stat. Phys.* **53** (1988): 531–541.
2. Fontanella, J. J., M. C. Wintersgill, C. S. Coughlin, P. Mazaud, and S. G. Greenbaum. "Application of the Bendler-Shlesinger Generalization of the Vogel Equation to Ion-Conducting Polymers." *J. Polymer Sci.* **29** (1991): 747–752.
3. Geisel, T., J. Nierwetberg, and A. Zacherl. "Accelerated Diffusion in Josephson Junctions and Related Chaotic Systems." *Phys. Rev. Lett.* **54** (1985): 616–620.
4. Hayot, F. "Levy Walk in Lattice Gas Hydrodynamics." *Phys. Rev. A* **43** (1991): 806–810.
5. Levy, P. *Theorie de l'Addition des Variables Aleatoires.* Paris: Gauthier-Villars, 1937.
6. Mandelbrot, B. B. *The Fractal Geometry of Nature.* San Francisco: W. H. Freeman, 1982.
7. Montroll, E. W., and G. H. Weiss. "Random Walks on a Lattice. II." *J. Math. Phys.* **6** (1965): 167–179
8. Montroll, E. W., and M. F. Shlesinger. "On the Wonderful World of Random Walks." *Studies in Statistical Mechanics* **11** (1984): 1–121.
9. Richardson, L. F. "Atmospheric Diffusion Shown on a Distance-Neighbour Graph." *Proc. Roy. Soc. London* **A110** (1926): 709–737.
10. Scher, H., M. F. Shlesinger, and J. T. Bendler. "Time-Scale Invariance in Transport and Relaxation." *Physics Today* **Jan.** (1991): 26–34.
11. Shlesinger, M. F., and B. D. Hughes. "Analogs of Renormalization Group Transformations in Random Processes." *Physica A* **109** (1981): 597–608.
12. Shlesinger, M. F., and E. W. Montroll. "On the Williams-Watt Function of Dielectric Relaxation." *Proc. Nat. Acad. Sci. (USA)* **81** (1984): 1280–1283.
13. Shlesinger, M. F., and J. Klafter. "Comment on Accelerated Diffusion in Josephson Junctions and Related Chaotic Systems." *Phys. Rev. Lett.* **54** (1985): 2551.
14. Shlesinger, M. F., B. J. West, and J. Klafter. "Levy Dynamics of Enhanced Diffusion: Application to Turbulence." *Phys. Rev. Lett.* **58** (1987): 1100–1103.
15. Shlesinger, M. F. "Fractal Time in Condensed Matter." *Ann. Rev. Phys. Chem.* **39** (1988): 269–290.
16. Todhunter, I. *A History of the Mathematical Theory of Probability.* Cambridge: Cambridge University Press, 1865.
17. Zaslavsky, G. "Stochastic Webs." *Chaos* **1** (1991): 1.

J. H. Lowenstein
Department of Physics, New York University, New York NY 10003

Dynamics of Web Maps: Parameter Dependence of Stochastic Layers

INTRODUCTION

Zaslavsky's web map[10,11,12] arises from the following idealized thought experiment. A charged particle moves in the xy-plane perpendicular to a uniform magnetic field. If undisturbed, it moves in a circle with uniform angular frequency Ω, and this imposes a linear relation between coordinate and velocity components:

$$\dot{x} = \Omega(y - y_c), \qquad (1)$$
$$\dot{y} = -\Omega(x - x_c), \qquad (2)$$

where (x_c, y_c) is the center of the circle. Now suppose the particle is subjected to instantaneous kicks, q times per revolution, by an electric field

$$\mathbf{E} = \mathbf{e}_y E_0 \sin ky \sum_n \delta\left(t - \frac{2\pi n}{\Omega q}\right). \qquad (3)$$

Since each kick leaves y and $\dot{x}$ unchanged, Eq. (1) is valid for all times, with y_c a constant of the motion, which we choose to be zero without loss of generality. The appearance of a typical orbit is suggested in Figure 1. We note that, with the aid of

Eq. (1), the relationship between the velocity vectors just prior to successive kicks can be expressed recursively, as an area-preserving mapping M (the web map) of the velocity plane,

$$M : \begin{pmatrix} u \\ v \end{pmatrix} \longrightarrow \begin{pmatrix} \cos \frac{2\pi}{q} & \sin \frac{2\pi}{q} \\ -\sin \frac{2\pi}{q} & \cos \frac{2\pi}{q} \end{pmatrix} \begin{pmatrix} u \\ v + a \sin u \end{pmatrix} \qquad (4)$$

where $u = k\dot{x}/\Omega$, $v = k\dot{y}/\Omega$, and a is a dimensionless parameter proportional to the kick amplitude.

Although the assumptions which went into the above model are unrealistic for practical applications in plasma physics, it is of some theoretical interest whether such a simple mechanism can lead to unbounded acceleration of particles. That is, are there initial conditions under which repeated application of the map M leads to ever increasing velocities? Or is there a dynamical obstacle that prevents orbits from marching out to infinity in the uv-plane? Such an obstacle would be a closed invariant curve surrounding the origin. Since the interior of such a curve is mapped into itself by the area-preserving map M, no orbit initially inside can cross to the outside. Because of their well-known role in the theorems[2,8] of Kolmogorov, Arnold, and Moser, we shall frequently refer to simple, closed, invariant curves as KAM curves.

The answer to our question is relatively simple in the cases $q = 3$, 4, or 6, where the KAM curves are restricted to the interiors of cells which tile the plane periodically, and there is an infinitely extended web of unbounded chaotic orbits. The

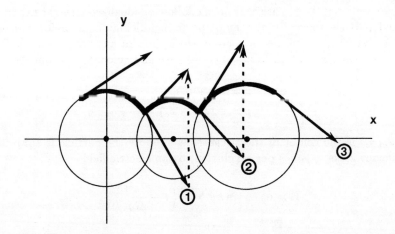

FIGURE 1 Motion of a periodically kicked charged particle in a uniform magnetic field. Here kicks in the y direction occur four times per revolution. Note the constancy of the y-component of the center of circular motion.

cases $q = 5, 7, 8$, etc. are far more subtle. Once again, for sufficiently large a, there is a web of chaotic orbits, this time with an apparent quasi-crystalline symmetry. Now periodicity no longer limits the size of KAM curves, and in fact numerical studies[3,7] show that in the $a \to 0$ limit there exist closed invariant curves of arbitrarily large radius. Based on perturbative calculations[1] and numerical experiments,[1,10,11,12] one expects an inexorable increase in the area occupied by chaotic orbits as the parameter a increases, so that all the KAM curves are eventually swallowed up by chaos. Beyond this broad picture, very little is known about the parameter evolution of KAM curves and quasi-crystalline stochastic webs. In this lecture we report on a small step toward gaining this understanding. We shall concentrate on a particular piece of the five-fold web, and try to follow graphically the evolution of one of its boundaries. We shall find that the behavior is a good deal more complicated than one might have guessed from studies of simpler maps.[1]

BASIC CONCEPTS

Before discussing our numerical explorations, it is important to make more precise some of the basic concepts. We restrict ourselves to the web map of Eq. (4) with $q = 5$.

SYMMETRIES

The geometry of typical orbits in the uv-plane (the so-called *phase portrait*) is characterized by some important symmetries. First of all, there is an invariance under the map M itself, which, for all points except those close to the origin, is approximately a clockwise rotation by $2\pi/5$. There are additional exact symmetries[6,9] of M-invariant objects, namely mirror reflections about the axes inclined at polar angles $3\pi/10$ and $4\pi/5$ (the product of the two reflections is just a total inversion). Exploiting these mirror symmetries is crucial to the effectiveness of numerical methods applied to long orbits.[6]

FIXED POINTS

There are infinitely many fixed points of the fifth-iterate map M^5 (this is the map, rather than M itself, which marches in small steps and traces out the "shape" of an invariant curve or stochastic layer), and they form an approximate quasi-crystalline array in the uv-plane. Since the map preserves areas, there are only two types of fixed points: *stable* (or *elliptic*) and *unstable* (or *hyperbolic*). The former are the centers of small-scale circulation and are not of particular interest to the present investigation. The hyperbolic fixed points, on the other hand, determine what we mean by the "shape" of long invariant curves. Intersecting at each such fixed point are two special M^5-invariant curves, its stable and unstable manifolds. Along the stable manifold, an orbit moves toward the fixed point, with the steps decreasing in size geometrically; along the unstable manifold, orbits move away from the fixed point, with steps increasing in size geometrically. Typical orbits in the vicinity of an unstable fixed point are "scattered" by it: they approach it along a stable direction and leave along an unstable one. A long orbit will visit the neighborhoods of many hyperbolic fixed points. Its shape will resemble a complicated polygon, with a rounded corner wherever it passes near a hyperbolic fixed point (see Figure 3 for some simple examples).

STOCHASTIC LAYERS

In the special case of an integrable model, the unstable manifold of a hyperbolic fixed point ends at another such point, forming part of a separatrix curve. On the other hand, a generic perturbation of an integrable model leads to breakdown of separatrices and their replacement by *stochastic layers* of nonzero thickness within which the hyperbolic fixed points are located. Almost all of the orbits within the layer are chaotic, and if one assumes ergodicity within the layer (islands of stability may be surrounded by the layer but are not considered part of it), one can define the layer to be the closure of any one of its chaotic orbits.

The web map is believed to be integrable only in the limit of vanishing a (the first-order approximation to M^5 is a Hamiltonian flow). Hence we expect to find stochastic layers associated with all the hyperbolic fixed points of the fifth-iterate map.

STOCHASTIC WEBS AND KAM CURVES

Being embedded in a common stochastic layer is clearly a connectivity equivalence relation for hyperbolic fixed points. The complete system of those connectivity components which surround the origin resembles, in structure, a spider web (at least if a is neither too large nor too small), and hence the name *stochastic web*. Our original questions concerning the possible trapping of orbits can be rephrased as follows: given a, does the stochastic web have a *finite component*? The alternative is that there is a connected web extending throughout the plane. If there is a finite component, then its outer boundary is a closed invariant curve C which traps all orbits with initial point interior to C. Typically, there will be a narrow annular region, between C and the web component surrounding C, within which quasi-periodicity reigns (in the form of closed invariant curves interlaced with island chains).

The relationship between hyperbolic fixed points, stochastic layers and webs, and KAM-dominated annuli is sketched in Figure 2. From our previous remarks about symmetries, it is obviously sufficient to restrict our attention to a sector of angle $\pi/5$.

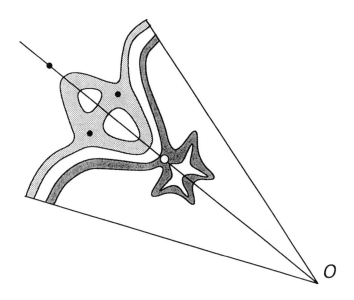

FIGURE 2 Sketch showing schematically two stochastic layers (shaded) which form connected components of a stochastic web. KAM curves, interspersed with island chains, are to be found in the annular region between the layers.

Ultimately we would like to know the connectivity structure of the stochastic web as a function of a. In particular, what is the smallest value a_0 such that for all $a > a_0$, the web has only a single component? As we shall see below, we are only beginning to make some headway toward answering such questions.

DETERMINATION OF STOCHASTIC LAYER BOUNDARIES

To investigate how stochastic layers evolve, it is crucial to have a reliable *operational* definition of the boundary of a stochastic layer. From its definition and the assumption of restricted ergodicity, one might try choosing a "random" assortment of initial conditions well within the layer and simply iterating away, thus defining the boundary as the limiting envelope of the chaotic orbits explored in this fashion. In many instances this can be misleading, due to the presence near the boundary of cantori: invariant fractal sets which can block the convergence of the envelope to the true boundary for unacceptably long times. More efficient is an approach from the quasi-periodic side, thinking of the boundary as the "last KAM curve" beyond which quasi-periodic orbits on closed curves are impossible.

A straightforward method of doing this has been given by Greene.[4,5] The main idea is to concentrate on a dense set of "noble" KAM curves. For a quasi-periodic orbit on such a curve, the average numbers of revolutions about the origin per map iteration (the *rotation number*) has a continued-fraction representation of the form

$$[m_0, m_1, m_2, \ldots, 1, 1, 1, \ldots] = m_0 + \cfrac{1}{m_1 + \cfrac{1}{m_2 + \cdots \cfrac{\cdots}{1 + \cfrac{1}{1 + \cfrac{1}{1 + \cdots}}}}} \tag{5}$$

In the neighborhood of a noble KAM curve, one expects to find a sequence of stable periodic orbits with rational rotation numbers in which the continued fraction representation (5) is truncated at n levels. According to Greene, there is a quantity R, the *residue*, whose scaling behavior as one proceeds down the sequence of approximating periodic orbits, gives a criterion for the existence of a KAM curve with the given irrational rotation number. If the limiting curve exists, R decreases to zero at a geometric rate; if the curve does not exist, the residue blows up geometrically. For the borderline case, applicable to the "last KAM curve," R approaches a constant close to 0.25.

Applying Greene's criterion requires an ability to locate long periodic orbits rapidly and precisely. Fortunately the existence of a symmetry axis greatly enhances our ability to make such calculations for the $q = 5$ web map.

EXPLORATION OF A SPECIFIC STOCHASTIC LAYER

We now turn to a computational exploration of a "typical" stochastic layer of the $q = 5$ web map. What I shall describe here is a graphical depiction of the parametric evolution of the selected layer. This is one of several approaches (including a direct application of Greene's criterion) which I have used to investigate this particular layer. The interested reader will find a detailed account in Lowenstein.[6]

The map M^5 has a hyperbolic fixed point (u_0, v_0) on the symmetry line near the point $(-05, 47)$ (its precise location depends on a, and is easily found by a one-dimensional search along the symmetry line). For values of a between about 0.21 and 0.36, the fixed point is immersed in a web component, one-tenth of which is shown in Figure 3(b). For at least part of the parameter range, this component is surrounded by the larger one shown in Figure 3(c), and the two are separated by a thin annular region. Figure 3(a) shows an orbit within the annulus. The relationships among the two stochastic layers, their associated fixed points, the annulus, and the origin are indicated schematically in Figure 2. Each of the pictures of Figure 3(a)–(c) is a plot of a single orbit with an arbitrarily selected initial point on the symmetry line. Although each of the orbits surrounds the origin, all points are mapped by symmetry transformations into the same sector of opening angle $\pi/5$.

To study the detailed evolution of the stochastic layer, we focus on a small neighborhood of the hyperbolic fixed point. This has the advantage of spreading out the orbits along the symmetry line, and allows a natural definition of the layer's

Dynamics of Web Maps

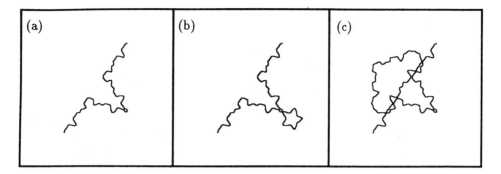

FIGURE 3 Plots of long orbits starting near the hyperbolic fixed point (u_0, v_0). Using symmetries, points are plotted in a single sector with opening angle 36°. The three orbits are (a) in the annulus between stochastic layers (close to a KAM curve), (b) in the inner stochastic layer (chaotic orbit), and (c) in the outer stochastic layer (chaotic orbit).

width, namely the distance of the boundary from the fixed point, measured along the symmetry line.

Figure 4 shows a sequence of twelve snapshots of representative orbits near the stochastic layer boundary, for equally spaced parameter values between 0.3030 and 0.3041. The plots are generated as follows:

i. For each a, the hyperbolic fixed point (u_0, v_0) is found by a search along the symmetry line. The field of view is set to be $u_0 - 0.06 < u < u_0 + 0.006$, $v_0 - 0.0054 < v < v_0 + 0.054$.

ii. Using the symmetry operations, we endow the angle $\pi/5$ sector straddling the symmetry line with periodic boundary conditions.

iii. Preliminary graphical explorations indicate that the primary periodic orbits (i.e., the minimal ones, corresponding to the largest island chains) near the layer boundary have periods between 494 and 510 in the selected parameter range. Note that these are periods for traversing the reentrant sector; the corresponding orbits circling the origin require approximately ten times as many iterations of M^5.

iv. For primary period n, we locate periodic orbits with rotation numbers

$$[n, 4, 1, 1, 1, 1] = n + \frac{5}{23},$$

$$[n, 2, 1, 1, 1, 1, 1] = n + \frac{8}{21},$$

$$[n, 1, 1, 1, 1, 1, 1, 1] = n + \frac{13}{21},$$

$$[n, 1, 3, 1, 1, 1, 1] = n + \frac{18}{23},$$

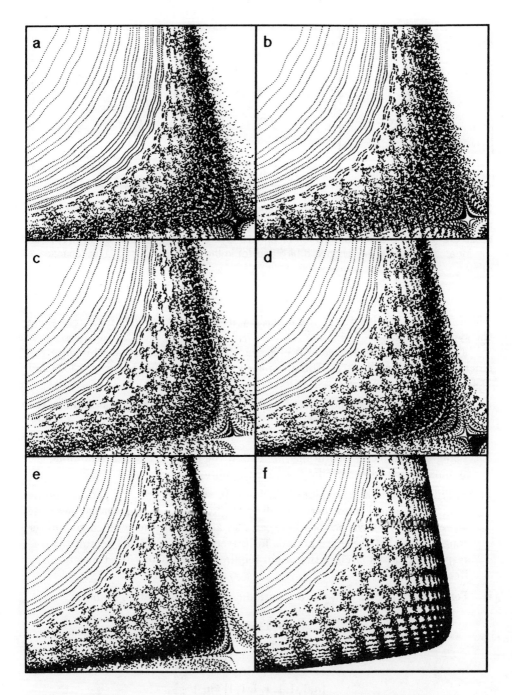

FIGURE 4 Snapshots of orbits near the inner stochastic layer boundary, with the parameter a increasing from 0.3030 to 0.3041 in steps of 0.0001. (continued)

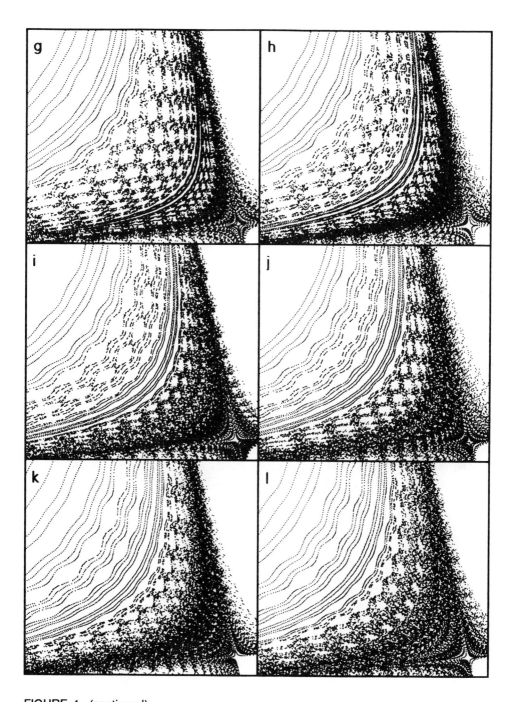

FIGURE 4 (continued)

using a search on the symmetry axis. Outside the stochastic layer, according to Greene's analysis, these orbits belong to island chains approximating KAM curves. Within the layer they are almost certainly unstable.

v. Starting near each of the periodic points (10^{-8} off the symmetry axis), we iterate M^5 15,000 times, plotting all those points which fall within the field of view. The near-KAM orbits will appear as dashed curves (actually lines of very narrow islands), while those well within the layer will soon show their chaotic character. Just inside the layer boundary, the orbits will spend a great deal of time near cantori, and the plot will show a fuzzy dashed curve.

A glance at a plot generated in this manner gives one a fairly good idea of the location of the stochastic layer boundary. This is confirmed by comparison with the more precise results obtained by systematic application of Greene's criterion.

What does our sequence of snapshots reveal about the parametric evolution of the stochastic layer in question? The first three frames show no dramatic changes in the stochastic layer, only an extremely gentle expansion. In Figure 4(d)–(f), a remarkable increase of stability occurs deep within the layer. The orbits are chaotic, but they spread out very little during the 15,000 iterations (this is particularly evident in Figure 4(f)). This is a prelude to the birth of a narrow channel of regularity which is barely visible in Figure 4(g) but grows rapidly and moves upward while the chaotic layer above it gradually dissolves. By the last frame we are more or less back to where we were at the beginning of the sequence.

The more comprehensive exploration of Lowenstein[6] shows that the cycle depicted in Figure 4 is repeated many times, with a generally increasing amplitude, as a increases from 0.30 to 0.36.

Perhaps the most interesting feature of Figure 4 is the discontinuous decrease, between frames (f) and (g), in the width of the stochastic layer. Finer subdivision of the interval[6] reveals that the collapse is at least by a factor of 4. Additional work will be needed to probe the details of the bifurcation which opens up an inner channel between $a = 0.3035$ and $a = 0.3036$, as well as to find the dynamical origin of the phenomenon.

ACKNOWLEDGMENTS

I would like to thank G. M. Zaslavsky, A. A. Chernikov, G. Schmidt, P. Cvitanovic, Ze Yang, and P. Kulesa for stimulating discussions.

REFERENCES

1. Afanasiev, V. V., A. A. Chernikov, R. Z. Sagdeev, and G. M. Zaslavsky. "The Width of the Stochastic Web and Particle Diffusion along the Web." *Phys. Lett. A* **144** (1990): 229–236.
2. Arnol'd, V. I. "Small Denominators and Problems of Stability of Motion in Classical and Celestial Mechanics." *Russian Math. Surveys* **18:6** (1963): 85–191.
3. Chernikov, A. A. Unpublished work.
4. Greene, J. M. "Two-Dimensional Area Preserving Mappings." *J. Math. Phys.* **9** (1968): 760–768.
5. Greene, J. M. "A Method for Computing the Stochastic Transition." *J. Math. Phys.* **20** (1979): 1183–1201.
6. Lowenstein, J. H. "Parameter Dependence of Stochastic Layers in a Quasicrystalline Web." *Chaos* **1** (1991): 473–481.
7. Lowenstein, J. H. Unpublished work.
8. Moser, J. K. "On Invariant Curves of Area-Preserving Mappings of an Annulus." *Nachr. Akad. Wiss. Göttingen, II Math. Phys.* **1** (1962): 1–20.
9. Piña, E., and E. Cantoral. "Symmetries of the Quasicrystal Mapping." *Phys. Lett. A* **135** (1989): 190–196.
10. Zaslavskii, G. M., M. Yu. Zakharov, R. Z. Sagdeev, D. A. Usikov, and A. A. Chernikov. "Stochastic Web and Diffusion of Particles in a Magnetic Field." *Zh. Eksp. Teor. Fiz.* **91** (1986): 500–516 (in Russian); *Sov. Phys. JETP* **64** (1986): 294–303 (in English).
11. Zaslavskii, G. M., M. Yu. Zakharov, R. Z. Sagdeev, D. A. Usikov, and A. A. Chernikov. "Generation of Ordered Structures and a Symmetry Axis from a Hamiltonian Dynamics." *Pis'ma Zh. Eksp. Teor. Fiz.* **44** (1986): 349–453 (in Russian); *JETP Lett.* **44** (1986): 451–456 (in English).
12. Zaslavskii, G. M., R. Z. Sagdeev, D. A. Usikov, and A. A. Chernikov. "Minimal Chaos, Stochastic Webs, and Structures of Quasicrystal Symmetry." *Usp. Fiz. Nauk.* **156** (1988): 193–251 (in Russian); *Sov. Phys. Usp.* **31** (1988): 887–915 (in English).

Wentian Li
Santa Fe Institute, 1660 Old Pecos Trail, Suite A, Santa Fe, NM 87501, and Box 167, Rockefeller University, 1230 York Avenue, New York, NY 10021

Non-Local Cellular Automata

Non-local cellular automata are fully discretized and uniform high-dimensional dynamical systems with non-local interactions. It is emphasized that although non-local interaction is not considered as a correct description of the physical world at its lowest level, at higher levels, it nevertheless is an important feature for systems in, for example, biology and economics. Many properties of non-local cellular automata are investigated in another publication.[7] In this lecture note, I will only highlight a few topics, including the analytic approximation of macroscopic dynamics, systems of coupled selectors, and group meeting problems.

FROM LOCAL TO NON-LOCAL DYNAMICAL SYSTEMS

One of the most important aspects of a complex system is its time evolution following a rule which does not change in time. A point of view, though perhaps extreme, is that since all physical laws are fixed (for example, there is no evidence that the gravitational force falls off as $1/r^2$ today—where r is the distance between two mass objects—but falls off as $1/r^3$ tomorrow), whatever has happened on the earth is a realization of a gigantic dynamical system with those fixed physical laws. With this point of view, the evolution of life as well as natural selection can also be modeled by complex dynamical systems with fixed rules, although this modeling will be extremely difficult because the evolution of life is much more complex than practically all model dynamical systems that we have known.

Since physical laws are local (there is no experimental evidence that physical interaction can be accomplished nonlocally), one may argue that we only need locally coupled high-dimensional dynamical systems to model everything. In other words, there is no need to introduce nonlocality in the model dynamical system.

However, in a more realistic modeling of the world around us, we do not come down to the very end of the microscopic description. The entities that interact with each other are not quarks, nucleus, atoms, or molecules, but things like neurons in a brain, animals in an ecological system, or agents in a stock market. As the level of description increases, two notions have been changed. First, the dynamics rule may not be fixed in time (they are not the golden, universal, time-invariant physical laws any more). Second, the interaction between entities may not be local.

If the dynamical rule is not time-invariant, it can be very hard to describe and to study the resulting dynamical system, unless the *dynamics of the rule* is describable. In other words, we need two sets of dynamical systems: at the higher level, there is a dynamics of the rules, and at the lower level, there is a dynamics of the entities. Many evolutionary models are of such nature. The complexity of the system results from the interplay between higher-level and lower-level dynamics. One can even imagine three or more levels of dynamics, in which the entity of the higher level is the rule of the level one step lower.

These multi-level dynamical systems are fascinating systems to study. But they are outside the realm of this lecture note. For the time being, to start from the simplest scenario, I will assume that the lower-level rules are not changed. An explanation for this assumption is that the higher-level rules function on a much longer time scale, so that during this time scale, the lower-level rules can be considered as unchanged.

The issue I want to address here is the following: what happens when the nonlocality is introduced to a dynamical system? One should know that the locality of interaction is a terrible assumption for many systems with a high-level description. For example, the transmission of signal from one neuron to another in the brain is through axons and dendrites. The distance between two neurons is an irrelevant piece of information concerning whether or not the two are connected to each other.

Note that the nonlocality at this level of the description (interaction between neurons) does not contradict the locality at the microscopic level: the traveling chemical signals do obey local physical and chemical laws. This fact, however, does not prevent us from explicitly incorporating the nonlocality into the modeling process when describing the interaction between neurons.

Similarly, two agents or brokers in a stock market can communicate via telephone line regardless of how far or close the two are to each other. Again, there is no contradiction with the locality of the physical laws. Admittedly, the electrical signal does take a longer time to travel for a longer telephone line than a short one, but the difference is so small compared with the time scale of the stock market activities, that this fact is irrelevant. The more important information is who makes phone calls to whom (whether the connectivity is one or zero) than the actual distance between them.

There are many, many other examples. What we have learned from this discussion is that when the level of description of a system is increased, one sometimes needs to explicitly introduce nonlocality to the modeling. This nonlocality does not violate the locality at the lowest level description—the physical description.

CELLULAR AUTOMATA AS A FULLY DISCRETIZED AND UNIFORM HIGH-DIMENSIONAL DYNAMICAL SYSTEM

What is a cellular automaton? With so many introductory articles and books existing on this topic, I will refer the reader to the original publications (e.g., Toffoli and Margolus,[12] and Wolfram[15,16]). To put it into simple terms, one can say that cellular automata are high-dimensional, fully discretized, synchronous, uniform, and locally coupled dynamical systems. There are high-dimensional dynamical systems that are not fully discretized, such as partial differential equations, coupled differential equations, and lattice maps (e.g., Crutchfield and Kaneko[2]). There are also high-dimensional, fully discretized dynamical systems that are not synchronous or uniform. The model systems I will introduce are high-dimensional, fully discretized, synchronous, and uniform dynamical systems with *nonlocal* connections.[7]

There exist other names that can describe nonlocally coupled, high-dimensional dynamical systems; for example, *automata networks*, or simply, *networks*. I will use the name "nonlocal cellular automata" to have a closer reference to the locally coupled cellular automata, in order to emphasize the uniformity and synchronousness of the system.

Suppose the state value of the component i at time t is x_i^t, and the total number of components in the system is N, then the state configuration of the system consists of state value for each component: $(x_1^t, x_2^t, \ldots x_N^t)$. An n-input nonlocal cellular automaton is defined by the rule $f(.)$:

$$x_i^{t+1} = f(x_{j_1(i)}^t, x_{j_2(i)}^t, \ldots x_{j_n(i)}^t), \qquad (1)$$

which says that each component i updates its state value by checking the state values of n other components, which have indexes $j_1(i), j_2(i), \ldots, j_n(i)$. Knowing the state values of these components, and knowing the rule $f(.)$ which is written as a *rule table* (a list of all possible n-component configurations as well as which state value they lead to), we are able to determine what the state value of the component i is at the next time step (x_i^{t+1}).

There are other types of networks previously studied. One of them, studied by Walker and Ashby, might be called "Ashby nets,"[13,14] is also uniform and synchronous, but not all inputs are randomly chosen—one input is always the component itself. More about Ashby nets will be discussed in the next section.

Another type of nets, that might be called the "Kauffman nets," is studied in Kauffman.[9,10] These nets are synchronous, but not uniform: the rule acting on one component may differ from that on another component. It is well known that different rules can lead to different dynamical behaviors, so mixing all of them into one system leads to rather poor statistics. If the number of inputs (n) and the number of components (N) are fixed, and we ask the question of what the "typical" transient and cycle times are, there will be no "good" answer. The median value (see, for example, Press et al.[11] for a definition of the median as well as the mean value) of a wide-spreading distribution of these statistical quantities, as used in Kauffman,[9] may not give a true "typical" value. Numerical results show that median cycle lengths for these nets are quite different from the mean cycle lengths, though I will not include these results nor discuss this type of net further here.

WIRING DIAGRAM

Besides the dynamical rule, the wiring diagram of a network also plays an important role in determining the dynamics. It could happen that with the same rule, some wiring diagrams lead to one type of dynamics, while other wiring diagrams lead to another. When we talk about dynamics of a nonlocal cellular automaton rule, there is an implication that almost all wiring diagrams ("typical" or randomly chosen) lead to the same dynamics.

It is in an analogous situation to local cellular automata. For local cellular automata, we also talk freely about the dynamics of a rule, without specifically mentioning the initial configuration. It is again implied that almost all typical or randomly chosen initial configurations lead to the same dynamics. This idea is essential to the concept of "attractor"; that is, whatever the initial conditions, they are all attracted to the same limiting behavior.

The wiring diagram dictates where to take inputs for each component. In some sense, it determines the direction of information flows. Obviously, wiring diagrams with different topological structures will transmit information in different ways, and dynamical behaviors can also be different.

For example, if one assumes that for each component i, one of its inputs is always itself:

$$\text{for all } i\text{'s } j_1(i) = i, \text{ but other } j_k(i)\text{'s are random } (k = 2, 3, \cdots, n), \qquad (2)$$

then the wiring diagram will not be completely random. I called this type of wiring diagram *partially local* or *partially nonlocal*.[7]

Some nonlocal cellular automata with partially local wiring diagrams were studied in Walker and Ashby.[13,14] These are 3-input, 2-state, nonlocal cellular automata with the second input being the component itself

$$\text{for all } i\text{'s, } j_2(i) = i, \text{ but } j_1(i) \text{ and } j_3(i) \text{ are random.} \tag{3}$$

It has been shown that for many 3-input, 2-state rules, fully nonlocal wiring diagrams and partially local wiring diagrams lead to different dynamics.[7]

Another issue related to the wiring diagram is the discussion on how dynamics are affected by changing the number of inputs. It is numerically shown that the number of inputs is important to determine the dynamical behavior.[6] If one randomly picks a rule, the more inputs one has, the more likely the dynamics are chaotic. For local cellular automata, the increase of the number of inputs will increase the percentage of rules that are chaotic. More detailed discussions are in Li, Packard, and Langton.[6]

Now back to the discussion of nonlocal cellular automata. Even though each component is supposed to receive n inputs, a particular realization of the random number generator may actually assign two inputs to be the same. If this happens, the rule as applied to that particular component will have one less number of input than it should have. And if many other components also have this *degeneracy* of inputs, it is more likely that the resulting dynamics acts as if the number of inputs is smaller. Some specific examples of the difference between the degeneracy-permitted and distinct-input diagrams are presented in Li.[7]

ANALYTIC APPROXIMATION OF MACROSCOPIC DYNAMICS

The ultimate method to study a dynamical system is to run the time evolution following the rule that updates the state value for each component. The simulation for 3-input, 2-state, nonlocal cellular automata has been carried out and the results are summarized in Li.[7]

If one is only interested in dynamics of *macroscopic* quantities, for example, the density of state 1, some alternative dynamical equations for that macroscopic quantity can be derived. These dynamical equations for macroscopic quantities are *not* equivalent to the original dynamics rules, but they will, in many cases, provide valuable information to the original dynamics.

The dynamical equation for the density of state 1 can be called *return map*:

$$d^{t+1} = F(d^t) \tag{4}$$

where d^t is the density of state 1 at time t, and the $F(.)$ is determined either by actually running the rule $f(.)$ or by some approximation schemes. Note that different original rules $f(.)$'s can give the same macroscopic dynamics $F(.)$.

One approximation scheme called *mean-field theory* assumes that all inputs are independent of each other, and the probability for having state 1 when the n inputs contain m state 1 and $n - m$ state 0 is estimated by counting the percentage of input configurations (containing m state 1 and $n - m$ state 0) that are mapped to state 1. For a general introduction, see Gutowitz.[4]

To illustrate this approximation scheme, let me use the following rule as an example (the triplet is the value of the three inputs, and the number below the triplet is the value to be updated to):

$$\begin{array}{cccccccc} 111 & 110 & 101 & 100 & 011 & 010 & 001 & 000 \\ 1 & 0 & 1 & 1 & 1 & 0 & 0 & 0 \end{array} \tag{5}$$

When all three inputs are 1, the state value will be 1; when two inputs are 1 and one input is 0, *two* out of three configurations will be mapped to 1; when one input is 1 and two inputs are 0, *one* out of three configurations will be mapped to 1; and when all three inputs are 0, the state value will never be 1. It is easy to show that one can approximate the return map by

$$d^{t+1} = (d^t)^3 + 2(d^t)^2(1 - d^t) + d^t(1 - d^t)^2. \tag{6}$$

Simple manipulation shows that it leads to

$$d^{t+1} = d^t; \tag{7}$$

that is, the density of state 1 does not change with time.

In fact, some important information can be extracted from this approximation of return maps. If the return map has a stable fixed-point solution equal to zero, the limiting density of state 1 should be zero or very low. That is the case when the original system has a fixed-point dynamics with zero-density or low-density spatial configurations.

If the return map has a non-zero, stable, fixed-point solution, there are two possibilities for the original dynamics: (1) the original system has a fixed-point dynamics with a spatial configuration about half filled with 0s and half with 1s, and (2) the original system is chaotic, with some kind of "thermal equilibrium states" being reached. Even though the state value for each component changes constantly, the density of 1s is nevertheless a constant.

I have yet to discover a return map with chaotic solutions. Generally speaking, it is very difficult for *macroscopic quantities to fluctuate chaotically*. Occasionally, numerical simulation shows that macroscopic quantities such as the density of state 1 do fluctuate irregularly. Nevertheless, it is always because these simulations are carried out for systems with finite sizes. The magnitude of these irregular fluctuation decreases as the system becomes larger. And, in principle, they will disappear in the infinite size limit. See, however, the discussions in Bohr et al.[1] and Kaneko.[8]

Although the return map is not equivalent to the original dynamical rule, it can provide valuable information. Because of the low dimensionality of the return maps, it is easier to study its own "bifurcation" phenomena (how dynamics of the return maps change with the parameter). From these studies, one can then understand some aspects of the bifurcation phenomena in the original system. A study of nonlocal cellular automata rule space following this strategy is carried out in Li.[7] In particular, it is partially understood why some bits in the rule table, which are called "hot bits" in Li and Packard,[5] are more important than others. It is because the hot bits change the form of the return map more drastically than other bits.[7]

SYSTEMS OF COUPLED SELECTORS

The cellular automaton rule defined in Eq. 5 can be written in another form:

$$x_i^{t+1} = \begin{cases} x_{j_1(i)}^t & \text{if } x_{j_2(i)}^t = 0 \\ x_{j_3(i)}^t & \text{if } x_{j_2(i)}^t = 1. \end{cases} \qquad (8)$$

In other words, if the second input is in state 0, the rule transmits the state value from the first input; if the second input is in state 1, the rule transmits the value from the third input. This rule can be called a *selector*, or a *multiplexer*, with the second input called a *control input*; that is, it decides which input to select.

This rule turns out to be the most intriguing 3-input 2-state nonlocal cellular automata. A typical spatial-temporal pattern for this rule is shown in Figure 1. Although the limiting dynamics is periodic, the transient dynamics looks chaotic. This combination of long chaotic transients and simple limiting dynamics is typical for systems on the "edge of chaos."

From Figure 1, we can see some dark and light horizontal stripes. In order for the dark stripe to form, if one component has state 1, other components also tend to have state 1 so that the total number of components with state 1 is increased. This seemingly simple fact implies the existence of certain cooperation among components. Indeed, other components do not have reasons to follow suit when one component switches from 0 to 1, unless they are dragged into doing so. The emergence of higher level structures is also a hallmark of the edge-of-chaos systems.

The transient time for systems of coupled selectors is observed to increase with the size of the system. More careful simulation shows that the increase is almost linear:

$$T_{av}(N) \sim N \qquad (9)$$

where $T_{av}(N)$ is the mean transient time for systems with size N. If we exclude all degeneracies in choosing inputs, it has been observed that the increase of transient time is more than linear[7]:

$$T_{av}(N) \sim N^{1.2}. \qquad (10)$$

time $T+1 \to 2T$

time $0 \to T$

FIGURE 1 A spatial-temporal pattern of the coupled selectors. The configuration of the system (horizontal string) consists either 1 (black) or 0 (white). And the updating of the configuration is represented by showing the configuration at each time step (time is increased going down).

There are many open issues concerning the transient behavior and I will not discuss them in length here. Briefly, there are questions on how large the system size should be in order to trust the scaling; what the distribution of transient times is for a fixed size; whether this distribution with respect to wiring sampling is different from that of initial configuration sampling; whether the mean transient time is a better quantity than the median transient time, and whether one should take the logarithm first, then do the average; how the permission of degenerate inputs change the result; and so on.

In Figure 1, the limiting configuration has a very low density of state 1. If we change the wiring and initial condition, and run the simulation again, it is possible that the limiting configuration can have a very high instead of a very low density of state 1. These two types of configurations are called *consensus states*. It is not clear before finishing the simulation which consensus state will be reached. In fact, it has been observed that the density could wander up and down in such a way that the system almost hits the high-density consensus state before turning the trend to eventually reach the low-density one!

GROUP MEETING PROBLEMS

Imagine a group of people having a meeting. The goal of the meeting is to find a consensus opinion: either most of the people vote yes, or most of them vote no. Requiring a 100 percent yes-vote or no-vote may not be realistic, so some compromise is made: we allow the meeting to finish whenever the density of yes or no is higher or lower than a certain threshold value.

The system of coupled selectors discussed in the last section can be recast into a group meeting problem. At the beginning of the meeting, each person votes yes or no randomly. Then each person chooses three other persons (he can also choose himself) as his or her "inputs." Each one of the three inputs is labeled as either the first, the second, or the third input. The second person is most important: whenever he or she votes no, the first person's vote will be followed; and whenever he or she votes yes, the third person's vote is followed.

This somehow bizarre way for a group meeting to proceed is, nevertheless, not as trivial as one might have thought. First of all, will the group meeting ever reach a consensus? By the result presented in the last section, the answer is yes. But this is true only when the three inputs are random chosen. There are examples where a consensus is never reached. For example, if the wiring is partially local, i.e., $j_2(i) = i$, it is almost always the case that the dynamics is chaotic and the density of state 1 is around 0.5.[7]

Second, even if a consensus is reached, do we know *which* one?

For the system of coupled selectors, we do not know beforehand whether it is an all-yes or all-no state that is reached. Both low- and high-density configurations are "traps" or "attractors" of the dynamics. If we consider the fluctuation of the density as a random walk (though it is a *deterministic* random walk because initial configuration, wiring diagram, and dynamical rule are fixed during the updating), it has a 50-50 chance to reach either the low-density or the high-density configuration.

The system of coupled selectors is not the only system to have two consensus states. Actually, there exists a large class of "unbiased" rules that behave similarly (by "unbiased," I mean that the rule does not have any reason to prefer either one of the consensus states). Interestingly, one such rule, a 7-input 2-state local cellular automaton, called Gacs-Kurdyumov-Levin rule, was proposed more than ten years ago,[3] defined as the following:

$$x_i^{t+1} = \begin{cases} \text{majority among } x_i^t, x_{i-1}^t, \text{ and } x_{i-3}^t & \text{if } x_i^t = 0; \\ \text{majority among } x_i^t, x_{i+1}^t, \text{ and } x_{i+3}^t & \text{if } x_i^t = 1. \end{cases} \quad (11)$$

It has been shown that Gacs-Kurdyumov-Levin rule (11) has two attractors: all-zero and all-one configurations. The all-zero consensus state will be reached if the initial density of state 1 is smaller than 0.5; and the all-one consensus state will be reached if the initial density is larger than 0.5 (this result is proved for Gacs-Kurdyumov-Levin rule (11) in the infinite size limit).

Similar to the systems of coupled selectors, consensus states may not be reached for Gacs-Kurdyumov-Levin rule if the wiring diagram is modified. For example, if the majority is chosen among x_i^t, x_{i-1}^t, and x_{i-2}^t when $x_i^t = 0$, and among x_i^t, x_{i+1}^t, and x_{i+2}^t when $x_i^t = 1$, then the limiting density will be more or less the same with the initial density. So, if the initial configuration is random, the limiting density will be around 0.5 instead of 0 or 1. We can see easily that the Gacs-Kurdyumov-Levin rule can also be translated to a group meeting problem. If the periodic boundary condition is used, we are going to have a "roundtable group meeting"!

With limited space and time, I can only introduce a few topics on nonlocal cellular automata. There are other major topics that are not covered here, for example, viewing nonlocal cellular automata as computers, and the structure of nonlocal cellular automata rule space. For interested readers, see my paper for more details.[7] I hope I have conveyed to readers the richness of dynamical behaviors for dynamical systems with nonlocal interaction, and I hope more people will share my excitement in studying these systems.

ACKNOWLEDGMENTS

The work at Santa Fe Institute was funded by MacArthur Foundation, National Science Foundation (NSF grant PHY-87-14918), and Department of Energy (DOE grant DE-FG05-88ER25054). This article was written at Rockefeller University where I was supported by DOE grant DE-FG02-88-ER13847.

REFERENCES

1. Bohr, T., G. Grinstein, Y. He, and C. Jayaprakash. "Coherence, Chaos, and Broken Symmetry in Classical, Many-Body Dynamical Systems." *Phys. Rev. Lett.* **58**(21) (1987): 2155–2158.
2. Crutchfield, J., and K. Kaneko. "Phenomenology of Spatial-Temporal Chaos." In *Directions in Chaos*, edited by B-L. Hao. Singapore: World Scientific, 1987.
3. Gacs, P., G. L. Kurdyumov, and L. A. Levin. "One-Dimensional Uniform Arrays that Wash Out Finite Islands." *Probl. Peredachi. Info.* **14**: (1978): 92–98.
4. Gutowitz, H. "Hierarchical Classification of Cellular Automata." *Physica D* **45**(1-3) (1990): 136–156.
5. Li, W., and N. Packard. "Structure of the Elementary Cellular Automata Rule Space." *Complex Systems* **4**(3) (1990): 281–297.
6. Li, W., N. Packard, and C. Langton. "Transition Phenomena in Cellular Automata Rule Space." *Physica D* **45** (1990): 77–94.
7. Li, W. "Phenomenology of Non-local Cellular Automata." *J. Stat. Phys.* **68**(5/6) (1992).
8. Kaneko, K. "Globally Coupled Chaos Violates the Law of Large Numbers but Not the Central-Limit Theorem." *Phys. Rev. Lett.* **65** (1990): 1391–1394.

9. Kauffman, S. A. "Metabolic Stability and Epigenesis in Randomly Constructed Genetic Nets." *J. Theor. Biol.* **22** (1969): 437–467.
10. Kauffman, S. A. "Emergent Properties in Randomly Complex Automata." *Physica D* **10** (1984): 145–156.
11. Press, W. H., B. P. Flannery, S. A. Teukolsky, and W. T. Vetterling. *Numerical Recipes in C.* Cambridge, MA: Cambridge University Press, 1988.
12. Toffoli, T., and N. Margolus. *Cellular Automata Machines—A New Environment for Modeling.* Cambridge, MA: MIT Press, 1987.
13. Walker, C. C., and W. R. Ashby. "On Temporal Characteristics of Behavior in a Class of Complex Systems." *Kybernetik* **3** (1966): 100–108.
14. Walker, C. C. "Behavior of a Class of Complex Systems: The Effect of System Size on Properties of Terminal Cycles." *J. Cybernetics* **1**(4) (1971): 57–67.
15. Wolfram, S. "Statistical Mechanics of Cellular Automata." *Rev. Mod. Phys.* **55** (1983): 601–644.
16. Wolfram, S., ed. *Theory and Applications of Cellular Automata.* Singapore: World Scientific, 1986.

Student Contributions

Jeremy John Ahouse,† Erhard Bruderer,‡ Angelica Gelover-Santiago,* Norio Konno,†† David Lazer,‡‡ and Stella Veretnik**

†Center for Complex Systems and Biophysics Program, Brandeis University, Waltham, MA 02254-9110, USA, to whom correspondence may be sent; ‡School of Business Administration, University of Michigan, Ann Arbor, MI 48109, USA; *Department of Complex Systems, Instituto de Fisica, UNAM, Apartado postal 20-364, C.P. 01000, Mexico D.F., MEXICO; ††Muroran Institute of Technology, Mizumoto-cho, Muroran, Hokkaido 050, JAPAN; ‡‡Department of Political Science, University of Michigan, Ann Arbor, MI 48109, USA; **Department of Genetics and Cell Biology, University of Minnesota, 356 Biological Science Center, 1445 Gortner Ave., St. Paul, MN 55108, USA

Reality Kisses the Neck of Speculation: A Report From the NKC Workgroup

During the 1991 Complex Systems Summer School, Stuart Kauffman lectured on the family of NKC models. We found these ideas intriguing and formed an NKC study group. This contribution to the lecture volume summarizes some of the ideas and musings of this group.

It has been suggested[26] that NKC is an acronym for "no known content." Whether or not this is the case, NKC models are (also) named for the three important parameters of the discrete fitness landscape models that are discussed in this paper.

This summary has three distinct parts. The first is a mathematically formal description of NKC models. The second is a list of critiques of current uses of NKC models. The third section suggests several new areas that NKC models may be useful. Space is limited so we will have to ask our readers to see other authors' treatments of the basic NKC model.[9,10,11,12,13,16,24,25]

1. A MATHEMATICAL FORMALISM

The formalism presented below is an attempt to add rigor to the NKC model. It does this in two ways: First, by introducing a mathematical formalism that can be used to find analytical results, and second, by specifying the model to a degree which allows for more specific critiques of the underlying assumptions. This formalism is based on our reading of the model as presented by Stuart Kauffman.[9,13] We believe that this is the first formalization of this kind, though several other approaches exist.[16,24,25]

In the description of the formalism, we will be applying the NKC model to genotypes. This allows us to use a familiar vocabulary. It is important to remind you that there is nothing inherent in this formalism that restricts us to this level of biological organization. It will also be immediately obvious that we are using caricature genotypes. They are simply binary strings; we will have more to say on the relationship between binary strings and biological genes in the second section of the paper.

One of the goals of this formalization is to derive analytical results for the NK family of models. Another is to formalize the operators that define the neighborhood of a given gene: one-mutant neighbor, inversion, and crossing over. These operators are defined below.

1.1 DEFINITIONS

Let N be a positive integer and K be a non-negative integer. N denotes the number of genes in the genotype of an organism (see Figure 1), and K denotes the number of other genes (see Figure 2) which depend on the fitness contribution (which will be specified later) of each gene, where $0 \leq K \leq N - 1$. Thus, K measures the richness of epistatic interactions among genes. $X_N = \{0,1\}^N$ corresponds to the configuration space of the genotype with N genes and $X_{K+1} = \{0,1\}^{K+1}$ is the collection of the $K+1$ genes on which the fitness contribution of each gene bears.

In a coevolutionary system, a positive integer S denotes the number of species (see Figure 3). Here we are using the binary string to represent the genes in a species, *not* the bases in a gene or amino acids in a protein. The genotype of each species is represented as a binary string, where each element stands for a gene.

1	2	3	...	...		N
1	0	1	...	...	1	0

FIGURE 1 An example of a genotype (binary string) of length N.

FIGURE 2 An example of several epistatic interactions; here $K = 4$.

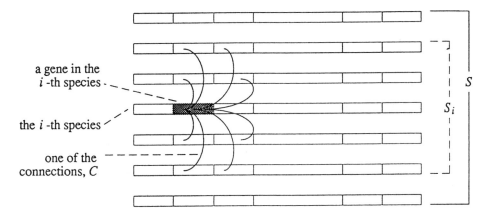

FIGURE 3 A coevolutionary system, where positive integer S denotes the number of species.

However, the model could be used at those levels. Often it is easier to visualize the relations if we think of the binary strings as analogous to DNA sequences. But Kauffman and Johnson's coevolution model describes the genes as being the elements of the binary string.[13] If this seems confusing, this arises from the fact that their descriptions move easily between different levels of organization. This shift may be unfortunate, but it illustrates the power of this model at different levels of organization. Remember that instead of species we could think of these as being chromosomes interacting or different genes on a chromosome interacting.

Each gene in the ith species depends on K genes internally and on C genes in each of the $S_i (\in \{1, \ldots, S\})$ species, S is the total number of species. S_i is a subset of S and represents all other species with which species i interacts. That is to say, a positive integer C is the number of other genes in other species which depend on the fitness contribution of each gene and S_i is the number of other species with which the ith species interacts. Let $X_{C \times S_i} = \{0, 1\}^{C \times S_i}$ denote the configuration space of the C genes in other species on which the fitness contribution of each gene bears.

We now introduce random variables to describe the elements of these interacting binary strings.

For each $j \in \{1, \ldots, S\}, \eta^{(j)}(\in X_N)$ is the genotype of the ith species with N_j genes, i.e.,

$$\eta^{(j)} = (\eta^{(j)}(1), \eta^{(j)}(2), \ldots, \eta^{(j)}(N)) \in X \text{ with } \eta^{(j)}(i) = \{0, 1\}.$$

We illustrate this in Figure 4.

In order to describe the interactions between elements of the string (i.e., interaction between genes within a chromosome), we define:

$$\Delta_i^{(j)}(K) = \{i_0^{(j)}, i_1^{(j)}, \ldots i_K^{(j)}\}$$
with $i_0^{(j)} = i$ and $i_r^{(j)} \neq i_S^{(j)}$ if $r \neq S$.

Here $\Delta_i^{(j)}(K)$ denotes the collection of indices which affect the fitness contributions of the ith gene in the jth species. And

$$(\eta^{(j)}(l): l \in \Delta_i^{(j)}(K)) = (\eta^{(j)}(i_0^{(j)}), \ldots, \eta^{(j)}(i_K^{(j)})) \in X_{K+1}$$

is the configuration which bears on the fitness contributions of the ith gene in the jth species (see Figure 5).

Similarly

$$\Delta_i^{(j)}(C) = \left\{ U_{S=S_1,\ldots,S_i} \left\{ i_{1,S}^{(j)}, i_{2,S}^{(j)}, i_{3,S}^{(j)}, \ldots, i_{C,S}^{(j)} \right\} \right\}$$

and $(\eta^{(j)}(l): l \in \Delta^{(j)}(C)) = (\eta^{(j)}(i_{1,1}^{(j)}), \eta^{(j)}(i_{2,1}^{(j)}), \ldots, \eta^{(j)}(i_{C,S_i}^{(j)}))) \in X_C$.

$\eta^{(1)}=$	η_1^1	η_2^1	$\ldots$	η_N^1
$\eta^{(2)}=$	η_1^2	η_2^2	$\ldots$	η_N^2
$\ldots$				
$\eta^{(j)}=$	η_1^j	η_2^j	$\ldots$	η_N^j
$\ldots$				
$\eta^{(s)}=$	η_1^s	η_2^s	$\ldots$	η_N^s

FIGURE 4 This diagram depicts several binary strings. Each of these strings represents a genotype.

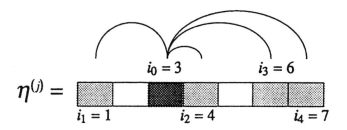

FIGURE 5 $\Delta_i(K) = \Delta_3(4) = \{i_0 = 3, i_1 = 1, i_2 = 4, i_3 = 6, i_4 = 7\}$.

Where $N = 7$ and $K = 4$, we would then write this as:

$$\eta^{(j),\Delta_3(4)} = \left(\eta^{(j)}(1), \eta^{(j)}(3), \eta^{(j)}(4), \eta^{(j)}(6), \eta^{(j)}(7)\right).$$

1.1.1 THE OPERATORS There are several operators that define which strings are accessible from a given string. These define the neighborhood of a given string. These operators are named "mutation," "crossover," and "inversion." These terms take their inspiration from looking at the NK model as describing DNA. This is a different organizational level from the one we have been using above.

New variations in the genotype can be introduced by mutation, crossing over, and inversion. These operations define *searchable* neighborhoods.

1. One-mutant neighbor operator:

$$G_m^{(j)}: N_j \to N_j, \text{ where } m \in \{1, \ldots, N_j\}$$

is defined by a change in state of the individual gene ($0 \to 1$ or $1 \to 0$) within the genotype. If $i \neq m$ (there is no change in the state of the gene),

$$\eta_m^{(j)}(i) = \eta_m^{(j)}(i);$$

if $i = m$ (invert the state of the mth gene),

$$\eta_m^{(j)}(i) = 1 - \eta_m^{(j)}(i).$$

Note that operation simply flips the bit position m.

2. Crossover operator:

$$G_a^{(j_1,j_2)}: (X_{N_{j_1}} \times X_{N_{j_2}}) \to X_{N_j}$$
$$G_a^{(j_1,j_2)}(\eta^{(j_1)}, \eta^{(j_2)}) = (\eta^{(j_1)}(1), \ldots, \eta^{(j_1)}(a), \eta^{(j_1)}(a+1), \ldots, \eta^{(j_2)}(N_j))$$

The other product of this crossing over is:

$$G_a^{(j_1,j_2)}(\eta^{(j_2)}, \eta^{(j_1)}) .$$

The point that the two genes are broken is a and we get two new genotypes by rejoining the substrings. In our description of binary strings as being genotypes belonging to different species it is not obvious how the crossing-over operator would be relevant. But as mentioned above the NKC model has different levels of applicability. In this case thinking of the binary strings as being analogs for bases in a gene on a chromosome would be more helpful.

3. Inversion operator:

$$G_{[a,b]}^{(j)}: N_j \to N_j, \text{ where } a \leq b \text{ and } a,b \in \{1,\ldots,N_j\}$$

is defined by:

$$G_{[a,b]}^{(j)}(\eta^{(j)}) = (\eta^{(j)}(1), \ldots, \eta^{(j)}(a), [\eta^{(j)}(b), \ldots, \eta^{(j)}(a+1)],$$
$$\eta^{(j)}(b+1), \ldots, \eta^{(j)}(N_j)) .$$

When $a = b$, then the inversion operators reduces to the identity operator. This operator is applied to a single string. Breaking and rejoining occurs immediately after the point a and b and the string of genes between points a and b gets inverted.

Furthermore, we can define the following subspaces of X corresponding to the above-mentioned operators.

1. One-mutant neighbor space:

$$X_{mu}(\eta^{(j)}) = \{G_m^{(j)}(\eta^{(j)}): m \in \{1,\ldots,N_j\}$$

2. Crossover space:

$$X_{cr}(\eta^{(j_1)}, \eta^{(j_2)}) = \{G_a^{(j_1,j_2)}(\eta^{(j_1)}, \eta^{(j_2)}): j_1 \neq j_2, a \in \{1,\ldots,N_j\}$$

3. Inversion space:

$$X_{inv}(\eta^{(j)}) = \{G_{[a,b]}^{(j)}(\eta^{(j)}): a \leq b, a,b \in \{1,\ldots,N_j\}\}$$

1.1.2 FITNESS The fitness contribution of the ith and jth genome (species) is a combination of all of the interaction internal to the genotype j and the interactions with genotypes of each of the other species, S_j. The fitness function assigns a value of fitness contribution to *each state of the set of genes that influence a given gene.* This may become a little convoluted so we have included some specific examples below.

We now define the fitness contribution of the ith gene and jth species:

$$W_i(\eta^{(j)}) = \frac{1}{1+S_j}\left\{W_i(\eta^{(j)}:\Delta_i^{(j)}(K)) + \sum_{l=1}^{S_j}\tilde{W}_i(\eta^{(l)}:\Delta_i^{(l)}(C))\right\}. \tag{1.1}$$

This is a combination of fitness contributions from internal interactions (as a function of K) and interspecies interactions (as a function of C). Where

$$W_i(\eta^{(j)}:\Delta_i^{(i)}(K)) = W_i(\eta^{(j)}(p):p\in\Delta_i^{(j)}(K))$$
$$\tilde{W}_i(\eta^{(l)}):\Delta_i^{(l)}(C)) = \tilde{W}_i(\eta^{(l)}(q):q\in\Delta_i^{(l)}(C))$$

and

$$\mathcal{R} = \begin{cases} W_i(\eta^{(j)}:\Delta_i^{(j)}(K)), \tilde{W}_i(\eta^{(l)}:\Delta_i^{(l)}(C)): \begin{cases} i=1,\ldots,N \\ j=1,\ldots,S \\ l=1,\ldots,S_j \end{cases} \\ (\eta^{(j)}(p):p\in\Delta_i^{(j)}(K))\in X_{K+1} \\ (\eta^{(l)}(q):q\in\Delta_i^{(l)}(C))\in X_C \end{cases}$$

$\mathcal{R}$ is the collection of (0,1)-valued random variables. In general, the cardinality of $\mathcal{R}$, $|\mathcal{R}|$, is a very large number. Commonly, for the sake of simple analysis, we may assume that $\mathcal{R}$ is the set of independent, identically distributed (IID) random variables.

The fitness of the jth species is defined as the average fitness contribution of each gene:

$$W(\eta^{(j)}) = \frac{1}{N}\sum_{i=1}^{N}W_i(\eta^{(j)}). \tag{1.2}$$

Hereafter, the above-mentioned NK family model will be called the $M = (N, K, C, S, S_i)$ model. Sometimes it is called the NKC model for short.

1.2 EXAMPLES OF THE $M = (N, K, C, S, S_I)$ MODEL

This section gives some examples of the NK family of models. We do this to help the reader tether some of the formalism to more concrete situations.

1.2.1 THE BASIC NK MODEL: $M(N, K, 0, 1, 0)$.

In this model, the number of species is one, so we will omit the superscript (j) so that:

$$\eta = \eta^{(j)}, W_i(\eta) = W_i(\eta^{(j)}),$$

and so on.

By definitions (1.1) and (1.2), we have

$$W(\eta) = \frac{1}{N} \sum_{i=1}^{N} W_i(\eta : \Delta_i(K)). \tag{1.3}$$

Furthermore, it is easily obtained that for $m = 1, \ldots, N$:

$$W(\eta_m) - W(\eta) = \frac{1}{N} \sum_{i : m \in \Delta_i(K)} \{W_i(\eta_m : \Delta_i(K)) - W_i(\eta : \Delta_i(K))\}. \tag{1.4}$$

Next, we will consider this model in more detail.

CASE A: $K = 0$ Equations (1.3) and (1.4) imply that for $m = 1, \ldots, N$,

$$W(\eta) = \frac{1}{N} \sum_{i=1}^{N} W_i(\eta(i)), \tag{1.5}$$

$$W(\eta_m) - W(\eta) = \frac{1}{N} \{W_m(1 - \eta(m)) - W_m(\eta(m))\} \tag{1.6}$$

where $W_i(\eta(i)) = W_i((\eta(i)))$

TABLE 1 In this table we calculate all of the fitnesses for all of the possible stringo for the case where $N = 3, K = 0, C = 0, S = 1,$ and $S_i = 0$.

η	W_1	W_2	W_3	W_η
000	$W_1(0)$	$W_2(0)$	$W_3(0)$	$\frac{1}{3}\{W_1(0) + W_2(0) + W_3(0)\}$
001	$W_1(0)$	$W_2(0)$	$W_3(1)$	$\frac{1}{3}\{W_1(0) + W_2(0) + W_3(1)\}$
010	$W_1(0)$	$W_2(1)$	$W_3(0)$	$\frac{1}{3}\{W_1(0) + W_2(1) + W_3(0)\}$
011	$W_1(0)$	$W_2(1)$	$W_3(1)$	$\frac{1}{3}\{W_1(0) + W_2(1) + W_3(1)\}$
100	$W_1(1)$	$W_2(0)$	$W_3(0)$	$\frac{1}{3}\{W_1(1) + W_2(0) + W_3(0)\}$
101	$W_1(1)$	$W_2(0)$	$W_3(1)$	$\frac{1}{3}\{W_1(1) + W_2(0) + W_3(1)\}$
110	$W_1(1)$	$W_2(1)$	$W_3(0)$	$\frac{1}{3}\{W_1(1) + W_2(1) + W_3(0)\}$
111	$W_1(1)$	$W_2(1)$	$W_3(1)$	$\frac{1}{3}\{W_1(1) + W_2(1) + W_3(1)\}$

For example, if $N = 3, K = 0, m = 2$.

$$\eta = (0,0,0) \text{ and } \eta_2 = (0,1,0),$$

then
$$W((0,1,0)) - W((0,0,0)) = \frac{1}{3}\{W_2(1) - W_2(0)\}. \quad (1.7)$$

For more details refer to Table 1.

CASE B: $K = N - 1$ Similarly, for $m = 1, \ldots, N$ we have

$$W(\eta) = \frac{1}{N}\sum_{i=1}^{N} W_i(\eta(i)) \quad (1.8)$$

$$W(\eta_m) - W(\eta) = \frac{1}{N}\sum_{i=1}^{N}\{W_i(\eta_m) - W_i(\eta)\}, \quad (1.9)$$

where
$$W_i(\eta) = W_i(\eta : \Delta_i(N-1)).$$

For example, if $N = 3, K = 2, m = 2, \eta = (0,0,0), \text{and} \eta_2 = (0,1,0)$, then

$$W((0,1,0)) - W((0,0,0)) = \frac{1}{3}\sum_{i=1}^{3}\{W_i((0,1,0)) - W_i((0,0,0))\}. \quad (1.10)$$

In general, see Table 2.

TABLE 2 In this table we calculate some of the fitnesses for all of the possible strings for the case: $N = 3, K = N - 1, C = 0, S = 1$, and $S_i = 0$.

η	W_1	W_2	W_3	W_η
000	$W_1(0,0,0)$	$W_2(0,0,0)$	$W_2(0,0,0)$	$\frac{1}{3}\{W_1(0,0,0) + W_2(0,0,0) + W_3(0,0,0)\}$
001	$W_1(0,0,1)$	$W_2(0,0,1)$	$W_2(0,0,1)$	$\frac{1}{3}\{W_1(0,0,1) + W_2(0,0,1) + W_3(0,0,1)\}$
⋮	⋮	⋮	⋮	⋮
111	$W_1(1,1,1)$	$W_2(1,1,1)$	$W_2(1,1,1)$	$\frac{1}{3}\{W_1(1,1,1) + W_2(1,1,1) + W_3(1,1,1)\}$

CASE C: $0 \leq K \leq N-1$ **(THE GENERAL CASE)** This case is more complicated than the previous $K = 0$ and $K = N - 1$ cases. For example, assume that

$$\Delta_i(K) = \{i = 1, \ldots, i, \ldots, i + r\}$$

with $l \geq 0, r \geq 0$ and $l + r = K$, and we adopt periodic boundary condition. Then for $m = 1, \ldots N$, we obtain

$$W(\eta) = \frac{1}{N} \sum_{i=N-r}^{N+1} W_i(\eta : \Delta_i(K)), \tag{1.11}$$

$$W(\eta_m) - W(\eta) = \frac{1}{N} \sum_{i=N-r}^{N+1} \{W_i(\eta_m : \Delta_i(K)) - W_i(\eta : \Delta_i(K))\}. \tag{1.12}$$

For example, if $N = 5, K = 2, (l = r = 1), m = 2, \eta = (0,0,0,0)$, and $\eta_5 = (0,0,0,1)$, then

$$\begin{aligned}W(\eta_5) - W(\eta) = &\frac{1}{5}[\{W_4((0,0,1)) - W_4((0,0,0))\} \\&+ \{W_5((0,1,0)) - W_5((0,0,0))\} \\&+ \{W_1((1,0,0)) - W_1((0,0,0))\}.\end{aligned} \tag{1.13}$$

In the above cases, $\Delta_i(K)$ is chosen deterministically. In particular, if $K = 2r(l = r)$, then $\Delta_i(K)$ is a symmetrically chosen epistatic set.

On the other hand, we can consider the case:

$$\tilde{\Delta}_i(K) = \Delta_i(K)\backslash\{i\}(= \{i_1, \ldots, i_k\})$$
$$\tilde{\Delta}_i(N) = \{1, \ldots, N\}\backslash\{i\},$$

where $A\backslash B = \{x : x \in A \cap B^C\}$ for sets A and B. That is the set of points that belong to A but not to B. (This is sometimes called the *difference*.)

In this case, $\tilde{\Delta}_i(K)$ is a randomly chosen epistatic set, i.e., there is a random variable Y such that

$$P[Y = \tilde{\Delta}_i(K)] = P_{\tilde{\Delta}_i(K)} \geq 0 \text{ and } \sum_{(\eta(l): l \in \tilde{\Delta}_i(K)) \in X_K} P_{\tilde{\Delta}_i(K)} = 1.$$

Notice that the random cases of $K = 0$ and $K = N - 1$ coincide with deterministic ones.

We hope that this formalism and our examples using it will clarify some of the descriptions of NK and NKC models. This formalism has already been used to achieve some analytical results.[14]

2. COMMENTS ON NKC MODELS

The comments in this section are a mixture of two forms. Some of our comments suggest areas where the NK/NKC models seem to diverge from the systems they intend to model so much that the model should be altered; others explore some of the details of the model structure.

As a backdrop for these comments, it is useful to review some of the stated[12] goals of NK models. Kauffman and Weinberger listed the following issues:

1. How many local optima exist in a landscape?
2. What is the distribution of optima in the landscape? Are they near one another in special subregions of the space or randomly scattered?
3. What are the lengths of uphill walks to local optima?
4. As an optimum is approached, the fraction of fitter neighbors must dwindle to 0. How rapidly does the fraction of fitter mutants dwindle?
5. Because the fraction of neighbors which are fitter dwindles to 0, there is some characteristic relation between the number of mutations "tried" and the number "accepted" on an adaptive walk. How are the two related?
6. How many alternative optima are accessible from a given starting point? Can a "low-fit" peptide typically climb to all possible local optima, or only a small fraction of those optima? Among the accessible alternative optima, how often will each be "hit" on independent adaptive walks from the same starting point?
7. How many of the possible peptides can climb to any specific optimum, including the global optimum? A small fraction? Almost all?
8. Since most adaptive walks end on local optima, what are the fitnesses of such optima and how do they compare with the global optimum in the space?
9. The one-mutant variants of a local optimum must be less fit than the optimum. But do all of the variants lead to nearly the same loss of fitness or is there high variance indicating precipitous cliffs and gentle ridges in different directions in the high-dimensional space?

This list indicates some of the initial goals. Each one of them is an opportunity to question the underlying assumptions. We offer some of those questions below.

2.1 SWARMS IN STATE SPACE

It should be clear that populations never settle down to the kind of equilibrium that allows them to be fairly represented as a *single* point in a high-dimensional binary state space. Each point in the space is a string of length D, the dimension of the space. Rather a population is really a swarm across this lattice, and the swarm is localized in part of the space with a hamming distance radius less than some e. (The hamming distance is a measure that counts the number of non-identical bits in two binary strings.) This approach allows us to imagine the bifurcation of a swarm and even imagine a population space interaction that excludes some strings.

There might be lethal strings that also lie within e and divide the space. We should like to see future modeling of NKC models grapple with this.

The NKC model, in effect, assumes that the population sizes do not change (they are monotypic and genes get fixed after each mutation). A more realistic assumption would be that populations change as a monotonic function of fitness. The importance of a particular species to the fitness of another species (through C linking genes) should be related to the population of that species. Thus, if a frog's sticky tongue at time t drives flies to near extinctions at time $t + 1$, that sticky tongue contributes less to the frog's fitness at $t + l$ than at t.

It is important that these population dynamics be incorporated into the model because it deals with populations whose fitness, and therefore sizes, would be fluctuating dramatically, and the resulting dynamics would affect whether and what types of equilibrium arise. It is exactly this long-term behavior that Kauffman is interested in (e.g., "frozen state" vs. "near frozen" vs. "liquid state").

2.2 WORD CHUNKING

One of the reasons for modeling with binary strings is the feeling that the results are easily portable to situations where more than a single bit is necessary at each location to describe the system. This can easily be seen when we apply the NK model to different biological levels. If the binary string represents the four bases in a strand of DNA, then we need two-bit *words* to specify the four bases (e.g., 00:adenine, 01:cystosine, 10:thymine, 11: guanine). If we were to be modeling the amino acids in proteins, we would need five-bit words (This would even leave a few extra redundant or meaningless words). If we imagine that we are looking at the whole set of proteins, we might need very long words indeed.

The question then arises: which of the manipulations that we apply to our binary strings (mutation, crossing over, inversion) are unchanged if we need to think with longer words? Are the conclusions claimed for NK models and especially NKC models complicated by the need to preserve behavior for a range of word sizes? We are particularly concerned about the meaning of the basic operators on different length words. Inversion and crossing over may be deleterious far more often when moe than one bit is used to represent a basic unit of information.

2.3 CHANGE IN N THROUGH EVOLUTION

An assumption of the NK models and (even more unlikely) in the NKC models is that there is a single value N and that it does not change in the course of the simulation. N is the length of the binary string. The length of the genome changes in evolution, and different species will not share the same N.

2.4 DISTRIBUTION OF K

Random distribution of fitness values do not reflect "realistic" fitness landscapes. The use of random assignments to model epistatic interactions was motivated by an admission of profound ignorance.[11] However, given that we do not know and possibly cannot know realistic values for the fitness contribution of an allele, it is also not clear that a random distribution of fitness represents the behavior of "real" systems well. It may be that conditions defining biological systems are just a small subset of these random assignments. Conclusions may thus be based on an atypical set of data. (It would be worthwhile examining the sensitivity of these conclusions to different assumptions about fitness landscapes.)

A closely related question is whether there are distributions that more closely fit our intuition of epistatic interactions. Clearly, using a single mean value (K, in the formalism) to represent the epistatic interactions is a caricature. Assuming a distribution on the number of interactions would be a better approximation. This distribution would likely be skewed, so that most genes are unlinked and a few would be highly linked (see Figure 6).

A similar argument can be made for the parameter C; however, it is more difficult to recommend a shape for the distribution. Clearly some members of a community are linked in many ways while many members are peripherally linked if they are linked at all.

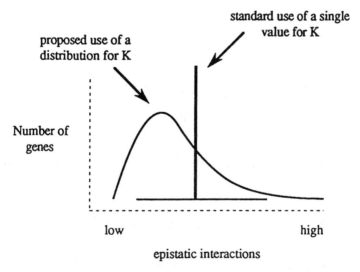

FIGURE 6 Current models use a single value for K and one possible distribution of K values.

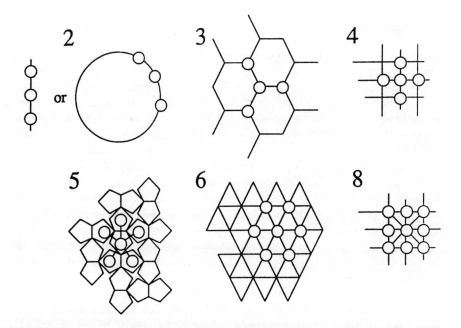

FIGURE 7 This diagram represents six different two-dimensional embeddings (tilings) of neighbors that preserve the notion of neighborhood for $C = 2, 3, 4, 5, 6, 8$. We have not found one for 7 yet.

2.5 WHAT ARE THE VALUES OF K AND C FUNCTIONS OF?

A species faces adaptive choices which affect its interconnectedness with the rest of the ecosystem. It is therefore possible that a system, rather than changing the ruggedness of its landscape (e.g., by changing K) to suit its interconnectedness with the ecosystem (C), would do exactly the opposite. More plausible is that both change.

2.6 REPRESENTING THE NEIGHBORHOOD OF INTERACTING SPECIES ON LATTICES

When looking at islands of chaos in seas of stability,[13] a rectangular grid is used. To have the intuitive feeling of neighborhood and also vary the number of interacting species, it is important to use an embedding in two dimensions that preserves the notion of neighbors. In Figure 7 we indicate some approaches to neighborhoods on planes that preserve our notion of closeness. (Notice that in the two-neighbor case a circle is a natural way to have periodic boundary conditions.) Previous work has dealt with the four-neighbor case (we think) because it was convenient to display and discuss since drawing it spans the plane.

2.7 THE ROBUSTNESS AND EXISTENCE OF K_{opt}

Kauffman suggests that there is an optimum K_{opt} towards which a species will evolve. This conclusion is derived under the assumption that no particular epistatic interaction or set of interactions is especially important. However, it is plausible that the epistatic interaction of *specific traits* might be much more important than a particular K value (i.e., K_{opt}). Thus, a successful large combination (much larger than K_{opt}) of epistatic interactions might overwhelm the disadvantages of a large K. Therefore, it is possible that $K \gg K_{opt}$. It would therefore be worth asking how the force of the "attraction" towards K_{opt} varies with distance from K. A further question is: how it would affect the dynamics of the system as a whole to have certain species "stuck" at $K \gg K_{opt}$?

It is not correct to infer the attraction of a K_{opt} from observations of higher fitness scores for species which have K's closer to K_{opt}. Presumably, the amount of epistatic interactions within a species is altered by the development of new traits (or changes in old ones) which interact with existing traits (or the disappearance of old traits which interacted with other traits). A change in K is not qualitatively equivalent to a single mutation in epistatic interaction with a single other gene. An incremental "change" in K (i.e., +1 or -1) results in *every* single gene increasing its epistatic interaction with a single other gene. Further, this is not a change in an evolutionary sense, since a species in the NK model cannot adapt by "changing its K." A test of the "K_{opt} conjecture" at the appropriate level of analysis would operate through "epistatic adaptation," one pair of genes at a time.

2.8 WHAT OTHER OPERATIONS DEFINE THE GENOTYPE NEIGHBORHOOD?

The one-mutant neighborhood is useful in that it defines what local genotype space can be explored. Evolution in the NK model is constrained to pass via 1-mutant fitter variants.[9,10,11,12,16] But there are other mechanisms, as important as mutation, that define the local neighborhood. (Here the neighborhood is the set of genotypes that can be *reasonably* explored in a single generation.)

The fraction of local genotype space to be explored is a function of population size, genome size, mechanisms of exploration, chance, etc. Since the part of the genotype space explored is not necessarily exhaustive, the movement of fitter neighbors rather than fittest neighbors seems a bit more accurate.

Mechanisms of exploration allow us to cross the discrete bit space. Mutation is one of those mechanisms and the one-mutant neighborhood is the "local" space that is explored in the standard NK model. If we combine inversions or diploid genotypes with crossing over, we can then jump far across the space in a single step. So, in our earlier discussion on formalism, we added the inversion and crossing-over operators.

These operators act as trap doors (worm holes if you wish) which connect parts of the space that may be far apart with respect to hamming distance. Population size acts as a constraint in as much as there are a finite number of offspring. Genome size interacts with population size in that a huge genome has a very large local neighborhood to explore and if the population size is small, only a small fraction of this space will be explored.

2.9 SHOULD ALL SPECIES HAVE THE SAME TIMINGS?

Genome size should be inversely proportional to the number of steps a population can take in a given iteration of the model. Large genomes should take fewer steps and so should be able to explore a smaller subset of the local genotype neighborhood. Here we see that a natural complication of using different N is that we now must worry about *relative* rates of evolution.

2.10 OTHER QUESTIONS

In the spirit of Kauffman and Weinberger's list (above), we will end this discussion with a list of questions that will need to be addressed in subsequent analyses of fitness landscapes within the NKC formulation.

1. What is the relationship between K and C?
2. Why is the evolution of K more important than evolution of C?
3. Why should the fitness contribution of K and C have the same magnitude? Could K and C be in a different "currency?"
4. How can the fitness of interacting genes be reflected in C?
5. What is the relationship between fitness components and the number of genes?
6. How do we introduce the long and convoluted series of events (development) that it takes from genes to traits? This is critical because it is traits that interact.
7. What is the mechanism of ecosystems tuning themselves via changing S's (the number of species)?
8. How important is it to model swarms across the discrete space instead of just populations as points?
9. Does it affect the model results if turns are taken synchronously or serially?
10. How does epistasis in NKC models correspond to the standard notions of epistasis?
11. How do we reflect the intuition that different members of an ecosystem are of different complexity?
12. At what levels of organization are the NK/NKC models most appropriate?

3. NKC MODELS AND SOCIAL SCIENCE

In this third part we want to broaden the horizon of NKC models to the social sciences. We examine the application of NKC models to the problem of coevolving complex strategies, such as firm diversification, and to coevolving complex belief systems, such as attitudes toward the government. By "coevolving" we mean that different actors adapt to each others actions over time; by "complex" we mean that individual actors interact with each other so that the outcome is more than the sum of individual actions. To demonstrate the usefulness of NKC models in the area of social sciences, we outline a few specific applications.

3.1 THE PROBLEM OF COEVOLVING STRATEGIES

Evolutionary models have been used to study the coevolution of strategies in the social sciences.[3,4] These models assume that effective behavior becomes more common either because of emulation of successful actors [20] or because of the physical replacement of actors by more successful innovators.[8] Coevolving strategies in NKC models differ from the evolutionary models above. Specifically, rather than assuming that unsuccessful behaviors are replaced, we assume that actors incrementally change their behavior over time. In this context NKC models would be interpreted as follows:

1. Each actor (species) has a set of actions (genes) that it may or may not engage in. There are N such actions which together form the strategy of an actor.
2. Each of those actions contribute to the success, or failure, of an actor. The overall success of an actor is called the performance (fitness) of an actor.
3. The contribution of the performance by a particular action is contingent on K other actions by the same actor, and C actions by each other actor.
4. Every actor chooses to change a single action during each time period. Identical to Kauffman, one can assume *fitter dynamics or fittest dynamics*.

One of the key assumptions of NKC models is that individuals choose only actions that improve their position locally. Such an assumption is consistent with the bounded rationality research tradition of human and organizational behavior.[6,17,20,22] The assumption of myopic behavior allows us to explain which of many possible Nash equilibria (or "local" Nash equilibria[13]) will be chosen from a given starting point. Such an approach not only allows us to explain why actors sometimes choose an inferior local optimum, but also why outcomes are history dependent and contingent on small events.[5]

3.2 COEVOLVING STRATEGIES OF FIRMS

NKC models can be used to study coevolving strategies of economic actors. A particular example would be a firm's decision to diversify into new markets. The decision to diversify is difficult because firms must take into account many interdependent dimensions, such as the profit potential of different markets, the synergy effects of operating in related markets, and the strategic responses of competitors.

For example, Michelin, a major European tire manufacturer, decided to enter the North American market because it expected high profits in that market due to its superior technology in producing high-quality radial tires. However, Michelin did not expect that Goodyear, a U.S. tire manufacturer, would retaliate by lowering its tire prices in major European markets. This example not only demonstrates the complexity of diversification, but also demonstrates the consequences of myopic behavior when strategic decisions are linked. Michelin tried to myopically improve its market position, but at the same time "deformed" Goodyear's profitability landscape, which in turn led Goodyear to myopically cut prices in Europe.

In a model of firm diversification, a single action would be to enter or exit a market. K would be interpreted as the number of synergies among different markets, and C the number of interdependent actions between each pair of competitors. Such a model of firm diversification can be used to study why firms "lock into" suboptimal positions (or "competency traps," as Levitt and March[15] call them). This model allows us to study how the effects of "lock in" varies with the technological complexity captured by K and C. Once the effects of "lock in" are understood, NKC models could be used to suggest improvement of strategic behavior, so that firms could "walk on rugged profitability landscapes" without falling into competency traps.

3.3 THE EMERGENCE OF BELIEF SYSTEMS

NKC models, applied to social psychology, can offer insights into the dynamics of changes in attitudes and beliefs within an organization or society. Further, it can explain the existence of particular types of reinforcing cleavages. For example, there is a strong correlation among attitudes towards government intervention in the economy, the necessity of a strong military, and the desirability of a social welfare system. We discuss an application below where people are the level of analysis, but these models of social conformity might also be applicable to "societies" of organizations (e.g., governments,[23] businesses, etc.).

Psychology offers the beginning of a solution. One's beliefs are affected by the beliefs of surrounding people.[7,19] Further, there tends to be a consistency among one's beliefs, such that the beliefs are mutually reinforcing ("cognitive consistency"[2]). For example, an individual who believes in the desirability of a large military budget would tend to believe that such expenditures come at a low-opportunity cost. However, these are results at the individual level. The question we are asking is about characteristics of aggregate opinion. The NKC model offers a means of examining what pattern in the aggregate would emerge out of individual

Reality Kisses the Neck of Speculation **349**

beliefs (for an early model see Abelson and Bernstein[1]; for more recent models see Novak et al.[21] and March[18]). In this context the NKC model would be interpreted as follows:

1. Each individual (species) has a number of beliefs (genes) which it may or may not hold. The total number of beliefs is N.
2. An index of cognitive frustration (fitness) can be constructed, based on the level of consistency among the individual's beliefs, and the congruence of those beliefs with those of surrounding individuals.
3. The contribution to cognitive frustration of each belief is contingent upon K other beliefs within that individual, and C other beliefs in each surrounding actor.
4. Each actor may change one belief each round, either under the assumption of "fitter dynamics" or "fittest dynamics" as outlined by Kauffman.

It would be useful to assume that all actors exhibit identical interactions among their own beliefs. For example, for *all* actors, believing "military spending is necessary to counter the 'Soviet threat'" and "military spending is good for the economy" results in a better cognitive frustration score than believing just one or the other. This model could be applied to the dynamics of opinion change. For example, until very recently the belief in the Soviet threat was the organizing principle of U.S. foreign policy. Beliefs about military spending, and about policies towards particular countries, among other things, were shaped by this overarching belief. Now that this belief has been exogenously changed, a critical question is how other beliefs about U.S. foreign policy will change.

3.4 SUMMARY

The above interpretation of NKC models as coevolving strategies shows that this family of models has a much broader range of application than just explaining coevolving species in biology. In particular, NKC models can be used in such diverse fields as economics, political science, organizational theory, and social psychology. NKC models can be applied to any research area that involves studying complex, coevolving behaviors at the individual, group, or organizational level.

4. CONCLUSIONS

We found NKC models to be stimulating and illuminating. We have tried to bring together three of the main directions that our group took. The first section of this paper summarized a mathematical formalism for the NK family of models. The second section focused one. Our third section explores possible applications of NK models to the social sciences.

We found the NK and NKC formulations to be a good intersection and basis for discussion for individuals whose areas of interest reached from immunology to evolution to physics and (as demonstrated in section 3 above) economics and political science. This is one of the strengths of this approach but may also be its downfall. We consistently found it difficult to rigorously apply the terms and relations of NKC models into the terminology and "facts" of a particular discipline. We look forward to future results in both the theoretical and specific applications of these models.

ACKNOWLEDGMENTS

We acknowledge Stuart Kauffman for living on the edge. Jeremy Ahouse would like to acknowledge Dr. Neil Simister for the time to pursue this project and Santa Fe for the inspiration to do so. Norio Konno would like to thank Lee Altenberg for useful discussions on the definitions of the NKC models. David Lazer would like to thank the University of Michigan for support from the Rackham Discretionary Fund.

REFERENCES

1. Abelson, R. P., and A. Bernstein. "A Computer Simulation Model of Community Referendum Controversies." *Public Opinion Quarterly* **27** (1963): 93–122.
2. Abelson, R. P. "Social Psychology and Rational Man." *Rationality and the Social Sciences*, edited by Mortimore and Benn. London: Rutledge, 1976.
3. Axelrod, Robert. "The Emergence of Cooperation Among Egoists." *Am. Pol. Sci. Rev.* **75** (1981): 306–318.
4. Axelrod, Robert. *The Evolution of Cooperation*. New York: Basic Books, 1984.
5. Axelrod, R., W. Mitchell, R. E. Thomas, S. Bennett, and E. Bruderer. "A Landscape Theory of Alliances with Application to Standard Setting." Working Paper 666, Graduate School of Business Administration, University of Michigan, 1991.
6. Cyert, Richard M., and James G. March. *A Behavioral Theory of the Firm*. Englewood Cliffs, NJ: Prentice-Hall, 1963.
7. Festinger, L., S. Schiter, and K. Back. *Social Pressure in Informal Groups*. Stanford: Stanford University Press, 1950.
8. Hannan, Michael T., and John Freeman. *Organizational Ecology*. Cambridge MA: Harvard University Press, 1989.

9. Kauffman, Stuart A., and Simon Levin. "Towards a General Theory of Adaptive Walks on Rugged Landscapes." *J. Theor. Biol.* **128** (1987)11–45.
10. Kauffman, Stuart A., Edward D. Weinberger, and Alan S. Perelson. "Maturation of the Immune Response via Adaptive Walks on Affinity Landscapes." In *Theoretical Immunology*, edited by Alan S. Perelson, 349–382. Santa Fe Institute Studies in the Sciences of Complexity, Proc. Vol. II. Reading, MA: Addison-Wesley, 1988.
11. Kauffman, Stuart A. "Adaptation on Rugged Fitness Landscapes." In *Lectures in the Sciences of Complexity*, edited by Daniel L. Stein, 527–618. Santa Fe Institute Studies in the Sciences of Complexity, Lect. Vol. I. Reading, MA: Addison-Wesley, 1989.
12. Kauffman, Stuart A., and Edward D. Weinberger. "The NK Model of Rugged Fitness Landscapes and Its Application to Maturation of the Immune Response." In *Molecular Evolution on Rugged Landscapes: Proteins, RNA, and the Immune System*, edited by Alan S. Perelson and Stuart A. Kauffman, 135–175. Santa Fe Institute Studies in the Sciences of Complexity, Proc. Vol. IX. Redwood City, CA: Addison-Wesley, 1990.
13. Kauffman, Stuart A., and Sonke Johnson. "Coevolution to the Edge of Chaos: Coupled Fitness Landscapes, Poised States, and the Coevolutionary Avalanches." *J. Theor. Biol.* **149** (1991): 467–506.
14. Konno, Norio. "Some Mathematical Results on the NK Model." This volume.
15. Levitt, Barbara, and James G. March. "Organizational Learning." *Annual Review of Sociology* **14(3)** (1988): 19–40.
16. Macken, Catherine A., and Alan S. Perelson. "Protein Evolution on Rugged Landscapes." *Proceedings of the National Academy of Science USA* **86** (1989): 6191–6195.
17. March, James G., and Herbert A. Simon. *Organizations*. New York: Wiley, 1958.
18. March, James G. "Exploration and Exploitation in Organizational Learning." *Organizational Science* **2(1)** (1991): 71–87.
19. Moscovici. "Social Influence and Conformity." In *The Handbook of Social Psychology*, edited by G. Linddrey, and E. Aronson, 347–412, 3rd ed. New York: Random House, 1985.
20. Nelson, Richard R., and Sidney G. Winter. *An Evolutionary Theory of Economic Change*. Cambridge MA: Belknap Harvard, 1982.
21. Novak, Andrejej, Jack Szamre, and Bob Latané.. "From Private Attitude to Public Opinion: A Dynamic Theory of Social Impact." *Psychological Review* **97(3)** (1990): 363–376.
22. Simon, Herbert A. "A Behavioral Model of Rational Choice." *Quart. J. Econ.* **69** (1955): 99–118.
23. Walker, Jack. "The Diffusion of Innovation Among American States." *Am. Pol. Sci. Rev.* **68** (1969): 880–899.
24. Weinberger, Edward D. "A More Rigorous Derivation of Some Properties of Uncorrelated Fitness Landscapes." *J. Theor. Biol.* **134** (1988): 125–129.

25. Weinberger, Edward D. "Correlated and Uncorrelated Fitness Landscapes and How to Tell the Difference." *Biol. Cyber.* **63** (1990): 325–336.
26. Weiner, M., and W. Tozier. "Complex Systems Lecture Style." 1991.

Paddy Andrews
The Physiological Laboratory, Cambridge University, Cambridge CB2 3EG, England;
e-mail: pra11@uk.ac.cam.phx

Complex Patterns, Simply Recognized

INTRODUCTION

Visual recognition may be taken to be the ability to respond specifically to a particular scene of view from among many similar ones. Images falling on our retinae also need somehow to be functionally related to our memories of previously seen images for identification to be of any use.

Creatures with a need to respond to the objects in their visual environment in unsophisticated ways can be preprogrammed with a repertoire of stimulus/response behaviors. With increasing environmental complexity, such programming becomes less practical, and a more flexible approach is required. Things seen need to be recorded and used to modify future action in order to survive.

The human visual system is powerful and anatomically complex. This raises the question: Is the process of recognition itself necessarily complex? What would be the minimum system requirement for human-like recognition in real visual environments? The following three major requirements may be put forward as probable prerequisites for the formation of useful, internal representations of visual images.

1. Selection: Whether it be via wavelength selectivity or some higher-level process, some choice about survival relevance has to be made about what should be extracted from the welter of information available to the retina.
2. Specificity: To avoid too much confusion, the system should display a high level of specificity so that even small changes in an image should result in perceptible differences in the resulting representation.
3. Relatedness: Things seen should be related in visual memory by a range of associations (...The shape of that hat reminds me of a car she was driving...).

A simple algorithm has been developed, with reference to the human visual system, which can associatively store and retrieve information about a large range of different images and thereby act as a visual recognition device.

The central question of how we respond to objects may be addressed as a problem of coding (to gain access to visual memory) and decoding (the subsequent execution of, for example, appropriate, voluntary muscle movements). This work addresses only the first part of the problem. The basic hypothesis is that images which look similar; a human being should result in the production of "similar" codes by any process which attempts to simulate human recognition performance.

It must be stated that such things as image movement, stereopsis, color vision, and level of attention have not been considered in this study. Neither was the objective to explain how efficient image transmission and reconstruction may be performed. The *sensation* of seeing itself is ignored. What is experienced when, for example, an (emotionally neutral) triangle is viewed may be just as much of an internal construct as is generally believed to occur when we see things in our mind's eye. This approach challenges Marr's[19] assertion that an explanation of vision must conform to the plain man's experience of it. It seems reasonable that no theory must contradict objective measurements of experience, but introspection is not necessarily a reliable test of any theory. The visual system must be able to deal with large numbers of combinations of sensory inputs (these are limited, in practice, by the finite human lifespan and the fact that our visual environment is actually much less than infinitely variable). Here we restrict the problem still further to the identification of monochrome images of objects (and parts of objects). Even so, the number of potential images is intimidatingly large.

As a familiar introductory example of the type of question for which an explanation if required, how can the following all be recognized as variants on the same theme, while still being seen as subtly different...

BACKGROUND

Seeing something is clearly necessary but not sufficient for it to be recognized (it might be the first experience of it, or it may be out of focus and thereby only identifiable as being within a certain general category, or it may be simply upside-down). Experiments on recognition performance require, ideally, the control of *a priori* visual experience, which is difficult with human subjects. Much of the work in the literature, therefore, does not take this into account or relies on limiting the recognition task to a discrimination between membership and non-membership of a designated set of such pictures. Demonstrations of the enormous capacity of the human visual system to store and retrieve pictorial material have been conducted. A few examples are given below.

Goldstein and Chance[10] presented their subjects, for three seconds each, with pictures in three categories: women's faces, magnified snowflakes, and inkblots (three seconds corresponds to no more than nine fixations). Subjects were able to achieve 71% success in distinguishing between those slides of faces which they had seen before and those which they had not. (The recognition rate corresponding to success by chance was 14%.)

Potter and Levy[23] found recognition accuracy for pictures varied from 15% (with a 125ms exposure) to 90% (for two seconds' exposure).

Recognition performance was found to be a positive function of the number of fixations on a given picture and is not dependent on viewing duration *per se* (if the number of fixations is restricted to being constant). Pictures viewed only peripherally are not remembered at all.[17]

Results such as this seem to show that at least some significant things in large numbers of pictures can be efficiently stored in memory, despite restricted access to their information content, if the pictures are looked at directly.

The history of attempts to achieve pattern recognition has included both efforts inspired by biological systems and those which ignored Nature. The list of ideas includes: whole-pattern templates, e.g., the "bug detectors" of Lettvin et al.[16]; feature (mini template) detectors, derived from interpretations of the work of Hubel and Wiesel[13]; Marr-Nishihara canonical elements[18]; massively parallel statistical sieves (neural networks); Fourier transforms motivated by the findings of Campbell and Robson[4]; and so-called structural models (lists of characteristic properties) from work on artificial intelligence. Few have been successful by any standard. The work described here is different in that it involves viewing each scene a small area at a time and forming a unique representation of each successive, small "window" taken as a whole.

The cortices of cats, monkeys, and humans have been shown to perform analysis of visual images by the use of oriented local filters tuned to different spatial frequencies (spatial frequency bandwidths of one octave and orientation tuning bandwidths of 15–20° are typical[7]). It is not clear what, if any, significance these cells have for the recognition of patterns. A very large part of the visual cortex of these species is devoted to treatment of signals from the fovea—a tiny area near the center of the retina. The area with diameter subtending, in humans, the central 20 minutes of arc of the fovea was designated the foveola by Polyak.[22] A fingernail seen at arm's length subtends about one third of a degree at the eye. The foveola, therefore, subtends an angle equal to one third of a fingernail. This tiny region seems to have great significance for the recognition of patterns.

The foveola is particularly suitable subject in the study of vision because:

- it has a large cortical representation;
- it contains comparatively few, regularly arranged anatomical elements (only cones are present, providing dichromacy[6]);
- it forms a direct link to the visual cortex;
- it contains a relatively low ratio of cones to ganglion cells (0.3)[28]; and the optics which produce faveolar images have been extensively investigated.[3]

In addition, cells tuned to a very wide range of spatial frequencies are found in the foveal region of the visual cortex.

Harmon[11] showed faces could be recognized using a coarse pixellation of 16×16 with eight grey levels. More recent work by Campbell[5] has confirmed that only a few hundred activated groups of seven cones at a time in the foveola of the human retina are required to identify most everyday objects. This has the result that faces can be identified at a distance of 35 m.

Consider constancy of recognition; it is commonly accepted that in order for a system to be able to recognize an object at a distance, for example, it should be able to manipulate a scale-invariant internal representation of that object so as to equate it to the current view and permit recognition. Such invariants actually seem not to occur in a foveola, however. Size constancy fails below 1/2 degree.[24]

- Rotation invariance: It is very important that, for example, a right-angled diamond and a square of equal side length are perceived as different.
- Position invariance: The threshold for displacement detection in an unstructured field is near 1.5 min arc[15] or about three foveolar cone diameters.

The freedom of the eye to move makes the notion of position invariance vague.

What about local and global changes of illumination and occlusion? Recognition under these circumstances is actually rather hard to do. We are not particularly good at spotting camouflaged wildlife or reading an eye chart on which a mixture of sunlight and the shadows of a leafy branch have been superimposed. By viewing a scene as a sequence of small areas, problems of figure/ground segmentation (such as looking for a particular bolt in a box of engineering components) can be rendered tractable.

OPERATION OF THE PRESENT MODEL

Every white dot in a black and white picture screen is assumed to spread, according to a simplified simulation of diffraction, which causes a group of cones to become activated. This is shown for a pattern consisting of two stars or dots in Figure 1 (natural scenes or hand-drawn images can be accommodated).

As the model eye moves from fixation to fixation, each successive small area of an image falling on the central few hundred receptors of a simulated retina is analyzed. The image of an object or part-object on this central 1/3 degree of the retina is assumed to be moved relative to its original position by signals communicated between retinal cell in the three principle axes of the regular, hexagonal cone mosaic (Figure 1). This activation pattern may then be sampled and processed by idealized cells which are each sensitive to a narrow range of orientation and spatial frequency. The differencing operator indicated in Figure 2 fulfills this function and generates results which are broadly consistent (C1, C2, C3...) with the kinds of responses actually recorded from complex cells in the cortices of mammals.[7] Attempts to explain the significance which these cells may have for the recognition of objects have hitherto been unsuccessful. For each of the three principal orientations of the receptor mosaic, the outputs of these simulated cells are summed, giving rise to the three-element code. It is of particular interest that these simulated cells generate relatively small responses to "meaningless" random dot patterns (visual noise). This is believed to be related to the physiological finding that no real long-term memories are formed from such images, thus avoiding potentially massive waste of memory capacity, which might result from combinatorial explosion.

The asymmetric, local transmission and adding of activation values in the plane of this simulated retine, shown in Figure 1, has the effect of specifically labelling edges within an image according to their orientations. When this resultant activation matrix is analyzed, by the oriented receptive fields, the (x, y, z) code produced is characteristic of the original image in the sense that any change in the image (other than adding uniform noise) must affect at least one of the (x, y, z) components. This type of process has been previously discussed in connection with the well-developed visual system of the octopus.[26]

A single receptive field width of three cones and three orientations has been used. This is the computationally simplest selection which avoids errors of orientation, etc., which is still capable of surprisingly effective recognition performance.

Resulting codes have been generated and plotted as the coordinates of points in a three-dimensional representation space, each of which uniquely stands for a particular view.

Each view of an object produces a slightly different coding so that similar views "clump together" in representation space. This results in automatic, nonrigid perceptual categorization. Novel objects are automatically classified by virtue of their proximity in representation space relative to those of known images. Known objects are coded and can reactivate their existing representation and its associates; i.e., they are recognized.

As we manipulate an object or move around it, we foveate many successive views. Hochberg[12] has said that perception depends on integration of the parts seen foveally in each of several glimpses. The continuity of these trajectories has the effect that objects presented in the continuously varying sizes and orientations of everyday experience can still be recognized.

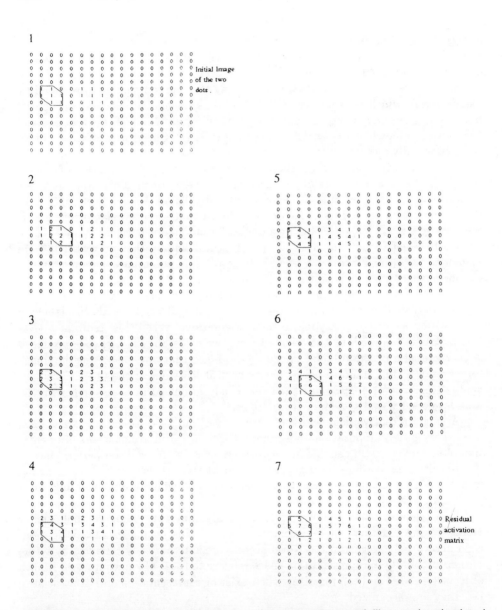

FIGURE 1 Activation levels, caused by an image of two "stars" falling on the simulated retina, are shown being transmitted asymmetrically along the axes of the hexagonal mosaic so as to form a residual activation matrix (mosiac is shown distorted to square for computational ease).

Warrington and Taylor[27] found that certain neurological patients were capable of recognizing objects only when seen from "conventional" viewpoints (suggesting

Complex Patterns, Simply Recognized

that they were unfamiliar with this view and had, therefore, no stored representation of it). Similarly, Palmer, Rosch, and Clare[20] reported that, for example, an unusual view of a horse was not easily recognizable. This suggests that, rather than computing what the plain view of a horse actually was from stored, conventional-view "coordinates," failure ever to have seen and recorded this view meant that it was simply not associated with other, conventional views labelled "horse." Yin[29] found that his subjects were poor at recognizing inverted faces. Diamond and Carey[8] found that this was true for a wide range of other objects, too.

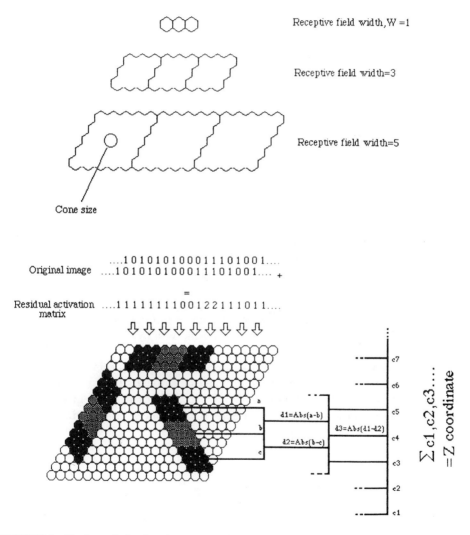

FIGURE 2 Design of simulated receptive fields.

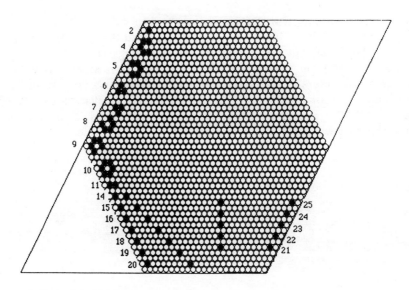

FIGURE 3 Some of the simplest possible pattern (see also Figures 7 and 8).

Faces (the relevant literature is reviewed in Bruce[2]) may be a special case in that they form a category, the members of which clearly differ from each other somewhat less than from the members of different groups (e.g., motor vehicles). For example, there have been many reports of patients who demonstrated an inability to recognize whole categories of objects (not just faces, as reported in Bodmer[1]) after *local* lesions. Some of these results seem to suggest that local lesions produce memory loss for objects of a very specific category.

Images consist of the spatial relationships between their elements (pixels). Any attempt to analyze an image and form its representation by counting up the number of edges in different orientations, for example, is likely to fail to encode the relationships which occur at corners, etc., or requires the *a priori* specification of a wide range of unwieldy "elements" of which all images may be assumed to be composed. An unspecific (resulting in confusions like "L" for "7") or insufficiently applicable code is produced. In the work described here, more of the spatial interrelationships are encoded, leading to an ability to form representations of a wide range of image types.

RESULTS

Results are presented for a small range of different images (Figures 3 through 6) as three-dimensional plots of their representations (Figures 7 through 10). Figure 3

Complex Patterns, Simply Recognized

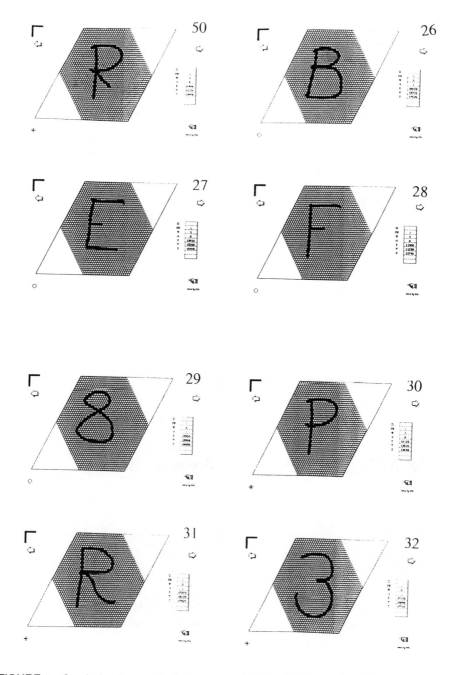

FIGURE 4 Symbols chosen for their apparent "similarity" (see also Figures 7 and 9).

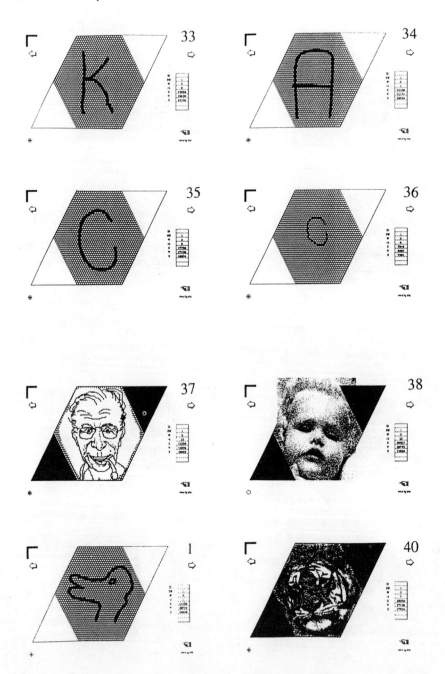

FIGURE 5 For a graphical description of the representations formed from these diverse images see Figures 7, 9, and 10.

Complex Patterns, Simply Recognized

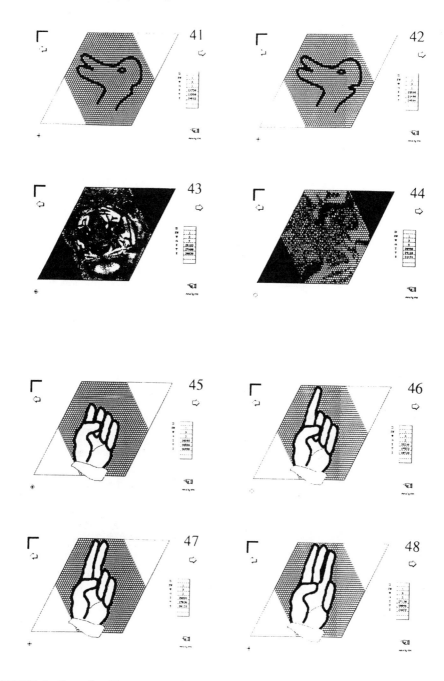

FIGURE 6 See also Figures 7 and 10.

364 Paddy Andrews

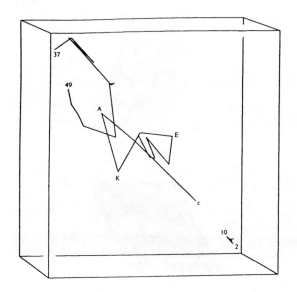

FIGURE 7 Overall view of "representation space."

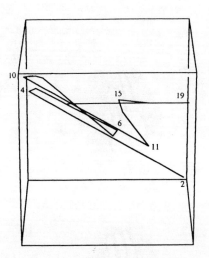

no.	Description
2	Single dot
4	5-dot c
5	5-dot c rot'd about vertical
6	three dots: one above two
7	three dots: two above one
8	small c: open at 10 o'clock
9	small c open at 4 o'clock
10	six dot hexagon
11	two dots touching horizontal
14	two dots..one space between
15	two dots..two..spaces between
16	two dots..three..spaces between
17	two dots..four..spaces between
18	two dots..five..spaces between
19	two dots..six..spaces between
20	two dots..seven spaces between
21	two dots..eight spaces between
22	two dots..nine spaces between
23	two dots..ten spaces between
24	two dots..eleven spaces between
25	two dots..twelve spaces between

FIGURE 8 Representations of some very simple patterns.

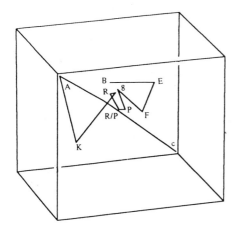

no.	Description
26	B
27	E
28	F
29	8
30	P
50	R/P
31	R
32	3
33	K
34	A
35	C
36	c

FIGURE 9 Representations of some handwritten capital letters.

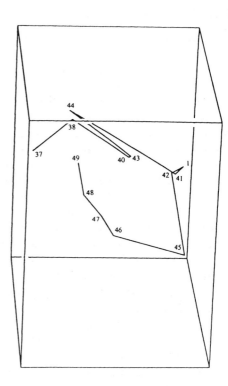

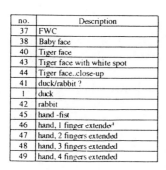

no.	Description
37	FWC
38	Baby face
40	Tiger face
43	Tiger face with white spot
44	Tiger face..close-up
41	duck/rabbit ?
1	duck
42	rabbit
45	hand -fist
46	hand, 1 finger extended
47	hand, 2 fingers extended
48	hand, 3 fingers extended
49	hand, 4 fingers extended

FIGURE 10 Some more complicated images, including a course temporal sequence.

is a composite of the simpler patterns, all of which appear as representations in the bottom right-hand corner of Figure 7. The numbers in bold type in Figures 4, 5, and 6 are used to indicate the relative positions of representations in Figures 7 through 10. Figure 7 shows the overall layout of representation space and indicates that a wide variety of different types of images can be accommodated within this scheme. There is a general increase in image "complexity" from the origin (single-point representations) outwards toward the representations of faces. Figures 8, 9, and 10 look in more detail at regions of the space shown in Figure 7.

Similar images do, indeed, result in similar codes and this, in turn, causes clustering of their representations. A short trajectory is shown for the image of a fist opening its fingers. Also note that the perceptually ambiguous image of the duck/rabbit is between that of the duck and that of the rabbit. This is true also for the ambiguous R/P image which lies between the points for P and R.

DISCUSSION AND CONCLUSIONS

These ideas may relate to the work on the inferotemporal cell ensembles known to be selectively responsive to complex visual stimuli. Perhaps an ensemble might correspond to a knot of trajectories in representation space, each signalling the presence of a view similar to those to which its neighbors are sensitive. Perret et al.,[21] for example, reported the apparent storage of face information in the inferotemporal cortex. Sakai and Miyashita[25] reported that IT cells recorded the temporal sequence of unfamiliar visual images. It has been shwon that IT cells have receptive fields centered on the fovea and that adjacent cells have similar response properties.

The number of foveations in a human lifetime (3 per second $\rightarrow 5 \times 10^9$) is, surprisingly, orders of magnitude less than the number of neurons in the visual centers of the brain, making it hard to dismiss these ideas purely on the grounds of "capacity." It is possible, it would seem, for recognition to be performed on the basis of stored representations of every single foveation in a human lifetime. This view of the visual system regards the brain as essentially a simple image analyzer linked, by a potentially simple coding process, to an enormous data bank of efficiently associated visual memories of shape information.

The postulated trajectories, if they exist in reality, could give an insight into prediction of what is about to appear. It may also be that linkage strength, between locations forming a trajectory, is related to probability of recall by some Hebb-like rule. Indeed, this system could be thought of as having the capacity to jump to the wrong conclusion: when asked to identify a church steeple, it may respond "rocket nose cone." This illustrates an ability to generalize and make errors.

The system described here is very simple; yet does seem to have some useful properties:

- It is potentially fast—one trial learning.

- Segmentation can be achieved by looking at small enough areas at a time—local shadows and occlusion can be accommodated.
- It is general and flexible.
- It is specific and accurate as long as images are not composed of regions which are separated by empty spaces; larger than 3W (see Figure 8).
- Only simple "technology" is required.
- Two ways to associate things are incorporated—visual "similarity" and experience of sequences stored as trajectories.

Kohonen[14] estimated in 1988 that there had been 30,000 papers published on pattern recognition and that the performance of artificial methods fell far short of that of biological sensory systems. Perhaps we fail to recognize that although biological systems are dauntingly complex at first sight, they often have an underlying simplicity of principle.

ACKNOWLEDGMENTS

Thanks are due to Professor F. W. Campbell, FRS; Professor Igor Aleksander; Aileen Briggs; the U.S. Department of the Environment; the National Science Foundation; the Office of Naval Research; a consortium of universities and laboratories, including the Santa Fe Institute; and Apple Computer Inc. This work was partly supported by a grant from the Kenneth Craik Fund of St. John's College, Cambridge.

Note: a copy of the software used in these investigations (which runs on the Apple Macintosh computer and makes use of MicroSoft's QuickBasic compiler and Apple's Hypercard software) is available from the author at the above address.

REFERENCES

1. Bodmer, J. "Die Prosop-Agnosie." *Archiv fur Psychiatrie und Nervenkrankheieten* **179** (1947): 6–53.
2. Bruce, V. *Recognizing Faces*. London: Lawrence Erlbaum, 1989.
3. Campbell, F. W., and R. W. Gubisch. "Optical Quality of the Human Eye." *J. Physiol.* **186** (1966): 558–578.
4. Campbell, F. W., and J. G. Robson. "Application of Fourier Analysis to the Visibility of Gratings." *J. Physiology* **197** (1968): 551–566.
5. Campbell, F. W., and Y. E. Shelepin. "The Mechanics of the Foveola and Its Role in Defining an Object." Paper presented at the Fergus Campbell Symposium on the Temporal and Spatial Domain, Twelfth ECVP Abstracts, A50.

6. Curcio, C. A., K. R. Sloan, R. Kalina, and A. Henrickson. "Human Photoreceptor Topography." *J. Comp. Neurology* **292** (1990): 497–523.
7. De Valois, E. L., and K. K. De Valois. *Spatial Vision*. Oxford: Oxford University Press, 1988.
8. Diamond, P., and S. Carey. "Why Faces Are and Are Not Special: An Effect of Expertise." *J. Exp. Psych.* **115** (1986): 107–117.
9. Ditchburn, R. W. *Eye Movements and Visual Perception*. Oxford: Clarendon Press, 1973.
10. Goldstein, A. G., and J. E. Chance. "Recognition of Complex Visual Stimuli." *Perception & Psychophysics* **9** (1971): 237–241.
11. Harmon, L. D. "The Recognition of Faces." *Sci. Am.* **229** (1973): 71–82.
12. Hochberg, J. "Levels of Perceptual Organization." In *Perceptual Organization*, edited by M. Kubovy and J. R. Plomerantz. Hillsdale, NJ: Erlbaum, 1981.
13. Hubel, D. H., and T. N. Wiesel. "Receptive Fields and Functional Architecture of Monkey Striate Cortex." *J. Physiology* **195** (1968): 215–243.
14. Kohonen, T. "The Role of Adaptive and Associative Circuits in Future Computer Designs." *Neural Computers*, edited by R. Eckmiller, and C. H. von der Malsburg. Heidelburg: Springer Verlag, 1988.
15. Legge, G., and F. W. Campbell. "Displacement Detection in Human Vision." *Vision Research* **21**: 205–213.
16. Lettvin, J. Y., H. R. Maturana, W. S. McCulloch, and W. H. Pitts. "What the Frog's Eye Tells the Frog's Brain." *Proc. Inst. Rad. Engrg.* **47** (1959): 1940–2051.
17. Loftus, G. R. "Eye Fixations and Recognition Memory for Pictures." *Cognitive Psych.* **3** (1972): 525–551.
18. Marr, D., and H. K. Nishihara. "Representation and Recognition of the Statial Organization of Three-Dimensional Shapes." *Proc. Roy. Soc. London B* **200** (1978): 269–294.
19. Marr, D. *Vision*. San Francisco, CA: Freeman, 1982.
20. Palmer, S. E., E. Rosch, and P. Chase. "Canonical Perspective and the Perception of Objects." In *Attention and Performance*, edited by J. Long and A. D. Baddeley, vol. IX. Hillsdale, NJ: Lawrence Erlbaum, 1981.
21. Perret, D. I., P. A. J. Smith, D. D. Potter, A. J. Mistlin, A. S. Head, A. D. Milner, and M. A. Jeeves. "Neurones Responsive to Faces in the Temporal Cortex: Studies of Functional Organisation, Sensitivity, and Relation to Perception." *Human Neurobiol.* **3** (1984): 197–211.
22. Polyak, S. L. *The Retina*. Chicago: University of Chicago Press, 1941.
23. Potter, M. C., and E. I. Levy. "Recognition Memory for a Rapid Sequence of Pictures." *J. Exp. Psych.* **82** (1969): 10–15.
24. Ross, J., B. Jenkins, and J. R. Johnstone. "Size Constancy Fails Below Half a Degree." *Nature* **283** (1980): 473–474.
25. Sakai, K., and Y. Miyashita. "Neural Organization for the Long-Term Memory of Paired Associates." *Nature* **354** (1991): 152–155.

26. Sutherland, N. S. "Visual Discrimination of Orientation and Shape by the Octopus." In *Perceptual Processing, Stimulus Equivalence and Pattern Recognition*, edited by P. C. Dodwell. New York: Meredith Corporation, 1971.
27. Warrington, E. K., and A. M. Taylor. "Two Categorical Stages of Object Recognition." *Perception* **7** (1978): 695–705.
28. Wassle, H., and B. B. Boycott. *Physiological Reviews* **71** (1991): 2.
29. Yin, R. K. "Face Recognition by Brain-Injured Patients: A Dissociable Ability?" *Neuro-Psychologia* **8** (1970): 395–402.

Antonio C. Roque Da Silva Filho
School of Cognitive and Computing Sciences, University of Sussex, Falmer, Brighton BN1 9QH, United Kingdom

Dynamical Behavior of a Pair of Spatially Homogeneous Neural Fields

Various types of dynamical behaviors are studied for a neural network made of excitatory and inhibitory neurons arranged separately in two layers under the assumption of uniform activity throughout the layers. The layers are treated mathematically as continuous one-dimensional fields, and the neurons have binary outputs. The analysis of the system follows one done previously by Amari, but now the synaptic strengths can vary with time according to two versions of Hebb's rule. The general existence conditions of the dynamical behaviors for the two versions are investigated, and the allowed cases presented.

1. INTRODUCTION

Neural networks are very complex dynamical systems, posing enormous difficulties for theoreticians to treat them mathematically. So far, only very simple kinds of networks could have been satisfactorily analysed in mathematical terms (for a review see, e.g., Levine[5]), leaving many questions about more general networks still unanswered. However, simple systems can sometimes give us valid insights about the behavior of more complex ones, and one can always have the hope that by adding some small new features to simple systems, one can find new types of behaviors and gain deeper understandings about more complex systems. The purpose of the present work is to try to show this for neural networks made of excitatory and inhibitory neurons.

Earlier studies on dynamical behaviors of neural networks consisting of neurons which are either excitatory or inhibitory, and have connections of the so-called lateral inhibition type, were done in the 1970s by Wilson and Cowan,[6,7] Ellias and Grossberg,[2] and Amari.[1]

In particular, Amari modelled the neurons as being arranged in a continuous fashion along a pair of one-dimensional neural fields, one made of excitatory neurons and the other made of inhibitory neurons. The states of the points on the fields were described by functions $u_i(x,t), i = 1,2$, giving, for each instant of time t, the average membrane potentials of the neurons around x on the excitatory and inhibitory fields respectively. The u's were assumed to have a rate of change with time proportional to the weighted integrals over the fields of the outputs of the excitatory and inhibitory neurons, including self-excitation. The outputs of the neurons were assumed to be given by the step function

$$f[u] = \begin{cases} 0 & \text{if } u \leq 0; \\ 1 & \text{if } u > 0. \end{cases}$$

This was done for mathematical convenience, since it was claimed that the results obtained would be valid for a monotonically increasing output function of u with saturation. The synaptic strengths were assumed to be time invariant and dependent only on the distances between the neurons, $\omega(x,x') = \omega(x-x')$. The inhibitory neurons did not have connections among themselves, and the connections from excitatory to inhibitory neurons had a very narrow fan out, so that only the inhibitory neurons imediately below a given point on the excitatory field would receive connections from it. Besides, the strengths of the excitatory-excitatory synapses were stronger than the strengths of the inhibitory-excitatory synapses at short distances, but weaker than them at longer distances (characterizing the lateral inhibition kind of connections).

Amari studied the dynamics of his field equations for two special cases, namely when the solutions are spatially homogeneous, $u(x,t) = u(t)$, and when the solution is a stationary travelling wave of a fixed shape, $u(x,t) = g(x - vt)$. He showed that solutions of both kinds are possible, and gave some examples of them. In particular, for the spatially homogeneous case, he found stable, oscillatory, and transient behaviors. Assuming spatial homogeneity one can represent the state of the system by the vector $\mathbf{u} = (u_1, u_2)$ in the u_1-u_2 plane. In the first quadrant u_1 and u_2 are positive, and because $f[u]$ is the step function, $f[u_1] = f[u_2] = 1$ in this quadrant. In the same way for the second, third, and fourth quadrants, one has $f[u_1] = 0$ and $f[u_2] = 1$, $f[u_1] = f[u_2] = 0$, and $f[u_1] = 1$ and $f[u_2] = 0$ respectively. In the u_1-u_2 plane, a stable state was identified as a constant vector in a certain quadrant towards which **u** would tend when having its initial position in a different quadrant. A transient behavior was identified as a situation in which the initial vector **u₀** happened to be in the same quadrant as the constant vector, so that the system would never get out of that quadrant, decaying quickly towards the constant vector. On the other hand, the oscillatory behavior was characterized by a constant jump of the state vector from a quadrant to the next one, and from this one to the next one, etc.

In this work, Amari's analysis is extended by incorporating into his model the following features:

- The inhibitory neurons have connections among themselves.
- The excitatory-inhibitory connections have a larger fan out, so that excitatory neurons at a point x can make synapses to inhibitory neurons located at points other than x.
- The two fields receive an external excitatory input v.
- All the synaptic strengths can vary with time according to rules defined in the next section.

2. THE FIELD EQUATIONS

The general field equations obeyed by the excitatory and inhibitory membrane potentials $u_1(x,t)$ and $u_2(x,t)$ are

$$\tau\frac{\partial u_1(x,t)}{\partial t} = -u_1(x,t) + \int \omega_1(x,x',t)f[u_1(x',t)]dx' \\ - \int \omega_2(x,x',t)f[u_2(x',t)]dx' + \\ s_1(x,t)v(x,t) - h_1, \quad (1)$$

and

$$\tau\frac{\partial u_2(x,t)}{\partial t} = -u_2(x,t) + \int \omega_3(x,x',t)f[u_1(x',t)]dx' \\ - \int \omega_4(x,x',t)f[u_2(x',t)]dx' \\ + s_2(x,t)v(x,t) - h_2, \quad (2)$$

where τ is the time constant of neuronal dynamics, assumed to be the same for both kinds of neurons; $\omega_i(x,x',t), i = 1,\ldots,4$ are the synaptic strengths of excitatory-excitatory, inhibitory-excitatory, excitatory-inhibitory, and inhibitory-inhibitory synapses respectively; $s_i(x,t), i = 1,2$ are the synaptic strengths of the connections between the external input $v(x,t)$ and neurons in the two fields; and $h_i, i = 1,2$ ($h_i > 0$) are the resting potentials towards which u_1 and u_2 decay in the absence of stimuli.

The assumption of spatial homogeneity permits one to rewrite these equation as:

$$\tau \frac{\partial u_1(t)}{\partial t} = -u_1(t) + f[u_1(t)] \int \omega_1(x,x',t)dx' - f[u_2(t)] \int \omega_2(x,x',t)dx' \\ + s_1(x,t)v(x,t) - h_1, \quad (3)$$

and

$$\tau \frac{\partial u_2(t)}{\partial t} = -u_2(t) + f[u_1(t)] \int \omega_3(x,x',t)dx' - f[u_2(t)] \int \omega_4(x,x',t)dx' \\ + s_2(x,t)v(x,t) - h_2. \quad (4)$$

Regarding the equations governing the time variation of the synapses, they will be assumed to be of a Hebbian type.[4] Two possible versions of the Hebbian rule will be considered as a way of comparing their implications for the system's behavior:

A. The first version is the one adopted by most of the authors in the literature. It assumes that the synapses vary proportionally to the product of the outputs of the pre- and post-synaptic neurons, denoted here by u_{pre} and u_{post},

$$\tau' \frac{\partial \omega_i(x,x',t)}{\partial t} = -\omega_i(x,x',t) + c_i(x,x')f[u_{post}(t)]f[u_{pre}(t)]. \quad (5)$$

B. The second version is the one proposed recently by the author.[3] In it the synapses where the pre-synaptic neuron is excitatory obey the same rule as Eq. (5), but the synapses where the pre-synaptic neuron is inhibitory obey the following rule:

$$\tau' \frac{\partial \omega_i(x,x',t)}{\partial t} = -\omega_i(x,x',t) + c_i(x,x')\left[1 - f[u_{post}(t)]\right]f[u_{pre}(t)]. \quad (6)$$

Let us call the first version type A, and the second one type B. In the type A version, the synaptic strength of both excitatory and inhibitory synapses will always increase when the pre- and post-synaptic neurons are firing in synchrony. In the type B version, this will happen as well for excitatory synapses, but inhibitory synapses will increase only when the pre-synaptic neuron is firing and the post-synaptic one is not firing. The constant τ' appearing in the above equations is the time constant characteristic of the synaptic dynamics and is the same for both versions and for the four synaptic types. The quantities $c_i(x,x'), i = 1,\ldots,4$ are assumed to be given by

$$c_i(x,x') = \begin{cases} c_i & \text{for } |x-x'| \leq \ell_i, \\ 0 & \text{for } |x-x'| > \ell_i, \end{cases}$$

where ℓ_i is the maximum distance within which the ith synaptic type can have non-zero strength.

As the external input is excitatory, the synapses between it and the neurons in the fields will always change with time according to the type A version of the Hebbian rule,

$$\tau'' \frac{\partial s_i(x,t)}{\partial t} = -s_i(x,t) + b_i f[u_i(t)]v(x,t), \qquad (7)$$

where $b_i, i = 1, 2$ are constants, and τ'' is the time constant characteristic of the dynamics of the input synapses. It is assumed to be different than τ'.

Since $f[u]$ is the step function, one can analyse the system's behavior in the u_1–u_2 plane, as Amari did. For each quadrant of this plane, the $f[u]$'s are constant, so that Eqs. (5) and (6) can easily be solved having solutions decaying exponentially with time as $\exp(-t/\tau')$,

$$w_i(x, x', t) = w(x, x')e^{-t/\tau'}(+c_i),$$

where the constants c_i were put in between brackets because they appear depending on the quadrant and the version of the Hebbian rule adopted.

For simplicity, we are going to assume that the external input is the same for all positions, and is kept constant up to a certain time t_0 and silenced immediately after that,

$$v(x,t) = \begin{cases} v & \text{for } 0 < t \le t_0; \\ 0 & \text{for } t > t_0. \end{cases}$$

This implies that the strengths of the input synapses have also to be spatially homogeneous and to decay exponentially with time, according to $\exp(-t/\tau'')$,

$$s_i(t) = s_i e^{-t/\tau''}(+b_i v), \qquad \text{for } t < t_0,$$

and

$$s_i(t) = s_i e^{-t/\tau''}, \qquad \text{for } t > t_0.$$

As it is possible to determine the temporal behavior of the synaptic strengths w_i and s_i for each quadrant, the equations for the membrane potentials u_i are reduced to the general type

$$\tau \dot{u}_i = -u_i + f(t),$$

where the $f(t)$'s are known functions of time, one for each quadrant. Equations of this type can be solved using the integrating factor $e^{t/\tau}$,

$$e^{t/\tau}(\dot{u} + u/\tau) = e^{t/\tau} f(t) \rightarrow u(t) = \int f(t') e^{(t'-t)/\tau} dt' + k e^{-t/\tau},$$

where k is a constant.

Hence, representing the global state of the network by the vector $\mathbf{u} = (u_1, u_2)$, one can write the solutions of Eqs. (3) and (4), one for each quadrant, as

1st Quadrant

$$\mathbf{u}(t) = \frac{\tau\tau'}{\tau' - \tau}\mathbf{W}^I e^{-t/\tau'} + \frac{\tau\tau''}{\tau'' - \tau} vse^{-t/\tau''} + \mathbf{k}^I e^{-t/\tau} + v^2\mathbf{b}^I + \mathbf{L}^I + \mathbf{h}, \tag{8}$$
for $t < t_0$,

$$\mathbf{u}(t) = \frac{\tau\tau'}{\tau' - \tau}\mathbf{W}^I e^{-t/\tau'} + \mathbf{k}^I e^{-t/\tau} + \mathbf{L}^I + \mathbf{h}, \quad \text{for} \quad t > t_0; \tag{9}$$

2nd Quadrant

$$\mathbf{u}(t) = \frac{\tau\tau'}{\tau' - \tau}\mathbf{W}^{II} e^{-t/\tau'} + \frac{\tau\tau''}{\tau'' - \tau} vse^{-t/\tau''} + \mathbf{k}^{II} e^{-t/\tau} + v^2\mathbf{b}^{II} + \mathbf{L}^{II} + \mathbf{h}, \quad \text{for } t < t_0, \tag{10}$$

and

$$\mathbf{u}(t) = \frac{\tau\tau'}{\tau' - \tau}\mathbf{W}^{II} e^{-t/\tau'} + \mathbf{k}^{II} e^{-t/\tau} + \mathbf{L}^{II} + \mathbf{h}, \quad \text{for } t > t_0; \tag{11}$$

3rd Quadrant

$$\mathbf{u}(t) = \frac{\tau\tau''}{\tau'' - \tau} vse^{-t/\tau''} + \mathbf{k}^{III} e^{-t/\tau} + \mathbf{h}, \quad \text{for } t < t_0, \tag{12}$$

and

$$\mathbf{u}(t) = \mathbf{k}^{III} e^{-t/\tau} + \mathbf{h}, \quad \text{for } t > t_0; \tag{13}$$

4th Quadrant

$$\mathbf{u}(t) = \frac{\tau\tau'}{\tau' - \tau}\mathbf{W}^{IV} e^{-t/\tau'} + \frac{\tau\tau''}{\tau'' - \tau} vse^{-t/\tau''} + \mathbf{k}^{IV} e^{-t/\tau} + v^2\mathbf{b}^{IV} + \mathbf{L}^{IV} + \mathbf{h},$$
for $t < t_0$,

$$\tag{14}$$

and

$$\mathbf{u}(t) = \frac{\tau\tau'}{\tau' - \tau}\mathbf{W}^{IV} e^{-t/\tau'} + \mathbf{k}^{IV} e^{-t/\tau} + \mathbf{L}^{IV} + \mathbf{h}, \quad \text{for } t > t_0. \tag{15}$$

The interesting point about these solutions is that they are the same for both versions of the Hebbian rule considered in this paper; only some of the constant vectors, defined below, are different. The constant vectors appearing in the above solutions are the following (apart from the $\mathbf{L}^i$'s, all of them are the same for the two versions of the Hebbian rule considered):

$$\mathbf{W}^I = (\omega_1 - \omega_2, \omega_3 - \omega_4), \quad \mathbf{W}^{II} = (-\omega_2, -\omega_4), \quad \mathbf{W}^{IV} = (\omega_1, \omega_3),$$

where the ω_i's are defined as

$$\omega_i \equiv \int_{-\infty}^{\infty} \omega_i(x,x')dx';$$

$\mathbf{s} = (s_1, s_2);\quad \mathbf{b}^I = (b_1, b_2);\quad \mathbf{b}^{II} = (0, b_2);\quad \mathbf{b}^{IV} = (b_1, 0);\quad \mathbf{h} = (-h_1, -h_2);$
$\mathbf{L}^I = (2\ell_1 c_1 - 2\ell_2 c_2, 2\ell_3 c_3 - 2\ell_4 c_4);\quad \mathbf{L}^{II} = (0, -2\ell_4 c_4);\quad \mathbf{L}^{IV} = (2\ell_1 c_1, 0)$ (Type A) ;
and
$$\mathbf{L}^I = (2\ell_1 c_1, 0);\quad \mathbf{L}^{II} = (0, -2\ell_4 c_4);\quad \mathbf{L}^{IV} = (2\ell_1 c_1, 0) \quad \text{(Type B)};$$

where

$$2\ell_i c_i \equiv \int_{-\infty}^{\infty} c_i(x,x')dx' = \int_{-\ell_i}^{\ell_i} c_i(x,x')dx'.$$

Thus, one has now the conditions of predicting the system's behavior for a given set of constants and an initial value for $\mathbf{u}$. Instead of doing that in this paper, which would involve the (quite arbitrary) stipulation of all the constants appearing in the equations, we are simply going to verify what sorts of dynamical behaviors are compatible with Eqs. (8)–(15).

3. DYNAMICAL BEHAVIORS ALLOWED BY THE EQUATIONS

There are two important times which enable one to determine the behavior of the system, namely t_0 when the external input stops being applied, and $t = \infty$ which gives the asymptotic value of $\mathbf{u}$ in the absence of external inputs. Eqs. (8)–(15) allow us to calculate the values of $\mathbf{u}$ at those times, depending on the initial quadrant in which $\mathbf{u}$ is. We will assume that the time t_0 during which the external signal is applied is much larger than the time constant of neuronal dynamics τ, but much smaller than the two time constants of synaptic dynamics, which will be assumed to have values of the same order,

$$\tau \ll t_0 \ll \tau' \sim \tau''.$$

Hence, one can write

$$e^{-t_0/\tau} \sim 0;\quad \text{and } e^{-t_0/\tau'} \sim e^{-t_0/\tau''} \sim 1 - t_0/\tau' \sim 1 - t_0/\tau'',$$

which leads us to rewrite Eqs. (8)–(15) as (the superscripts labelling the $\mathbf{u}$'s indicate the quadrant)

1st Quadrant

$$\mathbf{u}^I(t_0) \simeq \tau\left(1 - \frac{t_0}{\tau'}\right)\mathbf{W}^I + \tau\left(1 - \frac{t_0}{\tau''}\right)v\mathbf{s} + v^2\mathbf{b}^I + \mathbf{L}^I + \mathbf{h}; \qquad (16)$$

$$\mathbf{u}^I(\infty) = \mathbf{L}^I + \mathbf{h}; \qquad (17)$$

2nd Quadrant

$$\mathbf{u}^{II}(t_0) \simeq \tau\left(1 - \frac{t_0}{\tau'}\right)\mathbf{W}^{II} + \tau\left(1 - \frac{t_0}{\tau''}\right)v\mathbf{s} + v^2\mathbf{b}^{II} + \mathbf{L}^{II} + \mathbf{h}; \qquad (18)$$

$$\mathbf{u}^{II}(\infty) = \mathbf{L}^{II} + \mathbf{h}; \qquad (19)$$

3rd Quadrant

$$\mathbf{u}^{III}(t_0) \simeq \tau\left(1 - \frac{t_0}{\tau''}\right)v\mathbf{s} + \mathbf{h}; \qquad (20)$$

$$\mathbf{u}^{III}(\infty) = \mathbf{h}; \qquad (21)$$

4th Quadrant

$$\mathbf{u}^{IV}(t_0) \simeq \tau\left(1 - \frac{t_0}{\tau'}\right)\mathbf{W}^{IV} + \tau\left(1 - \frac{t_0}{\tau''}\right)v\mathbf{s} + v^2\mathbf{b}^{IV} + \mathbf{L}^{IV} + \mathbf{h}; \qquad (22)$$

$$\mathbf{u}^{IV}(\infty) = \mathbf{L}^{IV} + \mathbf{h}. \qquad (23)$$

Hence, the $\mathbf{u}^i(t_0)$ and $\mathbf{u}^i(\infty)$ do not depend on the initial state $\mathbf{u}^i(0)$. Irrespective of the point where $\mathbf{u}(t)$ starts off or enters in the ith quadrant, it will always go towards $\mathbf{u}^i(t_0)$ or $\mathbf{u}^i(\infty)$, depending on t being smaller or greater than t_0.

The first type of dynamical behavior to be checked against these equations is the full oscillatory one, where by full oscillation one means the system vector passing through all four quadrants. This is only possible for $t < t_0$.[1] There are two possible types of full oscillations, clockwise and anti-clockwise (see Figure 1).

---

[1] One can clearly see this, because $\mathbf{u}^{III}(\infty) = \mathbf{h} = (-h_1, -h_2) \in$ 3rd quadrant, so that $\mathbf{u}$ never gets out of the 3rd quadrant once it enters there after t_0.

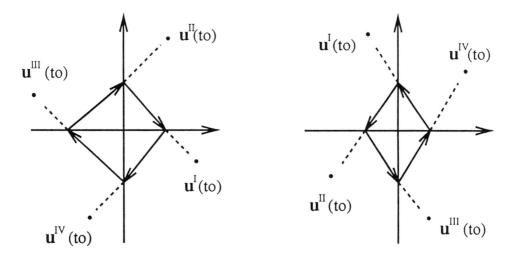

FIGURE 1 The two possible types of full oscillatory behavior. Only the anti-clockwise one is compatible with the equations.

TABLE 1 Types of Full Oscillations

Clockwise	Anti-clockwise
$\mathbf{u}^I(t_0) \in IV$;	$\mathbf{u}^I(t_0) \in II$;
$\mathbf{u}^{II}(t_0) \in I$;	$\mathbf{u}^{II}(t_0) \in III$;
$\mathbf{u}^{III}(t_0) \in II$;	$\mathbf{u}^{III}(t_0) \in IV$;
$\mathbf{u}^{IV}(t_0) \in III$;	$\mathbf{u}^{IV}(t_0) \in I$;

To test whether or not a full oscillation is compatible with Eqs. (16), (18), (20), and (22), one can decompose $\mathbf{u}^I(t_0)$–$\mathbf{u}^{IV}(t_0)$ into their components along vectors $(1,0)$ and $(0,1)$ and check whether or not the conditions (Table 1) can be simultaneously satisfied:

Performing the above-described analysis, one finds out that only the anti-clockwise behavior is allowed by the equations, and that this is the case for both type A and type B Hebbian rules. To show this here would involve writing down many algebraic inequalities, and this was not done for reasons of conciseness. For the type B Hebbian rule, the algebraic inequalities imply that the anti-clockwise oscillation is possible only for $\omega_2 > \omega_1$, but this does not happen for the type A

Hebbian rule, where both $\omega_1 > \omega_2$ and $\omega_2 > \omega_1$ are allowed; the condition for $\omega_1 > \omega_2$ being

$$2\ell_2 c_2 > \tau\left(1 - \frac{t_0}{\tau'}\right)(\omega_1 - \omega_2) + \tau\left(1 - \frac{t_0}{\tau''}\right)vs_1 + 2\ell_1 c_1 + v^2 b_1 - h_1.$$

It is interesting to mention here Amari's result concerning full oscillatory behavior.[1] In his paper Amari mentioned only the anti-clockwise oscillation and found that its existence condition is $\omega_2 > \omega_1$. However, any comparisons between the two results would be precipitated, because Amari's ω_i's are not the same as our ω_i's. Amari's ω_i's are the full strengths of the time-invariant synapses,

$$\omega_i^A = \int_{-\infty}^{\infty} \omega_i(x - x')dx',$$

where the superscript A indicates Amari, while in our case the full strengths of the synapses are time dependent and are not given by integrals of the ω_i's solely,

$$\omega_i^{full}(t) = \int_{-\infty}^{\infty} \omega_i(x, x', t)dx' = e^{-t/\tau'} \int_{-\infty}^{\infty} \omega_i(x, x')dx' \left(+ \int_{-\infty}^{\infty} c_i(x, x')dx'\right),$$

where the integral of $c_i(x, x')$ was put in between brackets because its presence in the above equation depends on the quadrant being considered.

Another interesting dynamical behavior whose possibility of existence can be checked with the use of Eqs. (16)–(23) is an oscillation between only two quadrants, which will be called a two-quadrant oscillation (see Figure 2).

For each of the four kinds of two-quadrant oscillations, there are nine possible dynamical cases. A list of them for the oscillations between the first and the second quadrants is given in Table 2.

Each pair of conditions in the above table have to be satisfied together with the conditions in its heading. Similar tables exist for the other three classes.

TABLE 2

$\mathbf{u}^I(t_0) \in II$ and $\mathbf{u}^{II}(t_0) \in I$		
$\mathbf{u}^{III}(t_0) \in III;$	$\mathbf{u}^{III}(t_0) \in III;$	$\mathbf{u}^{III}(t_0) \in III;$
$\mathbf{u}^{IV}(t_0) \in IV;$	$\mathbf{u}^{IV}(t_0) \in III;$	$\mathbf{u}^{IV}(t_0) \in I;$
$\mathbf{u}^{III}(t_0) \in II;$	$\mathbf{u}^{III}(t_0) \in II;$	$\mathbf{u}^{III}(t_0) \in II;$
$\mathbf{u}^{IV}(t_0) \in IV;$	$\mathbf{u}^{IV}(t_0) \in III;$	$\mathbf{u}^{IV}(t_0) \in I;$
$\mathbf{u}^{III}(t_0) \in IV;$	$\mathbf{u}^{III}(t_0) \in IV;$	$\mathbf{u}^{III}(t_0) \in IV;$
$\mathbf{u}^{IV}(t_0) \in IV;$	$\mathbf{u}^{IV}(t_0) \in III;$	$\mathbf{u}^{IV}(t_0) \in I;$

Dynamical Behavior of a Pair of Spatially Homogeneous Neural Fields

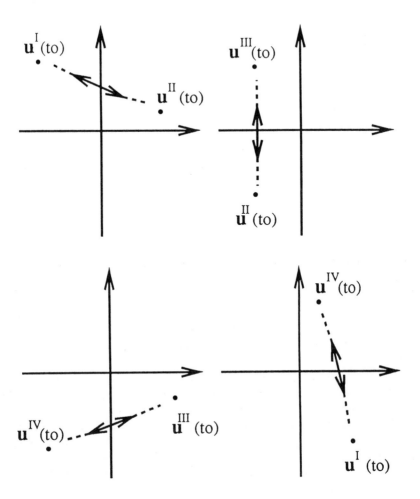

FIGURE 2 This figure shows schematically the four possible classes of two-quadrant oscillation, each class comprising nine types given by all possible behaviors in the two quadrants left. Notice that because, as soon as the system vector enters one of the "coupled" quadrants, it starts moving linearly towards the other one in the couple, it ends up doing small oscillations around the intersection of the line joining the two vectors in the "coupled" quadrants and the coordinate axis separating them.

Combining the conditions for two-quadrant oscillations with Eqs. (16)–(23), one obtains lots of algebraic inequalities. Analogously to the full oscillatory case, the inequalities reveal that two-quadrant oscillations are only possible while $t < t_0$, and most of the possible cases are ruled out by then. Only the three cases shown in Figure 3 are allowed.

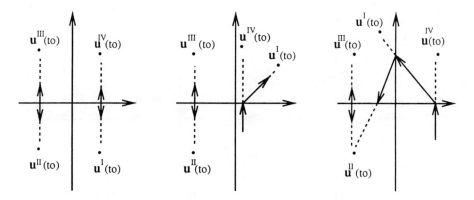

Type B Rule: $\omega_2 > \omega_1$

FIGURE 3 The only possible kinds of two-quadrant oscillations allowed by the equations. They are allowed for both types of Hebbian rules, but the third one can only exist for the type B rule if $\omega_2 > \omega_1$.

The three cases shown in Figure 3 are allowed to exist for both versions of the Hebbian rule considered in this paper. However, the case in which $\mathbf{u}^I(t_0) \in II$ and $\mathbf{u}^{IV}(t_0) \in I$ can only exist for type B if $\omega_2 > \omega_1$. An interesting case is the one in which oscillations between the second and the third, and between the first and the fourth quadrants are allowed to exist. Then, depending on the system starting off in the left or the right side of the u_1–u_2 plane, it will stay there and do small oscillations between the two quadrants of the initial half without jumping to the other half.

The cases left to analyse are the ones in which the system does not have any oscillatory behavior. For those cases the system can have only stable states, and there are four possibilities then, namely having one stable state, two stable states, three stable states, and four stable states. Obviously, for $t < t_0$ the system does not have a strict stable state because it will decay from its state at $t = t_0$ towards the allowed states for $t > t_0$. As we have seen above, for $t > t_0$ the system cannot have any oscillatory behavior, and then it can have only stable states. The possible stable states for $t > t_0$ are shown in Figure 4.

As we said before, for $t < t_0$ the system cannot have any real stable states, but we can define "stability prior to the vanishing of the external input," i.e., the system having one, two, three, or four stable points while the input is being applied,[2] and use this definition for the oscillatory cases to find out what stable behaviors of this kind are allowed by the equations.

---

[2] Notice that one of the possible cases of two-quadrant oscillations shown in Figure 3 has a stable state in this sense in the first quadrant.

The author has analyzed the algebraic inequalities for all 48 possibilities of stability before t_0, and found out that there are four monostable cases allowed (shown in Figure 5), six bistable cases allowed (shown in Figure 6), three tri-stable cases allowed (shown in Figure 7), and one (the only possible one) case having four stable states allowed (shown in Figure 8).

All stable states shown in Figures 5–8 are allowed for both versions of the Hebbian rule considered. Only for three of them (indicated in the figures), the condition $\omega_2 > \omega_1$ follows as an existence condition for type B rules. The cases in which the system vector starts off in a quadrant which contains a stable state, i.e., it cannot leave the quadrant, are Amari's transient states.[1]

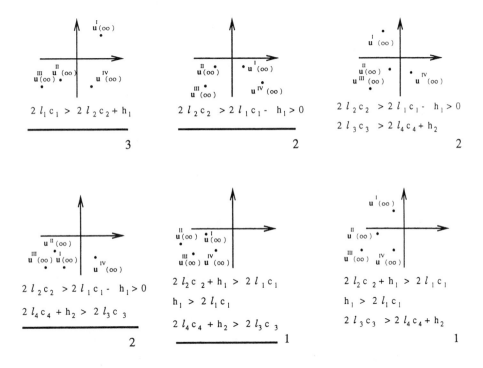

FIGURE 4 This figure shows the possible behaviors of the system for $t \to \infty$. The type B Hebbian rule allows only the four behaviors underlined, while type A allows all six. The number of stable states and the conditions for each behavior are given below each graph.

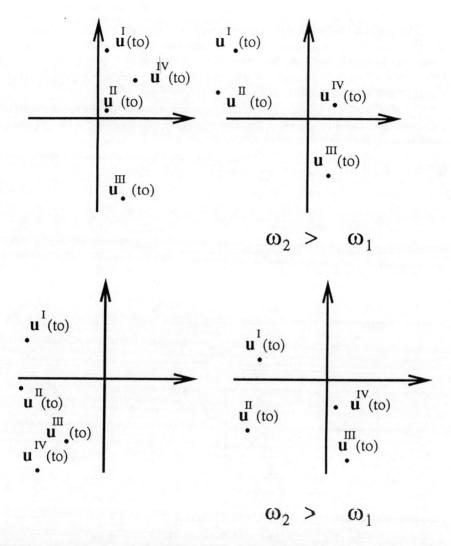

FIGURE 5 The four monostable cases before $t = t_0$ allowed by the equations. They are allowed for both versions of the Hebbian rule, but two of them (indicated below the graph) are only possible for type B if $\omega_2 > \omega_1$.

Dynamical Behavior of a Pair of Spatially Homogeneous Neural Fields

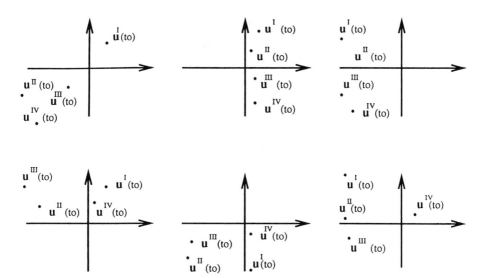

FIGURE 6 The six bistable cases before $t = t_0$ allowed by the equations. They are allowed for both versions of the Hebbian rule, without any restrictions on the relative values of ω_1 and ω_2.

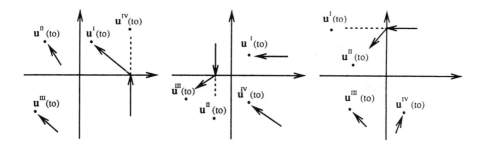

Type B Rule: $\omega_2 > \omega_1$

FIGURE 7 The three tri-stable cases before $t = t_0$ allowed by the equations. They are allowed for both versions of the Hebbian rule, but one of them (indicated below the respective graph) is only possible for type B if $\omega_2 > \omega_1$.

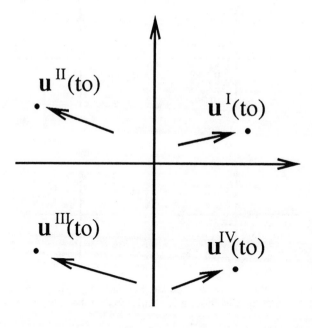

FIGURE 8 The only case having four stable states allowed is the only possible one. The system vector stays in the quadrant where it is initially while the external input is being applied.

4. CONCLUSIONS

This paper presented an example of a simple neural network which can be mathematically modeled and have its behavior fully understood analytically. Very few dynamical systems have this property, but the ones that have it can be used to give us some feeling about the behavior of more complex ones. In the case of the network studied in this work, it can be considered as an approximation for a network which receives uniform stimulation over a large part of it so that that part has roughly homogeneous activity.

We have shown, by adopting and extending Amari's approach to a similar network, that this network can have a variety of dynamical behaviors: full oscillations, two-quadrant oscillations, monostable states, bistable states, etc. The existence conditions found for these behaviors are the most general and, therefore, the weakest possible, in the sense of not assuming any particular set of values for the constants and parameters of the network (apart from the condition involving the time constants). For a given set of constants and parameters, especially for the biologically plausible ones, the mathematical inequalities to be satisfied would become tighter and the number of possible behaviors very much reduced.

ACKNOWLEDGMENTS

The author's Ph.D. studies in the U.K. are being supported by a grant from CNPq-Conselho Nacional de Desenvolvimento Científico e Tecnológico, from Brazil. The author would like to thank Harry Barrow for his comments and suggestions while this paper was being written, and Pepê for some helpful comments on the final layout of this paper.

REFERENCES

1. Amari, S. "Dynamics of Pattern Formation in Lateral-inhibition Type Neural Fields." *Biol. Cybern.* **27**(1977): 77–87.
2. Ellias, S. A., and S. Grossberg. "Pattern Formation, Contrast Control, and Oscillations in the Short Term Memory of Shunting On-Center Off-Surround Networks." *Biol. Cybern.* **20** (1975): 69–98.
3. Roque Da Silva Filho, A. C. "Analysis of Equilibrium Properties of a Continuous Neural Network Made of Excitatory and Inhibitory Neurons." *Network* (1992): submitted.
4. Hebb, D. O. *The Organization of Behaviour.* New York: Wiley & Sons,1949.
5. Levine, D. S. *Introduction to Neural and Cognitive Modeling*, chapter 4. Lawrence Erlbaum, Hillsdale, NJ: 1991.
6. Wilson, H. R., and J. D. Cowan. "Excitatory and Inhibitory Interactions in Localized Populations of Model Neurons." *Biophys. J.* **12** (1972): 1–24.
7. Wilson, H. R. , and J. D. Cowan. "A Mathematical Theory of the Functional Dynamics of Cortical and Thalamic Nervous Tissue." *Kybernetik* **13** (1973): 55–80.

Pedro Paulo Balbi de Oliveira*
School of Cognitive and Computing Sciences, The University of Sussex, COGS PG/ph, Falmer, Brighton BN1 9QH, E. Sussex, England; e-mail: pedrob@cogs.sussex.ac.uk; *On leave from the Computing and Applied Mathematics Laboratory, National Institute for Space Research, Brazil

A Cellular Automaton to Embed Genetic Search

A non-deterministic cellular automaton with periodic boundary conditions, whose temporal evolution resembles an artificial-life world, is presented. This artificial-life activity takes place in a two-dimensional world, where worm-like organisms roam around mating, reproducing, and being selected. Since the motivation for this work has been to embed some form of genetic search in cellular automata, the automaton is described in terms of its general capabilities to act as a framework within which genetic search problems can be defined. However, it is not an aim of the paper to discuss in detail any particular application. Although the concept of search has been traditionally associated with function optimization and with strategies for solving prespecified problems, these are not the connotations of search we mean here; rather, we refer to the process of exploring the space of possible genomes in particular universes, without any concern for optimization or preconceived evolutionary paths to be followed. Because of this, and also because the built-in selection process can be better seen as preserving the non-deleterious features of the organisms (in contrast to selecting for the most adapted ones), the nature of the evolutionary process eventually achieved should be seen as an instance of the exaptationist standpoint in evolutionary theory. The bridge between the activity of the organisms and the genetic search process is made by allowing that the main constituent of the organisms' bodies be the genomes that define the points in the search space under question. The fact that the cellular automaton relies upon only four states per cell allows for its use in a number of ways, some of which are discussed; indeed, the actual cellular automaton described is just one possible example of a large family. This flexibility opens up the possibility of the development of a new class of models to study emergence and self-organization in evolutionary processes, mainly from the standpoint of artificial life.

1. INTRODUCTION

Originally conceived as an abstract model of self-reproduction,[19] cellular automata are currently considered as models for complex natural systems that contain a large number of simple and locally interconnected elements. They can be thought of as mathematical or computational entities, as well as discrete dynamic systems. Cellular automata are made up of a set of elements (the *cells*) that are organized in an n-dimensional lattice (the *cellular space*), so that at any time, each cell can take on one among a set of discrete values (the cell *states*). The states of all cells in the lattice are updated (usually) synchronously, the new state of each cell being dependent upon the state of its *neighborhood*, i.e., its current state together with the states of a group of neighboring cells. The updating of each cell state is achieved by applying to the cell neighborhood a set of deterministic or non-deterministic *transition rules* which are the same for the entire cell space, providing a sort of underlying *physics* for the cellular automaton (see Wolfram[20] and Gutowitz[7] for extensive accounts of both theoretical and practical aspects concerning cellular automata).

By *genetic or evolutionary search* we mean a computational model of search gleaned from concepts in biological evolution, in which non-deterministic mechanisms provide *variability* and *selection* of "genome"-like structures that represent the points of the search space. As new genomes are created, the space is explored. New genomes are created through a sexual *reproduction* process involving already existing genomes, which implies that new genomes typically contain sequences of "genes" of their "parents." By calling these sequences *building blocks*, we can think of the search as a process in which building blocks are created and built upon, thus allowing the exploration of the search space.

Our aim in the paper is to present a two-dimensional cellular automaton with four states per cell, within which it is possible to embed a form of genetic search.[1] As a consequence, the characteristic feature of the automaton is that its temporal

---

[1]Since two out of the four possible states are indeed *classes* of states, in this sense it would be more appropriate to refer to an actual *family* of cellular automata.

evolution very strongly suggests an artificial-life-type world, where worm-like organisms roam around, mating, reproducing, and being selected. The bridge between the activity of the organisms and the genetic search process is made by allowing that the main constituent of the organisms' bodies be the genomes that define the points in the search space under question.

The most well-known forms of genetic search in the literature are the *genetic algorithms* and the *evolutionary strategies*,[8] although other methods also exist, such as Koza's *genetic programming* (see Koza,[10] for example) based on searching on a population of Lisp programs, the one used in MacLennan[14] in a study of the evolution of communication, and the so-called extended genetic algorithm used in Werner and Dyer[16] for the same kind of application; as far as we know no method has yet been devised for a cellular automaton. Although the concept of search has been traditionally associated with function optimization and with strategies for solving prespecified problems, these are not the connotations of search we mean here; rather, we refer to the process of exploring the space of possible genomes in particular universes, without any concern for optimization or preconceived evolutionary paths to be followed. Therefore, the usual characterization of genetic search in terms of creation of "useful" building blocks is not appropriate here; we will return to this point in subsection 2.4.

As far as genetic search is concerned, what we provide is a cellular automaton that, due to the features above, can be seen as a *framework* where a particular genetic search can be embedded. The emphasis of the presentation is on the description of the automaton itself. The discussions about how the framework can be used is made only in general terms; it is beyond the scope of the paper to discuss in detail any particular application. In the next Section we present the automaton by relying, whenever possible, on metaphorical concepts suggested by the artificial-life-type processes it supports, namely, movement, selection, mating, and reproduction; any details related to the actual state transitions involved can be found in the Appendices, which present the complete list of transitions being used. We then give details of the implementation, and discuss how to go about embedding genetic search within the framework. Finally, we sum up the main points raised in the paper, pinpoint some characteristics of the framework, and indicate directions that we are currently pursuing so as to extend it further.

2. THE CELLULAR AUTOMATON
2.1 A REMARK ON SEXUAL REPRODUCTION IN CELLULAR AUTOMATA

Considering the role of sexual reproduction in the provision of variability in nature, and the fact that the main genetic search methods rely upon sexual reproduction, it is appealing to have such a feature also appearing in the present case.

Although a number of cellular automata exhibiting the ability of self-reproduction have been discovered (see von Newmann,[19] Codd,[3] Banks,[1] Langton,[11] and Byl[2]), no cellular automaton capable of sexual reproduction has apparently been reported. The closest reference[18] in the literature seems to be where an abstract discussion is carried out on how to extend the cellular automaton described in von Newmann[19] so as to allow sexual reproduction; however, the complexity of the automaton renders it completely impractical for present purposes.

The complexity of those self-reproducing cellular automata, as expressed by the number of states in the initial configuration, as well as the number of possible states per cell, varies significantly and depends on the design constraints imposed on them. In particular, the imposition that an automaton should possess the abilities of universal computability and/or universal constructability implies an extreme complexity.[2] On the other hand, one wishes to create automata that are prevented from exhibiting a *trivial* self-reproduction whose oversimplification would preclude the modeling of any interesting issue involved in natural self-reproduction.[3]

The standpoint adopted here is somewhere between these extremes, since we have to satisfy a number of constraints such as the necessity of a mating configuration for the parental organisms, the necessity of having to cope with the movement of the parents and of the offspring as reproduction takes place, the premise of being able to describe the activity of the organisms from a high-level perspective, etc. These and other constraints will become clearer in the next sections.

2.2 THE GENERAL PICTURE

The simplest way to envisage our framework is by means of the metaphor of an artificial-life world in which worm-like organisms randomly roam around a two-dimensional world defined by the automaton's cell space. Each organism can have arbitrary length and is defined by a sequence of contiguous cells which constitute its body, as depicted in Figure 2. The two cells at both ends of an organism, the terminal cells, always take on a T-state, and can be intuitively thought of as its head and tail. The other cells between the terminal ones are the actual genomes which are the objects of the genetic search. Each cell of the genome represents a gene locus, while its state, represented here by a g-state, is one of the possible alleles for that particular gene. It should be noted that gene and terminal states represent *classes* of states. Throughout the paper, whenever we refer to a T-state or a g-state, we mean *any* member of the respective class; in the situations where it is necessary to distinguish between different states (as in Figure 2), a subscript is used.

Whenever possible each organism moves, each movement starting either leftwards or along the ascending diagonal on its left-hand side; as Figure 1 clarifies, we can say that the head of the organism can move either to the left or the top-left cells of the neighborhood. The top and right-hand edges of the cell space are wrapped

---

[2] As pointed out in Langton[11], as far as biological self-reproduction is concerned, neither of them seem to currently apply and it is very unlikely that they ever did.

[3] In Langton[11] it is also remarked that the self-reproduction of a 2-state cellular automaton performing addition modulo 2 fits into this category since it can be entirely described at the level of the automaton's underlying physics.

A Cellular Automaton to Embed Genetic Search

tl	t : top	tr	
l : left	c : center	r : right	$\Rightarrow c_{new}$
bl	b : bottom	br	

FIGURE 1 Moore neighborhood and the notation used according to the geographic position of the cells.

$$\text{Head} \mid \text{Genome} \mid \text{Tail}$$

| T | g_1 | g_2 | g_3 | $\cdots$ | g_n | T |

FIGURE 2 Example of an n-gene-long organism in the horizontal position. The g-states represent the states related to a particular genetic search being performed. Any well-formed organism must have its g-state cell delimited by the left and right terminals (T-states).

around, respectively, with the bottom and left-hand edges, giving the cellular space a toroidal geometry. With such periodic boundary conditions, these two movements are sufficient to ensure that the organisms have the ability to cover the entire world. In this way the organisms are able to approach any other in the world and, when two of them reach a predefined spatial configuration relative to each other, they mate and reproduce; after each mating, they begin wandering again, as do their offspring. Although the parental genomes can have different lengths, it can be seen that the newborn's length will not be more than one gene longer than the length of the longest parent.

All this artificial-life-type activity takes place over a *quiescent* background, that is, the inactive regions in the cell space that are not occupied by the cells of any organism, and that are represented here by 0-states.[4] The neighborhood we use for the state transitions of any cell is the "Moore" neighborhood, defined by the cell itself and the eight adjacent cells that surround it in a square lattice, as Figure 1 shows.

2.3 MOVEMENT OF THE ORGANISMS

The basic fact about movement is that either leftward or diagonal movement can only start towards a mostly quiescent region of the cell space, but, once started, it will always be completed even if the moving organism has started a reproduction process. In addition, a movement will never proceed if another organism enters the

---

[4]Quiescence means that an inactive cell that is surrounded just by other inactive cells will remain inactive.

neighborhood of its left-hand side terminal; this prevents organisms from "bumping" into each other.

As an organism moves, a special state comes into play so as to occupy the empty place of the cell that has just been "vacated." This movement state, represented here by an m-state, exists within an organism only while the movement is taking place, disappearing as soon as the organism stops. Figures 3 and 4 show organisms moving respectively to the left and diagonally, illustrating the action of the m-state. Although the figures show situations in which the organisms started their movement in one of the two possible directions and carried on in that direction through subsequent steps, in typical situations the organisms move in a composition of both.

Before starting a movement, the organism first "senses" a mostly quiescent neighborhood ahead in order to "check" whether the way ahead is "free." If that is the case, then it "casts" a movement state along the available direction, "trying" to start the movement. This situation can be seen in Figure 3 during the transitions from time t_0 to t_1 and from t_2 to t_3, and also in Figure 4 during the transitions from time t_2 to t_3 and from t_4 to t_5. If only one direction is available, only one movement state is cast, and the organism just moves in that direction. On the other hand, if both directions are available, a random choice is made among them, but also including the possibility that the organism just does not move, by simply "withdrawing" both movement states.

If an organism could not carry on its movement because of some obstacle in its way ahead, soon after all its m-states disappeared its body would remain in a position determined by the path it went through. To compensate for that, we allow an additional kind of movement which is an *upward movement of the body*, whose effect is, whenever possible, to set the body in the horizontal position; Figure 5 shows one such situation. As it will be clearer in subsection 2.5, the body's upward movement is relevant for the reproduction process, since it allows the increase of the rate of preservation of (non-deleterious) parental gene configurations in the offspring; in other words, its effect is to decrease the randomness associated with the process. But independently of this justification, the upward movement is interesting in itself due to the extra "realism" that it adds to the activity of the organisms without the need of any extra state.

t_0	·	·	·	T	g_i	g_j	T	·
t_1	·	·	m	T	g_i	g_j	T	·
t_2	·	·	T	m	g_i	g_j	T	·
t_3	·	m	T	g_i	m	g_j	T	·
t_4	·	T	m	g_i	g_j	m	T	·
t_5	·	T	g_i	m	g_j	T	·	·
t_6	·	T	g_i	g_j	m	T	·	·
t_7	·	T	g_i	g_j	T	·	·	·

FIGURE 3 Succession of snapshots of the same set of cells as a 2-gene-long organism moves 2 cells leftwards in successive iterations. The dots represent the quiescent state.

A Cellular Automaton to Embed Genetic Search

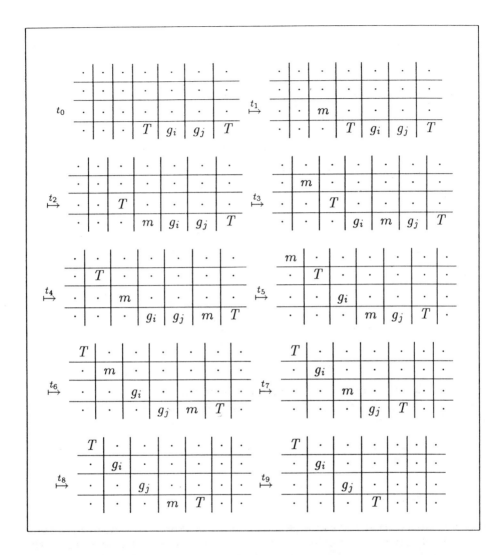

FIGURE 4 Successive snapshots of the same set of cells as a 2-gene-long organism moves 3 cells diagonally, from a horizontal initial position. The dots represent the quiescent state.

Note that, as a consequence of the three movements, the actual movement of the organisms is typically a composition of all of them, with different organism's cells moving in any of the three directions at the same time. The overall spatial disposition of the genomes in the cell space of the automaton is then always monotonically descending from the left, and, it is tempting to say, in a *worm*-like fashion. The complete list of the state transitions for movement can be found in Appendix A.

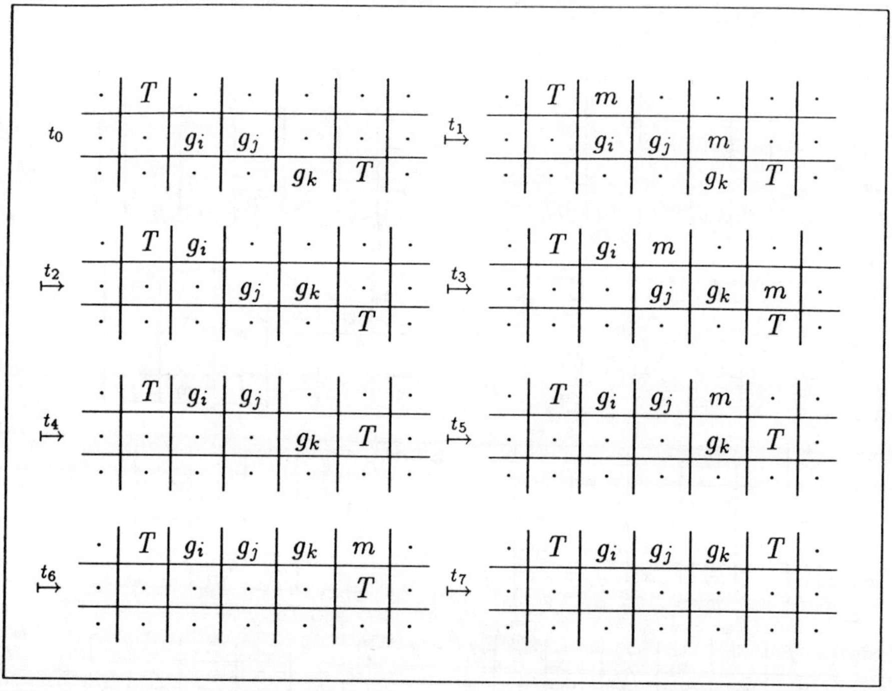

FIGURE 5 Subsequent snapshots of the same set of cells showing the body adjustment of a 3-gene-long organism, from an arbitrary initial position. The dots represent the quiescent state.

2.4 SELECTION

Selection takes place in the following way: if for some reason the state of a gene or a terminal cell changes to the quiescent state in a mostly quiescent neighborhood, the entire organism vanishes; the process occurs in a stepwise way, during the next set of iterations of the automaton. This feature is equivalent to saying that organisms which lose (at least) one terminal state and/or one g-state are not considered to be proper, well-formed organisms and then must die out. Appendix B presents the complete list of state transitions for selection; it is worth noting there what we mean here by a "mostly quiescent neighborhood."

As far as applications are concerned, it is necessary to design appropriate transition rules whose actions impose quiescence on at least one terminal or g-state in a particular neighborhood; as soon as this organism happens to be in a mostly quiescent neighborhood, it will eventually die out. For example, suppose one wishes all non-homogeneous genomes (i.e., genomes presenting two or more different types of

A Cellular Automaton to Embed Genetic Search

genes) to die out. In this case, a rule has to be added to the system so as to detect the presence of two different neighboring genes in the same organisms, so that, in this situation, the gene in the center cell of the neighborhood is *deleted*, i.e., that cell becomes quiescent. As a consequence, as soon as newborn genomes with this deleterious feature happened to be in a mostly quiescent neighborhood they would die out.

An important point here is that selection as described above is *not adaptationist*, i.e., the organisms are not *selected for* by some concept of fitness. The emphasis is not on preserving the fitter genomes, but on killing off the ones which have deleterious features in a particular situation. The emphasis thus is on concepts such as viability rather than fitness, and evolution by satisfying world constraints, rather than evolution towards solving predefined problems posed by the world. This way of looking at selection, *exaptationism* (a contraction for extra-adaptationism) is due to Gould and Vrba[6] and has increasingly gained support in evolutionary theory in recent years (see also Gould and Lewontin[5] and Piatelli-Palmarini[15]). Exaptationism is a generalization of traditional Darwinian adaptationism rather than an opposition to it, and its support has been due to the fact that exaptationist explanations in evolutionary theory have allowed clearer accounts of a number of genomal changes that are *neutral* in terms of their adaptive value but that are selected nonetheless. In order to keep coherence with the exaptationist standpoint, we should replace the concept of a "useful" building block for a *non-deleterious* one. In the current approach what is guaranteed is that any organism that is selected has some non-deleterious building block, even though it may be useless (note the contrast with the traditional parlance within the context of standard genetic search methods).

2.5 REPRODUCTION

Two organisms with any length will mate if they align their heads and their first gene, leaving a layer of quiescent states in between; the first state transition depicted in Appendix C clarifies this situation (the rest of the Appendix shows all the other transitions involved in reproduction). In the mating configuration, one of the parental organisms is on top of the quiescent layer and the other below, their heads being in the same column of the cellular space. Reproduction then goes on so that the new organism is produced in the quiescent layer, starting from the matching heads and stretching to the right. Born this way, the length of the newborn genome is never more than one gene longer than the length of its longest parental genome. Just after reproduction starts, as soon as the parent on the top find its way ahead "free," it restarts its movement; immediately after the way ahead is free for the newborn it too moves, even if its reproduction has not yet finished. Finally, the same thing happens to the parent on the bottom.

The cells of the newborn are created one at a time, both the genes and the terminal states. There are four basic classes of state transitions for reproduction: deterministic rules leading to a T-state, non-deterministic rules leading only to a

g-state or only to a T-state, and further non-deterministic ones leading to either of them. Reproduction starts by creating a head for the newborn whenever the situation described above takes place (although the actual T-state used is randomly chosen among the parental ones). Then it proceeds in a non-deterministic fashion by creating its genes. Finally, it creates the newborn's tail in a non-deterministic way, unless one of the following happens: first, the newborn has "moved too much" even before completely born (i.e., an m-state reached its right-hand extremity, as the last transition in Appendix C shows), or second, there is no more possibility for the newborn to acquire a gene from its parents (as shown in transitions 6, 7, and 8).

The fundamental point about reproduction is that it must be able to provide variability without being disruptive; i.e., it should allow for the preservation of the non-deleterious configurations of genes already existing in the neighborhood; in other words, the viable building blocks within the neighborhood should be preserved. Since in the current approach any genome that is able to exist in the cellular space has some viable building block, what we have to do is to allow the probability distribution of the non-deterministic rules to favor the reappearance of building blocks of the parental genomes, which are defined in the newborn by its most recently created cell and by the cell that is about to be created; this is accomplished by equally distributing the probability of the state transitions accordingly.

If there are no building blocks to be preserved, we just randomly choose any of the parental genes present in the neighborhood. Because reproduction is not prevented from taking place while the parental genomes are in movement, it may be the case that no parental gene is present in a neighborhood (see transition 3 in Appendix C for clarification). In this situation, the newborn gene to be created is randomly chosen from all currently possible genes. It should be mentioned that, even when there are building blocks to be preserved in the neighborhood, a gene can also be created through the latter process, thus giving a minimal uniform bias towards all possible g-states of the application concerned, equivalent to the maintenance of a residual background mutation. The g^*-state which appears in Appendix C refers to a g-state created in the newborn in the way we have just described.

We can now return to the motivation for having the upward movement of the body, as mentioned in subsection 2.3. According to the preceding paragraph, the emphasis of reproduction is on the preservation of the parental building blocks. So, if the organisms did not have the upward movement, the chance that a gene in the newborn was created from a neighborhood with few or no parental genes would be greater. The consequence would be that the rate of preservation of viable parental gene configurations would be smaller. Then, as hinted at earlier, the exploration of the search process would be more random, less oriented by the current state of the search.

Note that the transition above is fairly complex by normal standards in cellular automata applications. It should be clear, however, that our interest here is not on the emergence of reproduction, but on what can be developed assuming reproduction as a primitive we can rely on, and to a certain extent, manipulate.

3. DISCUSSION

The cellular automaton described was implemented in a Sun workstation using Cellsim 2.5, a public domain environment for cellular automata experiments[5]; the current implementation supports up to 256 different states in the world. Although the movement and the quiescent states are implemented as one state each, the terminal and the gene states are defined as ranges of state values specified by the user. The latter is important because it allows for the introduction of new features in the framework without necessarily creating conflicts with the existing transition; for example, it would be possible to add new kinds of heads, each of them with distinctive properties (we return to this point later on in this Section). Another feature of the implementation is that it is possible to control the non-determinism of the transitions by means of a set of parameters whose values are decided by the user; for example, it is possible to control the "amount" of each kind of movement, the rate of background mutation, etc.

In running experiments, even though selection is killing off organisms all the time, because the cellular space is finite sooner or later it gets overpopulated. As a consequence, we experience a *crowding effect* which implies that, after some degree of crowding is achieved, it becomes less likely that a reproduction involving long parents will be able to produce a similarly long offspring. The point is that less and less quiescent cells become available and so, once reproduction starts, it is normally curbed by a moving organism that gets into the quiescent layer in which the newborn is being created. But then the parental organisms start moving again, and similarly the newborn; as soon as the newborn's last gene also moves, reproduction necessarily stops, as mentioned earlier. The effect then is that, as the cellular space gets more and more crowded, an increasing bias towards shorter length genomes takes place.

Note however that the real agent of the bias is the transition (the last one in Appendix C) that adds the tail to the newborn as soon as it moves; in other words, there is an intrinsic selective pressure defined by the rule. It is worth observing that the crowding effect is due to the global behavior of the automaton, which "amplifies" the selective pressure already implicit in the rule. One way to minimize such an effect is to allow a background selective process which would randomly set cells to the quiescent state. This can be done by just adding a non-deterministic rule that leads to quiescence with a small probability, which would have to be worked out empirically, according to the domain concerned, as well as to the size of the cellular space being used.

The studies on cellular automata dynamics presented in Langton[13] suggest that, as far as the emergence of computation and life in natural and artificial systems is concerned, the "interesting" dynamics lies between order and disorder. Although the characterization of these dynamic regimes is not precise, there are some recurring patterns that have been accepted as necessary, such as the existence of very

---

[5] The C code that implements the automaton's state transitions is available from the author.

long transients, dependence on the size of the cellular space, high (but not maximal) temporal and spatial correlation between the cell states, and the existence of propagating structures. It happens that, provided that the overpopulation of the cell space is avoided, all these features have been captured in the cellular automaton described without having them as design constraints.

Although we are aware of the biological implausibility of the framework as it stands, there are a number of features that can be easily altered or added so that richer frameworks can be built, which could lead to models of some aspect of biological life as well as testbeds for artificial life. For example, mating here is, in principle, a matter of chance, not being driven by any characteristics of the domain (such as fitness). However, if one wishes to impose some selective mating among the organisms, it is enough to write a state transition, similar to the first one in Appendix C, with the difference that it would contain the specific parental genes that would allow reproduction to start. By placing this new transition before the equivalent, more general one, in the actual code of the automaton, the former would prevail over the latter without bringing any contradiction to the system's behavior. It should be clear that this example is absolutely general for any other aspect that one wishes to embed in the cellular automaton, and is indeed a central issue on the "programmability" of the framework; all that is needed is to satisfy the set of "hardwired" constraints defined by the existing state transitions and the kinds of states they involve. We refer to this important feature as the addition of *instantiated transitions*.

Through the same kind of reasoning, it would be very simple to allow the terminal states to be represented by distinct head and tail states. It is possible to go even further so as to allow the existence of different kinds of heads, which could be associated with the feature of specialization towards either of the directions of movement. A natural consequence would then be the addition of instantiated transitions to start reproduction so as to allow the movement specialization to be passed on to the newborn, according to various possible schemes, such as that the newborn of parents specialized in moving in the same direction would be more likely to move in that direction.

As far as reproduction is concerned, one could think of adding instantiated transitions that would change the distribution of probability of the non-deterministic transition rules so as to change the current bias towards the formation of building blocks according to some weighted, domain-dependent function of the number of building blocks that each candidate state defines. A trivial example would be just a weighted distribution according to the number of building blocks associated to each candidate state.

4. CONCLUSIONS AND PROSPECTS

The primary intention of this paper is to show a particular, non-deterministic cellular automaton that is able to embed genetic search. We pointed out that this automaton is just a member of a family, and also showed how its space of possible extensions can be explored. The framework provides flexibility to embed a number of features that could be used primarily for artificial-life experiments, and perhaps, also for some aspects of biological modeling. Then we showed that cellular automata can provide a distinctive framework to embed genetic search, which is meant here as a technique to explore a search space by means of non-deterministic mechanisms that provide variability and selection regarding the points of the space, in such a way that viable building blocks are created and built upon. As far as genetic search is concerned, the discussions were general, no attempt having been made to discuss any particular application in detail.

We stress that the possibility of defining the automaton in terms of the four primitive, "hard wired" concept states was an essential achievement, since the genetic search becomes dependent on this small set of state categories, ultimately rendering the programmability of the framework fairly simple, as we tried to show. As far as the general issue of cellular automata to embed genetic search is concerned, further developments can be directed to any aspect of their definition, bearing in mind for example, that only one (topological) species can exist in the automaton described.

As far as the artificial-life world embedded in the automaton is concerned, its major drawback has to do with the provision of interaction between organisms, which is currently very poor. Note that the only kinds of interaction provided are reproduction, and the ones derived from movement, as when an organism is in the way of another. However, a neat solution for this problem exists and is currently being worked out.[6] The definition of regions in the world composed of a new class of environmental E-states that could be "touched" by the organisms and resulting in mutual state modification would certainly solve the problem. The interactions among the organisms would then be made through the environmental states with virtually unbounded richness.

Another extension that is also being worked out refers to the introduction of the concept of an *intermediate* state, which would allow the organism to have two halves: the first half, representing the genotype as discussed in this paper, and the second half, representing the phenotype. The idea is that a newborn will be subjected to a developmental phase before it is fully created (in an egg-like fashion). So, after the reproductive process has created the newborn with only its genotype, as soon as its top parent leaves, a developmental process starts leading to the creation of the second half, where the genotype will be expressed.

The use of genetic search in cellular automata demands that a question being addressed be subjected to a formulation based on local constraints. This may be

---

[6]This is one of the topics in a forthcoming paper.[4]

difficult in a number of situations. Another source of difficulty that might even preclude particular applications is that the sort of framework we discussed usually demands a great deal of computational power.

The clear concepts of space and time that cellular automata embed, the strong, even fanciful sense of realism they get across, and their ability to support unification between operand and operator are just some of features they intrinsically carry with them, which are desired in artificial-life studies in general, as discussed in Langton.[12] As for aspects particularly relevant to genetic search, we can identify the fact that the state evolution of a cellular automaton according to local nonlinear rules provides a neat way to model phenotypic expression; the issue here is that there is a great deal of work done in cellular automata so that we can avoid an ad hoc dynamics, which would be unsound from a theoretical point of view, and whose analysis might be very difficult to perform. In addition, because within the context of cellular automata there is this well-defined concept of an underlying physics, it becomes natural to think of a unified process supporting the existence of both genotype and phenotype. A step further is the actual unification of evolution, development, and interactions of the organisms with their environment, which is in fact the direction we are currently pursuing.

We have tried to make the point about the applicability of the framework in addressing specific problems in artificial life; the question remains though, as to the effectiveness of the approach, mainly because the latter is beyond the scope of this paper. I believe however, that the most appropriate kinds of questions that should be addressed from the perspective we introduced have to do with using the organisms as probes into the emergence and self-organization of evolving systems that are not subjected to solving particular problems. What we have in mind here is the notion that in nature there are no problems being solved, but evolutionary paths being followed according to the constraints existing at each time (see Varela[17]). I think that such an appeal comes from two sources: first, the fact that cellular automata constitute a paradigmatic model for emergence, and second, the increasing support that the role of self-organization in the origins of order in evolution[7] has received recently, for example, as in Kauffman [9]

The exaptationist claim for looking at evolution from the point of view of constraint satisfaction clearly fits into the picture formed by the issues above. Now, even though one could also identify those ideas with adaptationism, the constraint that locality implies for selection in cellular automata seems to be much more in tune with the exaptationist standpoint. One might argue however, that since the notion of fitness function can certainly be interpreted either as a constraint to be satisfied, or as a specification of an evolutionary path to be followed, adaptationism and exaptationism are equivalent from an implementational point of view, and, as a consequence, the difference between them is "just" a matter of point of view of the experimenter. Although agreeing with the premises, I reject the conclusion drawn from it. I believe that the latter is exactly the crucial distinction between the two: if the issue at stake is self-organization, where the emphasis is on the ongoing process

---

[7] We could even say, order *in spite of* selection as contrasted to *due to* it.

rather than on the endpoint reached, the least biased and consequently, the most natural standpoint to take seems to be exaptationism. It is interesting to observe that, although this point of view is shared among many current evolutionists, it does not seem to be the case for many practitioners, say, in the genetic algorithm community, where adaptationism clearly prevails.

ACKNOWLEDGMENT

I thank Phil Husbands, Robert Davidge, Miguel Monteiro and, specially, Inman Harvey, for discussions about the work and comments on two previous versions of the paper. The second version was written during the Summer School and benefitted from useful conversations with Ken Rice about exaptationism; I'm grateful to the organizers of the School for making my participation possible, and to the University of Sussex for covering my travel costs. Thanks also to Chris Langton and Dave Hiebeler for making Cellsim public domain. Finally, I express my gratitude to the Brazilian institutions that jointly provide my financial support in the UK, the National Council for Scientific and Technological Development and the National Institute for Space Research.

APPENDIX

These Appendices present the complete list of state transition for the cellular automaton discussed. Unless otherwise stated, all the rules not shown are supposed to preserve the state of the centre cell (this applies in particular to the quiescent rule, which preserves a 0-state when all the surrounding cells are also 0). For all the Appendices the following holds:

- The symbol # is a *don't care* referring to either of the following states: T, g, or 0.
- When g/T appears in a neighborhood, it means that the corresponding cell can take on either of the two states. In addition, if the state transition also leads to g/T, the new state will follow the one that actually appears in the neighborhood.
- When more than one g-state appears in the neighborhood, no distinction is made between them, independently of their being equal or different to each other. Any case of ambiguity about which g-state of the neighborhood the transition leads to, is solved by subscripting the g-state by its geographic location in the neighborhood (according to Figure 1). Equivalent rationale applies for the neighborhoods which have more than one cell in a T-state or in a state represented by #.

- The number of a transition we sometimes mention in the text, refers to its position from the left to the right, and from the top to the bottom.
- The transitions characterized by the symbol $\stackrel{\bar{d}}{\Rightarrow}$ are non-deterministic; the ones with $\Rightarrow$ are deterministic.
- The neighborhoods showing both T-states of the same organism are due to the smallest well formed organism which has 3 cells.

A. STATE TRANSITIONS FOR MOVEMENT

$\begin{array}{|c|c|c|}\hline 0 & 0 & 0 \\ \hline 0 & 0 & T \\ \hline 0 & 0 & 0/T \\ \hline\end{array} \stackrel{\bar{d}}{\Rightarrow} 0/m$
$\begin{array}{|c|c|c|}\hline 0 & 0 & 0 \\ \hline 0 & 0 & 0 \\ \hline 0 & 0 & T \\ \hline\end{array} \stackrel{\bar{d}}{\Rightarrow} 0/m$
$\begin{array}{|c|c|c|}\hline 0 & m & 0 \\ \hline 0 & m & T \\ \hline \neq 0 & 0 & 0/T \\ \hline\end{array} \Rightarrow 0$

$\begin{array}{|c|c|c|}\hline 0 & m & 0 \\ \hline 0 & m & T \\ \hline 0 & 0 & 0/T \\ \hline\end{array} \stackrel{\bar{d}}{\Rightarrow} 0/m$
$\begin{array}{|c|c|c|}\hline \neq 0 & 0 & \# \\ \hline 0 & m & 0 \\ \hline 0 & m & T \\ \hline\end{array} \Rightarrow 0$
$\begin{array}{|c|c|c|}\hline \# & 0 & \neq 0 \\ \hline 0 & m & 0 \\ \hline 0 & m & T \\ \hline\end{array} \Rightarrow 0$

$\begin{array}{|c|c|c|}\hline 0 & 0 & 0 \\ \hline 0 & m & 0 \\ \hline 0 & m & T \\ \hline\end{array} \stackrel{\bar{d}}{\Rightarrow} 0/m$
$\begin{array}{|c|c|c|}\hline 0 & 0 & 0 \\ \hline 0 & m & T_r \\ \hline \# & 0 & 0/T_{br} \\ \hline\end{array} \Rightarrow T_r$
$\begin{array}{|c|c|c|}\hline 0 & 0 & T \\ \hline 0 & m & T \\ \hline \# & 0 & 0/T \\ \hline\end{array} \Rightarrow 0$

$\begin{array}{|c|c|c|}\hline \# & \# & \# \\ \hline \neq m & m & m \\ \hline \# & \# & \#_{br} \\ \hline\end{array} \Rightarrow \#_{br}$
$\begin{array}{|c|c|c|}\hline \# & 0 & \# \\ \hline \neq m & m & 0 \\ \hline \# & 0 & g_{br}/T_{br} \\ \hline\end{array} \Rightarrow g_{br}/T_{br}$
$\begin{array}{|c|c|c|}\hline \# & \# & \# \\ \hline \neq m & m & g_r \\ \hline \# & \# & \# \\ \hline\end{array} \Rightarrow g_r$

$\begin{array}{|c|c|c|}\hline \# & \# & \# \\ \hline g & m & T_r \\ \hline \# & \# & \# \\ \hline\end{array} \Rightarrow T_r$
$\begin{array}{|c|c|c|}\hline \neq 0 & \# & \# \\ \hline 0 & m & T_r \\ \hline \# & \# & \# \\ \hline\end{array} \Rightarrow T_r$
$\begin{array}{|c|c|c|}\hline \# & \# & \# \\ \hline m & g & \# \\ \hline \# & \# & \# \\ \hline\end{array} \Rightarrow m$

$\begin{array}{|c|c|c|}\hline m & 0 & \# \\ \hline 0 & g & \# \\ \hline \# & \# & \# \\ \hline\end{array} \Rightarrow m$
$\begin{array}{|c|c|c|}\hline \# & \# & \# \\ \hline m & m & \# \\ \hline \# & \# & \# \\ \hline\end{array} \Rightarrow 0$
$\begin{array}{|c|c|c|}\hline m & m & 0 \\ \hline 0 & g & \# \\ \hline \# & \# & \# \\ \hline\end{array} \Rightarrow m$

$\begin{array}{|c|c|c|}\hline \# & \# & \# \\ \hline m & T & 0 \\ \hline \# & \# & 0/T \\ \hline\end{array} \Rightarrow 0$
$\begin{array}{|c|c|c|}\hline \# & \# & \# \\ \hline m & T & 0 \\ \hline \# & \neq 0 & g/m \\ \hline\end{array} \Rightarrow 0$
$\begin{array}{|c|c|c|}\hline m & \# & \# \\ \hline 0 & T & 0 \\ \hline \# & \# & 0/T \\ \hline\end{array} \Rightarrow 0$

A Cellular Automaton to Embed Genetic Search

m	$\neq T$	$\#$
0	T	0
$\#$	$\neq 0$	g/m

$\Rightarrow 0$

0	0	$\#$
m	T	$\#$
$\#$	$\#$	$\#$

$\Rightarrow m$

m	0	$\#$
0	T	$\#$
$\#$	$\#$	$\#$

$\Rightarrow m$

$\neq 0$	$\neq 0$	$\#$
m	T	$\#$
$\#$	$\#$	$\#$

$\Rightarrow m$

$\#$	0	0
g/T	0	0
0	g	g/T

$\stackrel{\bar{q}}{\Rightarrow} 0/m$

$\#$	$\#$	$\#$
g/T	m	0
0	g_b	g/T

$\Rightarrow g_b$

g/T	m	0
0	g	g/T
$\#$	$\#$	$\#$

$\Rightarrow 0$

$\#$	0	0
g	0	0
0	T	0

$\stackrel{\bar{q}}{\Rightarrow} 0/m$

$\#$	$\#$	$\#$
g	m	0
0	T_b	0

$\Rightarrow T_b$

g	m	0
0	T	0
$\#$	$\#$	$\#$

$\Rightarrow 0$

B. STATE TRANSITIONS FOR SELECTION

0	0	0
$\neq 0$	g	0
0	0	0

$\Rightarrow 0$

$\neq 0$	0	0
0	g	0
0	0	0

$\Rightarrow 0$

0	0	$\#$
0	T	0
$\#$	$0/T$	0

$\Rightarrow 0$

T	$\#$	$\#$
0	T	0
$\#$	0	0

$\Rightarrow 0$

0	0	$\#$
0	T	0
$\#$	$\#$	T

$\Rightarrow 0$

0	0	$\#$
0	g	$\#$
$\#$	$\#$	$\#$

$\Rightarrow 0$

0	m	0
0	$\neq 0$	0
0	0	0

$\Rightarrow 0$

$\#$	0	$\#$
0	m	0
$\#$	0	0

$\Rightarrow 0$

0	0	$\#$
$\#$	m	0
0	0	0

$\Rightarrow 0$

0	0	$\#$
0	m	0
$\#$	$\#$	$\neq T$

$\Rightarrow 0$

0	0	$\#$
0	m	$\neq T$
$\#$	$\#$	0

$\Rightarrow 0$

0	0	0
0	m	0
0	g/T	$\#$

$\Rightarrow 0$

0	0	0
g	m	0
0	g	0

$\Rightarrow 0$

$\#$	$\#$	$\#$
m	g	0
$\#$	$\neq g$	0

$\Rightarrow 0$

0	g	$\#$
0	g	$\#$
$\#$	$\#$	$\#$

$\Rightarrow 0$

m	T	$\#$
0	g/m	$\#$
$\#$	$\#$	$\#$

$\Rightarrow 0$

0	m	g/T
0	g	$\#$
$\#$	$\#$	$\#$

$\Rightarrow 0$

C. STATE TRANSITIONS FOR REPRODUCTION

- The state g^* means a g-state that is non-deterministically generated according to the explanation in subsection 2.5.
- The index min used in some of the terminal states is just an implementation detail that defines the default terminal state used.

$$\begin{array}{|c|c|c|} \hline 0 & T_t & g \\ \hline 0 & 0 & 0 \\ \hline 0 & T_b & g \\ \hline \end{array} \overset{\bar{d}}{\Rightarrow} T_t/T_b \qquad \begin{array}{|c|c|c|} \hline T & g/m & \neq 0 \\ \hline T & 0 & 0 \\ \hline T & g/m & \# \\ \hline \end{array} \overset{\bar{d}}{\Rightarrow} g^* \qquad \begin{array}{|c|c|c|} \hline \# & g/m & \neq 0 \\ \hline g & 0 & 0 \\ \hline \neq 0 & g/m & \# \\ \hline \end{array} \overset{\bar{d}}{\Rightarrow} g^*$$

$$\begin{array}{|c|c|c|} \hline \# & T_t & 0 \\ \hline g & 0 & 0 \\ \hline \neq 0 & g/m & \# \\ \hline \end{array} \overset{\bar{d}}{\Rightarrow} g^*/T_t \qquad \begin{array}{|c|c|c|} \hline \# & g/m & \neq 0 \\ \hline g & 0 & 0 \\ \hline \neq 0 & T_b & 0 \\ \hline \end{array} \overset{\bar{d}}{\Rightarrow} g^*/T_b \qquad \begin{array}{|c|c|c|} \hline \# & T_t & 0 \\ \hline g & 0 & 0 \\ \hline \neq 0 & T_b & 0 \\ \hline \end{array} \overset{\bar{d}}{\Rightarrow} T_t/T_b$$

$$\begin{array}{|c|c|c|} \hline \# & T_t & 0 \\ \hline g & 0 & 0 \\ \hline \# & 0 & 0 \\ \hline \end{array} \Rightarrow T_t \qquad \begin{array}{|c|c|c|} \hline 0/T & 0 & 0 \\ \hline g & 0 & 0 \\ \hline \neq 0 & T_b & 0 \\ \hline \end{array} \Rightarrow T_b \qquad \begin{array}{|c|c|c|} \hline \# & g/m & \neq 0 \\ \hline g & 0 & 0 \\ \hline \# & 0 & 0 \\ \hline \end{array} \overset{\bar{d}}{\Rightarrow} g^*/T_{min}$$

$$\begin{array}{|c|c|c|} \hline 0/T & 0 & 0 \\ \hline g & 0 & 0 \\ \hline \neq 0 & g/m & \# \\ \hline \end{array} \overset{\bar{d}}{\Rightarrow} g^*/T_{min} \qquad \begin{array}{|c|c|c|} \hline \# & g/m & \neq 0 \\ \hline \# & g & 0 \\ \hline \# & 0 & m \\ \hline \end{array} \Rightarrow 0 \qquad \begin{array}{|c|c|c|} \hline \# & 0 & m \\ \hline \# & g & 0 \\ \hline \# & g/m & \# \\ \hline \end{array} \Rightarrow 0$$

$$\begin{array}{|c|c|c|} \hline \# & g/m & \# \\ \hline 0 & T & 0 \\ \hline 0 & T & q \\ \hline \end{array} \Rightarrow 0 \qquad \begin{array}{|c|c|c|} \hline \# & \# & \# \\ \hline g & m & 0 \\ \hline \# & \# & \# \\ \hline \end{array} \Rightarrow T_{min}$$

REFERENCES

1. Banks, E. R. *Information Processing and Transmission in Cellular Automata.* Ph.D. thesis, Massachusetts Institute of Technology, 1971.
2. Byl, John "Self-Reproduction in Small Cellular Automata." *Physica D* **34** (1989): 295–299.
3. Codd, E. F. *Cellular Automata.* ACM monograph series. New York: Academic Press, 1968.
4. de Oliveira, Pedro P. B. "Methodological Issues Within a Framework to Support a Class of Artificial-Life Worlds in Cellular Automata." Presented at the British Computer Society Workshop on Cellular Automata, Imperial College, London, Feb. 1992. Submitted to the Workshop Proceedings.
5. Gould, S. J., and R. C. Lewontin. "The Spandrels of San Marco and the Panglossian Paradigm: A Critique of the Adaptationist Programme." In *Conceptual Issues in Evolutionary Biology: An Anthology*, edited by E. Sober, 252–270. Cambridge, MA: Bradford Books/MIT Press, 1984.
6. Gould, S. J., and E. S. Vrba. "Exaptation—A Missing Term in the Science of Form." *Paleobiology* **8** (1982): 4–15.
7. Gutowitz, Howard, ed. Proceedings of the International Workshop "Cellular Automata: Theory and Experiment," held at CNLS (Los Alamos Laboratories), NM, Sept. 9–12, 1989. *Physica D* **45(1–3)** 1990.
8. Hoffmeister, F., and T. Bäck. "Genetic Algorithms and Evolutionary Strategies: Similarities and Differences". In *Proceedings of the First International Workshop on Parallel Problem Solving from Nature*, edited by Hans-Paul Schwefel, 455–470. Lecture Notes in Computer Science, vol. 496. Heidelberg: Springer-Verlag, 1991.
9. Kauffman, Stuart. "Principles of Adaptation in Complex Systems". In *Lectures in the Sciences of Complexity*, edited by D. L. Stein, 619–712. Santa Fe Institute Series in the Sciences of Complexity, Lectures Volume I. Reading, MA: Addison-Wesley, 1989.
10. Koza, J. R. "Genetic Evolution and Co-evolution of Computer Programs." In *Artificial Life II*, edited by C. G. Langton, C. Taylor, J. D. Farmer, and S. Rasmussen, 603–629. Santa Fe Institute Studies in the Sciences of Complexity, Proceedings Volume X. Redwood City, CA: Addison-Wesley, 1992.
11. Langton, C. G. "Self-Reproduction in Cellular Automata". *Physica D* **10** (1984): 134–144.
12. Langton, C. G. "Studying Artificial Life With Cellular Automata." *Physica D* **22** (1986): 120–149.
13. Langton, C. G."Computation at the Edge of Chaos: Phase Transitions and Emergent Computation." *Physica D* **42(1–3)** (1990): 12–37.
14. MacLennan, B. "Synthetic Ethology: an Approach to the Study of Communication." In *Artificial Life II*, edited by C. G. Langton, C. Taylor, J. D.

Farmer, and S. Rasmussen, 631–658. Santa Fe Institute Studies in the Sciences of Complexity, Proceedings Volume X. Redwood City, CA: Addison-Wesley, 1992.
15. Piatelli-Palmarini, M. "Evolution, Selection and Cognition: From 'Learning' to Parameter Setting in Biology and in the Study of Language." *Cognition* **31(1)** (1989).
16. Werner, G. M., and M. Dyer. "Evolution of Communication in Artificial Organisms." In *Artificial Life II*, edited by C. G. Langton, C. Taylor, J. D. Farmer, and S. Rasmussen, 659–687. Santa Fe Institute Studies in the Sciences of Complexity, Proceedings Volume X. Redwood City, CA: Addison-Wesley, 1992.
17. Varela, F. *Conaître: les Sciences Cognitives, Tendances et Perspectives*. Paris: Seuil, 1989.
18. Vitányi, Paul M. B. "Sexually Reproducing Cellular Automata." *Mathematical Biosciences*. **18** (1973): 23–54.
19. von Neumann, J. *The Theory of Self-Reproducing Automata*. Edited and completed by A. W. Burks). University of Illinois Press, Illinois, 1966.
20. Wolfram, S. *Theory and Applications of Cellular Automata*. Singapore: World Scientific, 1986.

Norio Konno
Department of General Education, Faculty of Engineering, Muroran Institute of Technology, Mizumoto 27-1, Muroran, Hokkaido, 050 Japan

Some Mathematical Results on the NK Model

The NK model is a simple biological model; however, it is hard to analyze mathematically in detail, except by simulation. In this paper, a relation between the NK model and the spin model is given in an explicit form. Furthermore, by calculating the correlation $W(\eta)$ and $W(\eta')$, a rigorous meaning of the ruggedness of the landscape is presented in a suitable assumption.

1. INTRODUCTION

We will consider the NK model which was introduced by Kauffman.[2] First, we present the rigorous definition of the NK model. Let N be a positive integer and K be a non-negative integer. The NK model is based on only two alleles at each of the N genetic loci. In general, the number of alleles at each locus can be extended $A \in \{2, 3, \cdots\}$.

Let η denote a configuration of genotype, i.e., $\eta = (\eta(1), \cdots, \eta(N))$ with $\eta(i) \in \{0, 1\}$ $(i = 1, \cdots, N)$. $X = \{0, 1\}^N$ is a configuration space of the genotype with N loci. Each genetic locus, i, has epistatic interactions from K other loci, $\{i_1, \cdots i_K\}$.

The configuration space of the above 2^{K+1} contributions of alleles is defined by $X_{K+1} = \{0,1\}^{K+1}$. It is noted that $0 \leq K \leq N - 1$.

The fitness contribution of each locus, $W_i^{(K)}(\eta(i), \eta(i_1), \cdots, \eta(i_K))$, is specified by the configuration of the alleles of the $K + 1$ loci, $\Delta_i(K) = \{i, i_1, \cdots, i_K\}$.

Remark that $\{W_i^{(K)}(\eta(i), \eta(i_1), \cdots, \eta(i_K)) : i = 1, \cdots, N, (\eta(l) : l \in \Delta_i(K)) \in X_{K+1}\}$ is the collection of $N \times 2^{K+1}$ random values. Then, for each genotype $\eta \in X$, the fitness of genotype, $W(\eta)$, is defined as the average of the fitness contribution of each locus;

$$W(\eta) = \frac{1}{N} \sum_{i=1}^{N} W_i^{(K)}(\eta(i), \eta(i_1), \cdots, \eta(i_K)).$$

The above-mentioned model is called the NK model.

The one-mutant neighbor $\eta_j \in X (j = 1, \cdots, N)$ with respect to $\eta \in X$ is given by; $\eta_j(i) = \eta(i)$ if $i \neq j$ and $\eta_j(j) = 1 - \eta(j)$.

In this paper, Section 2 will give a relation between the NK model and the spin model. Next, we will obtain a rigorous meaning of the ruggedness of fitness landscape of the NK model in Section 3. Finally, Section 4 is devoted to summary and discussions.

2. RELATION BETWEEN THE NK MODEL AND THE SPIN MODEL

Let $\Omega = \{-1,1\}^N$ and $\Omega_{K+1} = \{-1,1\}^{K+1}$. Following Palmer,[5] for each spin configuration $S \in \Omega$, we define the fitness function, $F(S)$, as a sum of N contributions, with the ith contribution depending on $S(i)$ and K other $S(j)$'s;

$$F(S) = \sum_{i=1}^{N} F_i^{(K)}(S(i), S(i_1), \cdots, S(i_K)). \tag{2.1}$$

Noting that from the following basic relation

$$S(i) = 2\eta(i) - 1 \ (i = 1, \cdots, N), \tag{2.2}$$

it is easily obtained

$$F(S) = W(\eta), \tag{2.3a}$$

and

$$F_i^{(K)}(S(i), S(i_1), \cdots, S(i_K)) = \frac{1}{N} W_i^{(K)}(\eta(i), \eta(i_1), \cdots, \eta(i_K)). \tag{2.3b}$$

In a similar fashion, the one-mutant neighbor $S_j \in \Omega (j \in \{1, \cdots, N\})$ with respect to $S \in \Omega$ is given by; $S_j(i) = S(i)$ if $i \neq j$ and $S_j(j) = -S(j)$.

Each $F_i^{(K)}(S(i), S(i_1), \cdots, S(i_K))$ takes 2^{K+1} values, then we have the next representation of it by simple calculation.

Some Mathematical Results on the NK Model

THEOREM 2.1 In the NK model,

$$F_i^{(K)}(S(i), S(i_1), \cdots, S(i_K))$$
$$= J_{i,0}^{(K)}(i) + \sum_{p_1=0}^{K} J_{i,1}^{(K)}(i_{p_1})S(i_{p_1}) + \cdots$$
$$+ \sum_{p_1,\cdots,p_r \in \{0,\cdots,K\},(p_1,\cdots,p_r) distinct} J_{i,r}^{(K)}(i_{p_1},\cdots,i_{p_r})S(i_{p_1})\cdots S(i_{p_r}) + \cdots$$
$$+ J_{i,K+1}^{(K)}(i_0, i_1, \cdots, i_K)S(i_0)S(i_1)\cdots S(i_K),$$

where $i_0 = i$ and

$$J_{i,r}^{(K)}(i_{p_1},\cdots,i_{p_r}) =$$
$$\frac{1}{2^{K+1}} \sum_{\tilde{S}(i_0)=\pm 1,\cdots,\tilde{S}(i_K)=\pm 1} \tilde{S}(i_{p_1})\cdots \tilde{S}(i_{p_r})F_i^{(K)}(\tilde{S}(i),\tilde{S}(i_1),\cdots,\tilde{S}(i_K)). \quad (2.4)$$

Then $J_{i,r}^{(K)}(i_{p_1},\cdots,i_{p_r})$ may be considered as a random coefficient of r points correlation of $\{S(i_{p_1}),\cdots,S(i_{p_r})\}$. For example, we consider $K = 1$. In this case, for each $i = 1,\cdots,N$ and $i \neq j$, we can write

$$F_i^{(1)}(S(i), S(j)) = J_{i,0}^{(1)}(i) + J_{i,1}^{(1)}(i)S(i) + J_{i,1}^{(1)}(j)S(j) + J_{i,2}^{(1)}(i,j)S(i)S(j),$$

where

$$J_{i_0}^{(1)}(i) = \frac{1}{2^2} \sum_{\tilde{s}(i)=\pm 1,\tilde{s}(j)=\pm 1} F_i^{(1)}(\tilde{S}(i),\tilde{S}(j)),$$

$$J_{i,1}^{(1)}(i) = \frac{1}{2^2} \sum_{\tilde{s}(i)=\pm 1,\tilde{s}(j)=\pm 1} \tilde{S}(i)F_i^{(1)}(\tilde{S}(i),\tilde{S}(j)),$$

$$J_{i,1}^{(1)}(j) = \frac{1}{2^2} \sum_{\tilde{s}(i)=\pm 1,\tilde{s}(j)=\pm 1} \tilde{S}(j)F_i^{(1)}(\tilde{S}(i),\tilde{S}(j)),$$

$$J_{i,2}^{(1)}(i,j) = \frac{1}{2^2} \sum_{\tilde{s}(i)=\pm 1,\tilde{s}(j)=\pm 1} \tilde{S}(i)\tilde{S}(j)F_i^{(1)}(\tilde{S}(i),\tilde{S}(j)).$$

Remark that the relation (2.4) implies that $J_{i,r}^{(K)}(i_{p_1},\cdots i_{p_r})$ and $J_{i,m}^{(K)}(i_{p_1},\cdots,i_{p_m})$ are independent for any distinct pair (r,m), even if $N \times 2^{K+1}\{F_i^{(K)}(S(i), S(i_1), \cdots, S(i_K))\}$ random variables are independent, identically distributed (IID).

3. RUGGED LANDSCAPE OF THE NK MODEL

In this section, we show a rigorous result corresponding to the ruggedness of the fitness landscape.

THEOREM 3.1 Assume that $N \times 2^{K+1} \{F_i^{(K)}(S(i), S(i_1), \cdots, S(i_K)) : i = 1, \cdots, N, (S(i), S(i_1), \cdots, S(i_K)) \in \Omega_{K+1}\}$ random variables are IID with mean $m(F)$ and variance $v(F)$. Then

$$E[F(S)F(S')] = N^2 m(F)^2 + \{N - (K+1)\}v(F), \qquad (3.1)$$

for any $S, S' \in \Omega$, where S' is one-mutant neighbor of S.

PROOF Since $\{F_i^{(K)}(S(i), S(i_1), \cdots, S(i_K)) : i = 1, \cdots, N, (S(i), S(i_1), \cdots, S(i_K)) \in \Omega_{K+1}\}$ is a collection of IID random variables, we have

$$\begin{aligned}
E[F(S)^2] &= E\left[\{\sum_{i=1}^{N} F_i^{(K)}(S : \Delta_i(K))\}^2\right] \\
&= \sum_{i=1}^{N} E\left[F_i^{(K)}(S : \Delta_i(K))^2\right] \\
&\quad + \sum_{i \neq j} E\left[F_i^{(K)}(S : \Delta_i(K))\right] E\left[F_j^{(K)}(S : \Delta_j(K))\right] \\
&= Nv(F) + N^2 m(F)^2.
\end{aligned}$$

where $F_i^{(K)}(S : \Delta_i(K)) = F_i^{(K)}(S(i), S(i_1), \cdots, S(i_K))$ for $i = 1, \cdots, N$.

On the other hand, there is an $l \in \{1, \cdots, N\}$ such that $S' = S_l$. Then, in a similar fashion, we get

$$\begin{aligned}
&E\left[\{F(S') - F(S)\}^2\right] \\
&= E\left[\{\sum_{i=1}^{N}\{F_i^{(K)}(S_l : \Delta_i(K)) - F_i^{(K)}(S : \Delta_i(K))\}\}^2\right] \\
&= \sum_{i : l \in \Delta_i(K)} E\left[\{F_i^{(K)}(S_l : \Delta_i(K)) - F_i^{(K)}(S : \Delta_i(K))\}^2\right] \\
&\quad + \sum_{i,j : i \neq j, l \in \Delta_i(K), \Delta_j(K)} E\left[F_i^{(K)}(S_l : \Delta_i(K)) - F_i^{(K)}(S : \Delta_i(K))\right] \\
&\quad \times E\left[F_j^{(K)}(S_l : \Delta_j(K)) - F_j^{(K)}(S : \Delta_j(K))\right] \\
&= 2(K+1)v(F).
\end{aligned}$$

Therefore,

$$E[F(S')F(S)] = \frac{1}{2}\{E[F(S')^2] + E[F(S)^2] - E[\{F(S') - F(S)\}^2]\}$$
$$= N^2 m(F)^2 + \{N - (K+1)\}v(F).$$

Note that Theorem 3.1 is also obtained by using Theorem 2.1; however, the proof is more complicated than the above one.

Furthermore, the following result is easily derived from relations (2.2) and (2.3), and Theorem 3.1.

COROLLARY 3.2 Assume that $N \times 2^{K+1}\{W_i^{(K)}(\eta(i),(\eta(i_1),\cdots,\eta(i_K))) : i = 1,\cdots,N,(\eta(i),\eta(i_1),\cdots,\eta(i_K)) \in X_{K+1}\}$ random variables are *IID* with mean $m(W)(=\frac{1}{N}m(F))$ and variance $v(W) = \frac{1}{N^2}v(F))$. Then

$$E[W(\eta)W(\eta')] = m(W)^2 + \frac{N-(K+1)}{N^2}v(W), \tag{3.2}$$

for any $\eta, \eta' \in X$, where η' is one-mutant neighbor of η.

For example, $\Delta_i(K) = \{i-l,\cdots,i,\cdots,i+r\}$ with $l+r = K, l \geq 0, r \geq 0$ and periodic boundary condition is one of the typical cases of above-mentioned results.

Equation (3.2) implies that K increases from 0 to $N-1$, then the correlation of $W(\eta')$ and $W(\eta)$ decreases monotonically. And it corresponds to the following Kauffmans statement[2]: Increasing the richness of epistatic interactions, K, increases the ruggedness of fitness landscape. In particular, when $K = N-1$, Eq. (3.2) is equal to

$$E[W(\eta')W(\eta)] = E[W(\eta')]E[W(\eta)]. \tag{3.3}$$

Hence, it shows that $\{W(\eta) : \eta \in X\}$ is the collection of uncorrelated random variables.

4. SUMMARY AND DISCUSSIONS

First, this paper presented rigorous definition of the NK model. Next, by using the definition of it, we got the relation between the NK model and the spin model. This result suggests that the NK model is more difficult than the spin model to analyze mathematically. Finally, in the simple case, we showed that if K increases from 0 to $N-1$, then the ruggedness of fitness landscape increases monotonically by the direct computation of $E[W(\eta)W(\eta')]$. The NKC model[3] is an extended model of the NK model in order to study coevolutionary processes. In the NKC model, rigorous clarification of relation between ruggedness of fitness landscape and the edge of chaos[4] is a future interesting problem. In connection with it, various problems of the NK/NKC models are discussed in Ahouse et al.[1]

ACKNOWLEDGMENTS

This work was partially financed by the Grant-in-Aid for Encouragement of Young Scientists of the Ministry of Education, Science and Culture (Japan). The author would like to thank Daniel L. Stein, Lee Altenberg, Jeremy J. Ahouse, and Erhard Bruderer for their useful comments and suggestions during the 1991 Complex Systems Summer School.

REFERENCES

1. Ahouse, J. J., E. Bruderer, A. Gelover-Santiago, N. Konno, D. Lazar, and S. Veretnik. "Reality Kisses the Neck of Speculation: A Report From the NKC Workgroup." This volume.
2. Kauffman, S. A. "Adaptation on Rugged Fitness Landscapes." In *Lectures in the Sciences of Complexity*, edited by D. Stein, 527-618. Santa Fe Institute Studies in the Sciences of Complexity, Lect. Vol. I. Redwood City, CA: Addison-Wesley, 1988.
3. Kauffman, S. A. "Principles of Adaptation in Complex Systems." In *Lectures in the Sciences of Complexity*, edited by D. Stein, 619-712. Santa Fe Institute Studies in the Sciences of Complexity, Lect. Vol. I. Redwood City, CA: Addison-Wesley, 1988.
4. Kauffman, S. A., and S. Johnson. "Coevolution to the Edge of Chaos: Coupled Fitness Landscapes, Poised States, and Coevolutionary Avalanches." *J. Theor. Biol.* **149** (1991): 467-505.
5. Palmer, R. "Optimization on Rugged Landscapes." In *Molecular Evolution on Rugged Landscapes*, edited by A. S. Perelson and S. A. Kauffman, 3-25. Santa Fe Institute Studies in the Sciences of Complexity, Proc. Vol. IV. Redwood City, CA: Addison-Wesley, 1990.

Masatoshi Murase
Research Institute for Fundamental Physics, Kyoto University, Kyoto 606, JAPAN

Complex Dynamics of Flagella

A flagellum swimming in a viscous medium is modeled by a one-dimensional array of opposed active elements. The resultant model is mathematically described by a fourth-order partial differential equation. In the model, the active element is characterized by both *hysteresis* and *excitability* with respect to the sliding motion between the filaments. Hysteresis means that the element is either turned "on" or "off," depending on the history of the sliding motion. Excitability is defined when *active* sliding is triggered by *passive* sliding over a threshold. The combination of these properties leads to a spatio-temporal sliding pattern within the flagellar system, which in turn causes a bending pattern. Numerical simulations for the present model reveal that (i) *intrinsic* instability arises from this model system, (ii) the direction of propagating waves is reversed, (iii) such direction-reversing propagating waves are replaced by unidirectional waves after the insertion of a passive region at one end, and (iv) the increase in the system size leads to chaotic behavior.

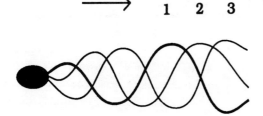

FIGURE 1 Propagating waves typical of "normal" flagella. Successive waves (1 → 3) propagate toward the tip of a flagellum as indicated by the arrow.

1. INTRODUCTION

Flagella are hair-like projections which are found on eukaryotic cells.[1] Their primary function is to move single cells through a fluid for locomotion. Most flagella show regular base-to-tip bend propagation[11] as illustrated in Figure 1. However, others show quite complex dynamical behavior such as the reversal of the direction of propagating waves,[10,16,1] collision of waves which travel in the opposite directions,[1,15] *intermittent* movements with *stopping* and *starting* transients,[9] and co-existence of different waves on different sections of a long insect flagellum.[26] Surprisingly, there is no essential difference in the structure of these flagella. The problem is, thus, to clarify the underlying mechanism leading to various modes of complex behavior. Although many theoretical studies have been performed, they have focused on the regular base-to-tip bend propagation only.[8-16] No attempt has been made to understand the potentially important complex behavior.

In the present paper, I will examine the above problem based on recent theoretical studies.[17-21]

2. THE SLIDING FILAMENT MECHANISM

It is now established that bending waves in flagella are caused by the sliding filament mechanism.[31,27,28] Although actual flagella have nine outer microtubules,[32] they are approximated by a two-filament system on the assumption that bending occurs in a single plane. As illustrated in Figure 2, bending does not occur when any part of the filaments slides equally (Figure 2(B)). If, however, sliding is restricted on local regions, bending is generated between the sliding and nonsliding region (Figure 2(C)). For such bending to be reversed, the direction of sliding must be reversed (Figure 2(D)). The flagellar system is, thus, modelled by a one-dimensional array of opposed active elements, each of which has its own "preferred" direction.

---

[1]Confusingly, bacterial flagella share the same name as those of eukaryotes. They are, however, completely different in structure and function.

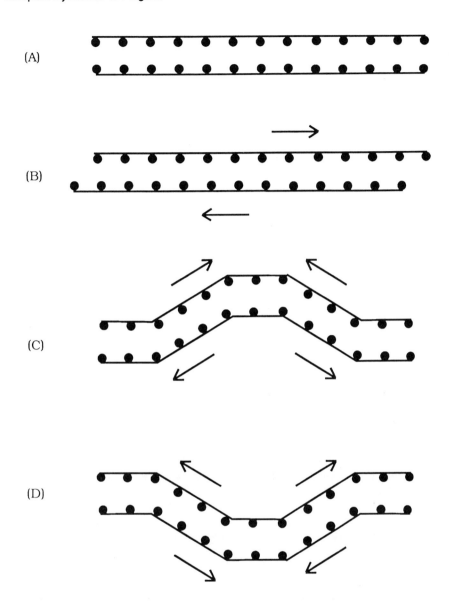

FIGURE 2 Diagrams showing how sliding motion causes bending motion in a two-filament system. (A) The flagellum is straight and no bending occurs without sliding motion. (B) No bending is initiated when sliding occurs equally throughout the length of the flagellum. (C) If sliding is localized, bending occurs between the sliding and nonsliding regions. (D) When the direction of sliding is reversed, the flagellum bends in the direction opposite to the previous direction as shown in (C). The arrows indicate the directions of relative sliding.

3. DERIVATION OF THE BASIC EQUATION

An arc length, s, is introduced to measure the distance along the flagellum from the base. Then, the sliding displacement, σ, is defined as a function of time, t, and space, s. Under the condition that sliding is restricted on local regions, we can assume that the sliding displacement, σ, is proportional to the bending angle, θ, between a horizontal axis and a line tangent to the flagellum. Once σ is specified, we can easily obtain the flagellar shape by simple integration (cf. Figure 4). For convenience, σ is defined as a dimensionless sliding displacement and is allowed to vary between 0 and 1.

The moment-balance equation for a flagellum is written by

$$M_V + M_S = M_E = 0 \tag{1}$$

where M_V, M_S, and M_E are the external viscous, internal shear, and internal elastic moments, respectively. To obtain the basic equation, let us specify each moment in Eq. (1).

First, the external viscous moment, M_V, is given by the external viscous force,[2] F_N:

$$\frac{\partial M_V}{\partial s} + F_N = 0. \tag{2}$$

The external viscous force, F_N, in turn obeys the following force-balance equation[12]:

$$\frac{\partial F_N}{\partial s} + C_N V_N = 0 \tag{3}$$

where C_N and V_N are normal components of the external viscous drag coefficient and the velocity, respectively. In Eqs. (2) and (3), inertial terms are ignored because the *Reynolds* number of flagella is extremely small. The normal component of the velocity, V_N, is, then, specified under the condition of continuation:

$$\frac{\partial V_N}{\partial s} = \frac{\partial \sigma}{\partial t}. \tag{4}$$

In Eqs. (3) and (4), translational movements of the flagellum as a whole are neglected based on the small-amplitude assumption.[26] This simplifies the algebra and the essential results should not be affected.[2]

Secondly, the internal shear moment, M_S, is defined by the internal shear force,[3] S:

$$\frac{\partial M_S}{\partial s} = S. \tag{5}$$

Lastly, the internal elastic moment, M_E, is proportional to the curvature:

$$M_E = E_B \frac{\partial \sigma}{\partial s} \tag{6}$$

Complex Dynamics of Flagella

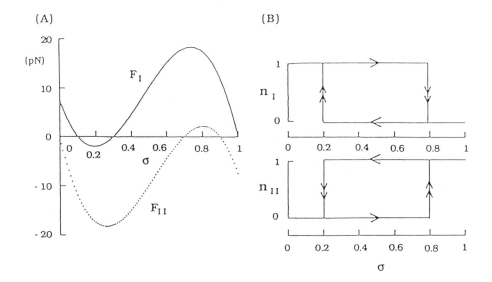

FIGURE 3 The cubic force-distance and hysteresis switching functions. (A) F_I and F_{II} are represented by solid and dotted lines, respectively. They are defined as a function of the sliding displacement, σ. The force constant, Q, is taken as 250 pN. (B) The binary function is defined in the region $0.2 < \sigma < 0.8$. n_I and n_{II} give either the discrete values 0 or 1 under the condition of $n_I + n_{II} = 1$.

where E_B is the bending resistance.

Combining the above equations, we obtain the following basic equation:

$$C_N \frac{\partial \sigma}{\partial t} + \frac{\partial^2 S}{\partial s^2} + E_B \frac{\partial^4 \sigma}{\partial s^4} = 0. \tag{7}$$

4. THE MODEL

The problem is how to specify the internal shear force, S, in such a way that Eq. (7) gives rise to various modes of wave phenomena. In the present model, the internal shear force, S, is defined as follows:

$$S = F_I n_I + F_{II} n_{II} - K_e(\sigma - 0.5) - \gamma \frac{\partial \sigma}{\partial t} \tag{8a}$$

$$F_I = Q(\sigma - 0.1)(\sigma - 0.3)(1 - \sigma) \tag{8b}$$

$$F_{II} = Q(\sigma - 0.9)(\sigma - 0.7)(-\sigma) \tag{8c}$$

$$n_I = \begin{cases} 1 & 0 < s \leq 0.2 \\ 0 & 0 < s < 0.8 \end{cases} \quad \text{(if initially } n_I = 0 \text{ for } \sigma > 0.2\text{)} \tag{8d}$$

$$n_I = \begin{cases} 1 & 0 < s < 0.8 \\ 0 & 0.8 \leq s < 1 \end{cases} \quad \text{(if initially } n_I = 1 \text{ for } \sigma < 0.8) \quad (8e)$$

where F_I and F_{II} are two opposing force-distance functions, n_I and n_{II} are two switching functions,[2] K_e is the force constant of the passive elastic component, and γ is the internal viscous resistance. In the following simulations, γ is taken to be zero except for Section 5.1 because it is negligible in experimental conditions.[2] Excitability is represented by Eqs. (8b) and (8c), where Q is their force constant. See Figure 3(A) for details. Hysteresis is represented by Eqs. (8d) and (8e). To avoid the competition between the two opposing elements, it is assumed that $n_I + n_{II} = 1$. See Figure 3(B) for details.

Equations (7) and (8) are solved on the assumption that moments and forces vanish at both ends. These free-end boundary conditions are:

$$\left.\frac{\partial \sigma}{\partial s}\right|_{s=0,L} = \left.\frac{\partial^2 \sigma}{\partial s^2}\right|_{s=0,L} = 0 \quad (9)$$

where L is a length of a model system.

5. SIMULATION RESULTS
5.1 INTRINSIC INSTABILITY

Although the internal viscous resistance, γ, has been considered to be negligible, large values of γ are empirically introduced to stabilize the wavelength of simulated waves in some models.[5,20] This section investigates the effect of changing the ratio between the internal viscous resistance, γ, and the external viscous drag coefficient, C_N, on the stability of solutions to Eqs. (7) and (8). For this purpose, three sets of values of γ and C_N are used: (i) $\gamma = 50$ pNms/24 nm, $C_N = 0$; (ii) $\gamma = 50$ pNms/24 nm, $C_N = 0.5$ pNms/μm^2; and (iii) $\gamma = 0$, $C_N = 5$ pNms/μm^2. A 50-μm-long model flagellum is set to be *homogeneous* along the length of the system except that forced periodic oscillations are applied at one end in order to generate propagating waves.

Figure 4 shows the simulation results. In each case, the sliding displacement, σ, is plotted against space, s, in the left, and the corresponding bending pattern is shown in the right. The time interval between the two successive patterns is 5 ms. As the ratio of γ/C_N is decreased, the sliding pattern is deformed in two ways (see left panels) though its corresponding bending pattern does not change as much

---

[2]Subscripts I and II indicate two subsystems I and II, respectively.

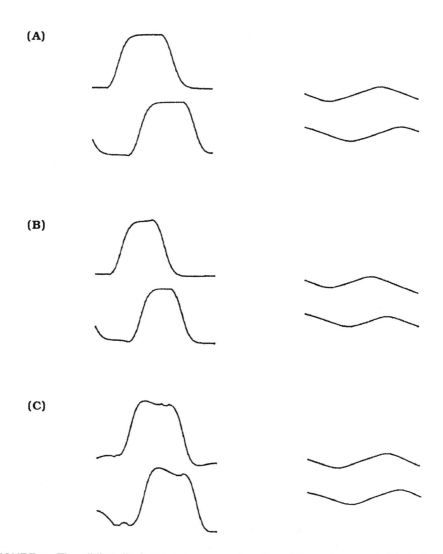

FIGURE 4 The sliding displacement, σ, as a function of the space, s, shown in the left, and the corresponding bending pattern shown in the right. The model flagellum is set to be homogeneous ($Q = 250$ pN and $K_e = 1$ pN/24 nm for $0 < s < 50$ μm) except that forced oscillations are applied. The period of the oscillations is 60 ms. The flagellar shapes in the (x, y) coordinate are obtained by: $x(s) = \int_0^s \cos(\sigma - 0.5)ds$, $y(s) = \int_0^s \sin(\sigma - 0.5)ds$. Two successive patterns in each panel are shown at 5-ms time intervals. Parameters are: (A) $\gamma = 50$ pNms/24 nm, $C_N = 0$; (B) $\gamma = 50$ pNms/24 nm, $C_N = 0.5$ pNms/μm^2; and (C) $\gamma = 0$, $C_N = 5$ pNms/μm^2.

(see right panels). First, the plateau phases of the sliding pattern become spiky at local regions. Since spiky regions are localized, they are caused by the second-order space derivative term in Eq. (7). Second, the plateau phases are globally inclined. These global changes result from long-range interactions which are described by the fourth-order space derivative term in Eq. (7).

The system described by Eq. (7) is subjected to intrinsic instability when $\gamma = 0$ and $C_N = 5$ pNms/μm^2 (see Figure 4(C)). In the following simulations, solutions to Eqs. (7) and (8) are obtained under these conditions as they correspond to the experimental conditions.[21] Because of the instability inherent in this model system, the dynamical behavior must be studied for a long time. For this purpose, two types of representations are used. One is the *energy dissipation* which is obtained by integrating $(\partial \sigma/\partial t)^2$ with respect to space, s. This simply indicates the intrinsic instability. The other is a space-time diagram of σ in which the regions for $\sigma > 0.5$ are plotted by bars against space, s, at 5-ms time intervals. This plot reflects the spatio-temporal sliding pattern.

5.2 REVERSAL OF PROPAGATING WAVES

A 50-μm-long model flagellum has a homogeneous structure, in which opposed active elements are arranged along the system from one end to the other. This model system is initially set to be straight except for the one end (i.e., the left end). Such an initial bend is developed and propagates toward the other end (i.e., the right end).

Figure 5(A) shows the energy dissipation. A number of spiky patterns exist which correspond to intrinsic instability. There are two minima in the time course of the energy dissipation: one is at $t = 1120$ ms and the other is at $t = 2340$ ms. Figure 5(B) shows the space-time diagram of σ. Waves which propagate toward the right are represented by successive bars moving in the rightward direction. As indicated by the first arrow at $t = 1120$ ms, the direction of propagating waves is reversed. This reversal occurs as follows. The trailing edge of the original wave first slows down, while the leading edge does not significantly change its propagating velocity. Then, the wave changes its form and the deformed part sends out a wave which propagates in the direction opposite to the original direction (i.e., wave splitting[19]). This new wave collides with the subsequent wave. Since the new wave is large enough, it can destroy the other. As a result, there are only waves which propagate toward the left. The next reversal of these propagating waves occurs at $t = 2340$ ms as indicated by the second arrow.

If two waves which propagate in the opposite directions are identical, they pass through on collision.[21] Non-annihilating propagating waves of this kind are known as *solitons*. Non-annihilating waves are also observed in real flagella.[1,15]

Complex Dynamics of Flagella

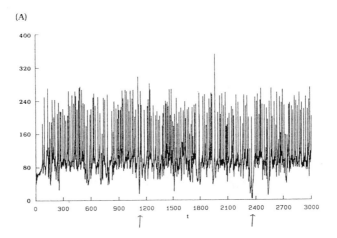

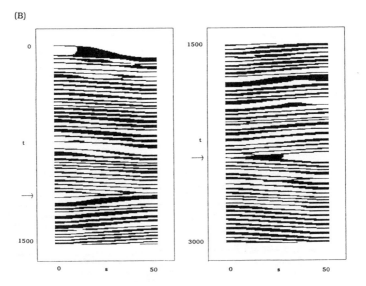

FIGURE 5 The energy dissipation (A) and space-time diagram of σ (B). The flagellum is set to be homogeneous. Parameters are: $\gamma = 0$, $C_N = 5$ pNms/μm^2, $Q = 250$ pN and $K_e = 1$ pN/24 nm for $0 < s < 50$ μm. Simulation results are shown up to $t = 3000$ ms.

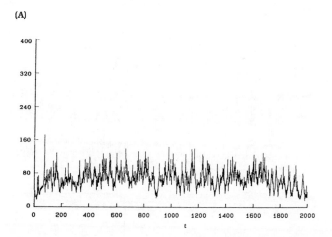

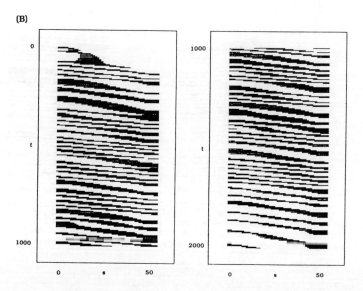

FIGURE 6 The energy dissipation (A) and space-time diagram of σ (B). The flagellum is set to be inhomogeneous. Parameters are: $\gamma = 0$, $C_N = 5$ pNms/μm^2, $Q = 250$ pN and $K_e = 50$ pN/24 nm for $s = 1$ μm, $Q = 250$ pN and $K_e = 1$ pN/24 nm for $1 < s < 40$ μm, and $Q = 0$ and $K_e = 1$ pN/24 nm for $40 < s < 50$ μm. Simulation results are shown up to $t = 2000$ ms.

5.3 INSERTION OF PASSIVE REGION AT ONE END

The model system examined in the previous section demonstrated the reversal of propagating waves and soliton-like behavior. The problem still remaining is how to demonstrate unidirectional waves typical of "normal" flagella. To solve this problem, let us consider the fine structure of sea urchin sperm flagella which show the regular waves. These flagell are 41–43 μm long. Each flagellum has an inert *terminal piece* of 5–8 μm long at the distal end[26] and has a *basal plate* at the basal end.[30] Based on these observations, opposed active elements are removed from the distal 10 μm of the 50-μm-long model flagellum, and a strong elastic component is placed at the base. Mathematically, this situation is modeled when $Q = 0$ for $40 < s < 50$ μm and $K_e = 50 pN/24$ nm for $s = 1$ μm.

Figure 6(A) shows the energy dissipation. The peaks of spiky patterns are reduced extensively. The passive terminal region works like a bulk system which can absorb the instability arising from the active region. Figure 6(B) shows the space-time diagram of σ. As a result of the reduction of the intrinsic instability, only unidirectional propagating waves are demonstrated.

5.4 INCREASE IN SYSTEM SIZE

The model system is set to be homogeneous again, but its length is set to be 100 μm. A single propagating wave is initially present in the system. It propagates to the right and two waves are reflected at the right end based on the wave splitting mechanism (see Section 5.2). The first one propagates slowly, while the second propagates quickly. Since the system size is doubled, the average value of the energy dissipation is almost doubled as indicated by Figure 7(A). Figure 7(B) shows the space-time diagram of σ. As indicated by the first arrow, the second wave collides with the first one at $t = 425$ ms. After the collision, they continue to propagate. Collision of two waves which propagate in the same direction is experimentally observed. Following the collision, the system shows unidirectional propagating waves for a while. However, as indicated by the second arrow, the spatio-temporal sliding pattern begins to be chaotic at $t = 1260$ ms. There are different sections which show quite different wave parameters such as the wavelengths and wave frequencies. This chaotic behavior may correspond to the wave patterns observed in a long insect flagellum.[26]

6. DISCUSSION

The most important problem is how to specify the internal shear force, S, in such a way that Eq. (7) gives rise to various types of wave phenomena. In the present paper, the shear force, S, was defined as a function of σ under the condition of $\gamma = 0$ in Eq. (8a) as in Sections 5.2–5.4:

$$S = S(\sigma). \tag{10}$$

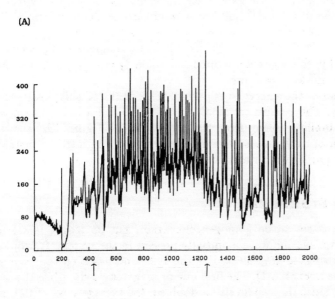

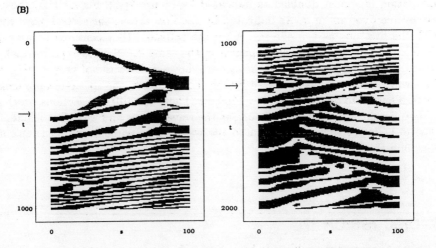

FIGURE 7 The energy dissipation (A) and space-time diagram of σ (B). The flagellum is set to be homogeneous. Parameters are: $\gamma = 0$, $C_N = 5$ pNms/μm^2, $Q = 250$ pN and $K_e = 1$ pN/24 nm for $0 < s < 100$ μm. Simulation results are shown up to $t = 2000$ ms.

It is very difficult to solve the above problem because the system described by Eqs. (7) and (10) is subjected to the intrinsic instability. To understand this situation, let us consider a simple case where the internal shear force, S, is proportional to the sliding displacement, σ. Then, the second term in Eq. (7) corresponds to the *negative* diffusion leading to *destabilization*, while the third term causes *stabilization*. The competition between the two properties leads to intrinsic instability. Furthermore, there are only even powers of the space derivatives. This means that symmetry holds with respect to space, s; that is, the equation is invariant under the spatial inversion $s \to -s$. As a result, both distally propagating and proximally propagating waves were equally developed.

To get unidirectional waves, the structural asymmetry such as the terminal piece without active elements was taken into account. The passive region absorbed instability arising from the active region. The passive region in isolation does not show any function. But it can work to control orders when it coexists with the active region. By analogy with this model behavior, it is important to study any network systems (e.g., gene network, immune network, and neural network) which involve non-active elements.

Besides the present model, two other types of models have been proposed in order to account for normal base-to-tip bend propagation: *curvature-controlled* models[8-16] and *self-oscillatory* models.[5] Curvature-controlled models assume that the shear force, S, is defined as a function of the curvature, $\partial \sigma / \partial s$:

$$S = S\left(\frac{\partial \sigma}{\partial s}\right). \tag{11}$$

To understand the meaning of Eq. (11), let us consider a simple case that the shear force, S, is proportional to the curvature, $\partial \sigma / \partial s$. Then Eq. (7) does not hold the symmetry with respect to space, s, because of the presence of an odd power of the space derivative. As a result, either distally or proximally propagating waves are present depending on the sign of the proportionality constant. However, once the sign of the constant is specified, these models cannot account for two waves propagating in the opposite directions. Furthermore, there is no direct experimental evidence which supports Eq. (11).

Self-oscillatory models assume high internal viscosity, γ, to get unidirectional propagating waves. Here, the shear force, S, is conventionally represented as follows:

$$S = S - \gamma \frac{\partial \sigma}{\partial t}. \tag{12}$$

Let us consider the extreme case of $C_N = 0$. Equation (7) can be reduced to the following reaction-diffusion equation:

$$\gamma \frac{\partial \sigma}{\partial t} = E_B \frac{\partial^2 \sigma}{\partial s^2} + S. \tag{13}$$

In this case, it is easy to get unidirectional propagating waves if an appropriate pace-maker is placed at one end of the system. However, the internal viscosity, γ, is generally considered to be negligible, which is inconsistent with Eq. (12). It is now clear that any model except for the present model is based on ad hoc assumptions to account for regular wave phenomena.

Intrinsic instability has not been discussed in the field of cellular motility. One reason for this is that theoreticians have focused on the regular behavior though there are experimental observations for irregular modes of wave phenomena. Another reason is that it is very difficult to grasp the deformed patterns from the flagellar shape only (see right panels of Figure 4). For these reasons, the observed irregularity has been ascribed to *random noise*. Equations similar to Eq. (7) have been discussed in different physical contexts. For example, the *Kuramoto-Sivashinsky* equation[17,29,18] and the *generalized reaction-diffusion* equation[25] have this class of intrinsic instability. Numerical simulations for these equations show complex dynamics. Despite the diversity of dynamical systems, it is very interesting to notice that there may be a common principle behind them. I hope that the present study stimulates the investigation of such a principle.

ACKNOWLEDGMENTS

I am particularly grateful to Dr L. M. Simmons, Jr., Vice President for Academic Affairs at the Santa Fe Institute, for his hospitality while I was visiting the institute during 1991 Complex Systems Summer School.

REFERENCES

1. Alexander, J., and R. G. Burns. "Differential Inhibition by Erythro-9-[3-(2-hydroxynonyl)] Adenine of Flagella Like and Cilia Like Movement of *Leishmania* Promastigotes." *Nature* **305** (1983): 313–315.
2. Blum, J., and M. Hines. "Biophysics of Flagellar Motility." *Quart. Rev. Biophys.* **12** (1979): 103–180.
3. Brokaw, C. J. "Bend Propagation by a Sliding Filament Model for Flagella." *J. Exp. Biol.* **55** (1971): 289–304.
4. Brokaw, C. J. "Computer Simulation of Flagellar Movement. I. Demonstration of Stable Bend Propagation and Vend Initiation by the Sliding Filament Model." *Biophys. J.* **12** (1972): 564–586.
5. Brokaw, C. J. "Molecular Mechanisms for Oscillation in Flagella and Muscle." *Proc. Nat. Acad. Sci. U.S.A.* **72** (1975): 3102–3106.
6. Brokaw, C. J. "Models for Oscillation and Bend Propagation by Flagella." *Symp. Soc. Exp. Biol.* **35** (1982): 313–338.

7. Brokaw, C. J. "Computer Simulation of Flagellar Movement. VI. Properties of an Oscillatory Two-State Cross-Bridge Model." *Biophys. J.* **48** (1985): 633–642.
8. Brokaw, C. J. "Operation and Regulation of the Flagellar Oscillation." In *Cell Movement*, vol. 1, edited by F. D. Warner, P. Satir, and I. R. Gibbons, 267–279. New York: Alan R. Liss, 1989.
9. Gibbons, I. R. "Transient Flagellar Waveforms in Reactivated Sea Urchin Sperm." *J. Musc. Res. Cell Motility* **7** (1986): 245–250.
10. Goldstein, S. F., M. E. J. Holwill, and N. R. J. Silvester. "The Effects of Laser Microbeam Irradiation of the Flagellum of *Crithidia (Strigomonas) oncopelti*." *Exp. Biol.* **53** (1970): 401–409.
11. Gray, J. J. "The Movement of Sea-Urchin Spermatozoa." *Exp. Biol.* **32** (1955): 775–801.
12. Gray, J. J., and G. J. Hancock. "The Propulsion of Sea-Urchin Spermatozoa." *J. Exp. Biol.* **32** (1955): 802–814.
13. Hines, M., and J. Blum. "Bend Propagation in Flagella. I. Derivation of Equations of Motion and Their Simulation." *Biophys. J.* **23** (1978): 41–57.
14. Hines, M., and J. Blum. "Bend Propagation in Flagella. II. Incorporation of Dynein Cross-Bridge Kinetics into the Equations of Motion." *Biophys. J.* **25** (1979): 421–442.
15. Holwill, M. E. J. "The Motion of *Strigomonas oncopelti*." *J. Exp. Biol.* **42** (1965): 125–137.
16. Holwill, M. E. J., and J. L. McGregor. "Control of Flagellar Wave Movement in *Crithidia oncopelti*." *Nature* **255** (1975): 157–158.
17. Kuramoto, Y. *Chemical Oscillations, Waves, and Turbulence*. Berlin: Springer-Verlag, 1984.
18. Manneville, P. "The Kuramoto-Sivashinsky Equation: A Progress Report." In *Propagation in Systems Far from Equilibrium*, edited by J. E. Westreid, H. R. Brand, P. Manneville, G. Albinet, and N. Boccara, 265–280. Berlin: Springer-Verlag, 1988.
19. Marek, M., and H. Sevcikova. "Electrical Field Effects on Propagating Pulse and Front Waves." In *Self-Organization, Autowaves and Structures Far from Equilibrium*, edited by V. I. Krinsky, 161–163. Berlin: Springer-Verlag, 1984.
20. Murase, M., and H. Shimizu. "A Model of Flagellar Movement Based on Cooperative Dynamics of Dynein-Tubulin Cross-Bridges." *J. Theor. Biol.* **119** (1986): 409–433.
21. Murase, M., M. Hines, and J. Blum. "Properties of an Excitable Dynein Model for Bend Propagation in Cilia and Flagella." *J. Theor. Biol.* **139** (1989): 413–430.
22. Murase, M. "Simulation of Ciliary Beating by an Excitable Dynein Model: Oscillations, Quiescence and Mechano-Sensitivity." *J. Theor. Biol.* **146** (1990): 209–231.
23. Murase, M. "Excitable Dynein Model With Multiple Active sites for Large-Amplitude Oscillations and Bend Propagation." *J. Theor. Biol.* **149** (1991): 181–202.

24. Murase, M. *Dynamics of Cellular Motility.* Manchester: Manchester University Press, 1992.
25. Murray, J. D. *Mathematical Biology.* Berlin: Springer-Verlag, 1989.
26. Rikmenspoel, R. "Movement of Sea Urchin Sperm Flagella." *J. Cell Biol.* **76** (1978): 310–320.
27. Sale, W. S., and P. Satir. "Direction of Active Sliding of Microtubules in *Tetrahymena* Cilia." *Proc. Natl. Acad. Sci. U.S.A.* **74** (1977): 2045–2049.
28. Satir, P., J. W. Steider, S. Lebduska, A. Nasr, and J. Avolio. "The Mechanochemical Cycle of the Dynein Arm." *Cell Motility* **1** (1981): 303–327.
29. Sivashinsky, G. I. "Nonlinear Analysis of Hydrodynamic Instability in Lamina Flames—I. Derivation of Basic Equations." *Acta Astronautica* **4** (1977): 1177–1206.
30. Sleigh, M. A. *The Biology of Cilia and Flagella.* Oxford: Pergamon Press, 1962.
31. Summers, K., and I. R. Gibbons. "Adenosine Triphosphate-Induced Sliding of Tubules in Trypsin-Treated Flagella of Sea-Urchin Sperm." *Proc. Nat. Acad. Sci. U.S.A.* **68** (1971): 3092–2096.
32. Warner, F. D. "The Fine Structure of the Ciliary and Flagellar Axoneme." In *Cilia and Flagella*, edited by M. A. Sleigh, 11–37. London: Academic Press, 1974.

Jan D. van der Laan and Maarten C. Boerlijst
Theoretical Biology/Bioinformatica, University of Utrecht, Padualaan 8, 3584 CH Utrecht, The Netherlands

Cellular Automata with Non-Uniform Rules: An Illustration of Kauffman's Boolean Network Theory

INTRODUCTION

Kauffman[6,7,8,9] introduced random Boolean networks in order to study the phenomenon that every multi-cellular organism has a (limited) number of different cell types, although the genetic material in all cells is identical. Each node in the network is a binary automaton, which can be either on or off (true or false). Every automaton in the network is connected to K other automata. An automaton will change its state according to a transition function, which is based upon the states of the connected automata. If $K = 2$, there exist $2^{(2^2)}(= 16)$ different transition functions, each determining in a different way the effect of the two connected automata. Out of these $2^{(2^K)}$ possible transition rules, one rule is randomly assigned to every automaton in the network. For the rest of its "lifetime," the automaton will obey this transition rule. The assignment of the transition rules is done by filling in a look-up table of all possible input configurations with ones and zeros; every position in this table will have a probability p to become one, and $(1 - p)$

to become zero (usually p = 0.5). The model constructed in this way models the assumed network of interacting genes in a cell.

These Boolean networks have been analyzed for different K and p, and they show interesting behavior.[6,22,23,24] They possess a limited number of attractors, but also exhibit chaotic behavior if $K > 2$ (at $p = 0.5$). The set of attractors in a Boolean network could be linked to the set of cell types, by assuming that epistatic interactions between genes (e.g., gene regulation) do not allow all possible configurations, but rather force the network of connected genes to a strongly limited number of patterns of genetic activity.

Kauffman's Boolean networks have been further analyzed by using the cellular automaton formalism, thus putting local, spatial constraints upon the interactions.[2,3,4,5,14,15,16,17,18] These studies mainly focussed upon the phase transition between frozen and chaotic behavior, damage spreading, and fractal dimension in relation to the percolation threshold.[25,14,16,17,18] In most cases the parameter p has been used in order to study these phenomena. For two-dimensional cellular automata and $K = 4$, the transition between fixed-point or periodic behavior and chaotic behavior is at approximately $p = 0.31$. Above this critical p chaos will arise, and damage does not remain localized.[17]

The aim of this paper is to exploit the cellular automaton formalism to illustrate the main results of Kauffman's Boolean network theory and to show the beauty of the patterns that arise. We believe that the work on this subject is lacking a visualization of the rich dynamics of these networks. So, we will show the dynamics of one- and two-dimensional cellular automata with non-uniform rules.

The patterns of behavior of two-dimensional cellular automata can be difficult to grasp, even when displayed as a movie. However, by using one-dimensional cross sections apparent chaotic two-dimensional behavior displays an amazing amount of structure (compare Poincaré sections). We already applied this technique successfully in the study of cellular automata in another context,[26,10] and the present study is another example of its usefulness.

Furthermore, we will discuss the shortcomings of p as a parameter to characterize the system. For reasons of simplicity we will start with one of the simplest cellular automata: one-dimensional with $K = 2$.

FIXED-POINT AND PERIODIC BEHAVIOR

Starting with an even mixture of all possible rules in a cellular automaton of 100 cells, we observe an amazing variety of localized fixed-point and periodic behavior, which emerges after only a few generations (Figure 1).

Cellular Automata with Non-Uniform Rules

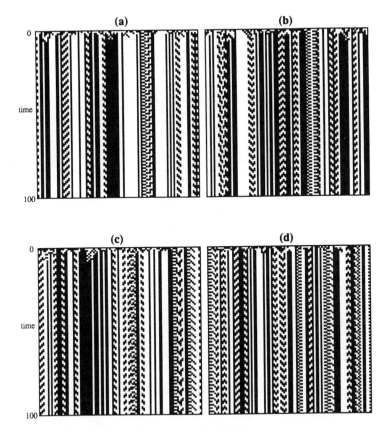

FIGURE 1 Behavioral patterns in a non-uniform one-dimensional cellular automaton ($K = 2$). Four different time series (a-d) of a one-dimensional cellular automaton of 100 cells ($K = 2$). Each replicate uses a different initial configuration as well as a different pattern of rules.

We examined a very small part of the "genome" ($N = 10$) in order to determine the number of attractors and the extent of their basins of attraction. Therefore, the outcome of all 2^{10} (= 1024) possible initial configurations has been studied. The results, presented in Figure 2, show that there are only four different types of behavior. This number is of the order of magnitude of $\sqrt{N}$ as observed in Kauffman's Boolean networks and also sometimes estimated to be the number of cell types in multi-cellular organisms.[9] The four behavioral patterns are all periodic, with periods 3, 4, 6, and 12. The basins of attraction are 112, 128, 272, and 512.

rules

1	2	3	4	5	6	7	8	9	10
8	6	4	12	5	5	1	7	2	1

attractors

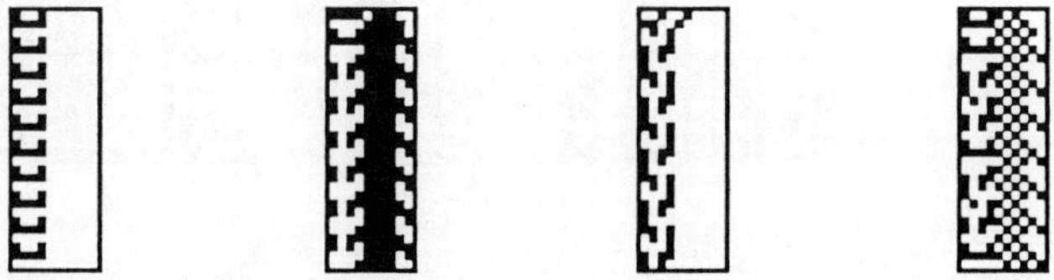

basin of attraction	112	128	272	512
cyclelength	3	4	6	12

FIGURE 2 Attractors in a one-dimensional CA of length 10.

FORCING AND NON-FORCING FUNCTIONS

Kauffman attributed his results to the special properties of certain rules: certain rules exhibit a forcing (or canalizing) effect.[6,7,8,9] This means that a cell obeying one of these particular rules, can be forced to a state by only one of the neighbors, regardless of the state(s) of the other neighbor(s). In Table 1 all rules of the one-dimensional ($K = 2$) cellular automaton are listed together with their characterization as forcing, half-forcing, non-forcing, or immune. The clearest example of forcing is formed by the rules that effect a copy of the state of one of the two neighbors to the one that obeys the rule (rule 3 and 5). Another example is the logical AND function (rule 1); if one of the neighbors is zero, it does not matter what the state of the other neighbor is; the outcome will be zero. However, if one neighbor is one, then the outcome is determined by the state of the other neighbor. This is the reason why we call this rule "half-forcing." The logical function exclusive OR (XOR, rule 6) is an example of the opposite of forcing: in all cases both neighbors will determine the outcome together. Another rule with special properties is the rule which keeps a cell clamped to a state, regardless of the states of the neighborhood (rule 0 and 15); this is what we call an "immune" rule.

TABLE 1 Rule Space with $K = 2$

rule [1]	0	1	2	3	4	5	6	7	8	9	10	11	12	13	14	15
L R	I	H	H	F	H	F	N	H	H	N	F	H	F	H	H	I
0 0	0	0	0	0	0	0	0	0	1	1	1	1	1	1	1	1
0 1	0	0	0	0	1	1	1	1	0	0	0	0	1	1	1	1
1 0	0	0	1	1	0	0	1	1	0	0	1	1	0	0	1	1
1 1	0	1	0	1	0	1	0	1	0	1	0	1	0	1	0	1

[1] L=Left, R=Right, I=Immune, F=Forced, H=Half-forced, N=Non-forced

Closer examination of the results of the experiment presented in Figure 2 reveals that if the input pattern at positions 6, 7, and 8 is 1-1-1, the system will always end in the cycle of period 4, regardless of the states of the other positions (since this happens in 12.5% of the cases, the extent of the basin of attraction is easy to understand $0.125 \times 1024 = 128$). The rules 5 (F), 1 (H), and 7 (H) at these positions force the entire system into this pattern.

A parameter often used in the analyses of uniform, simple one-dimensional cellular automata with $K = 2$ or 3 is λ (the proportion of non-zeros in a rule).[11,12,13,19,20,21] For one-dimensional cellular automata with five states and $K = 4$, Langton[11] showed that by varying λ between 0.0 and 1.0, one goes from fixed point to periodic to chaotic behavior and backwards. However, several studies have shown that λ alone is not capable of characterizing the rule space sufficiently if either the number of states or K is small.[11,12,13] Other parameters have been suggested, among which is the so-called dependency.[21] This dependency parameter is analogous to the extent of forcing (a forcing rule has a low value of dependency, whereas a non-forcing rule has the highest value of dependency). We also see a correspondence between λ and the parameter p, as used in the analyses of cellular automata with non-uniform rules. We therefore followed Hartman,[5,6] who studied two-dimensional cellular automata with non-uniform rules (combinations of AND and XOR), and by making a series of runs of the one-dimensional ($K = 2$) cellular automaton with different proportions of the non-forcing rules (rule 6 and 9). The results are presented in Figure 3 in which transition from fixed-point and periodic behavior to chaotic behavior can be observed. The effect of forcing rules is dramatic: at a proportion of 80- or 85-percent non-forcing rules, highly structured regions appear in the time plots if the local density of forcing rules is high, whereas the intermediate "non-forcing regions" still show localized chaotic behavior. The emergent chaotic behavior in these simulations seems to contradict the results of Stauffer, who concluded that chaotic behavior does not occur in one-dimensional cellular automata with $K = 2$. However, in his simulations the proportion of non-forcing rules remained fixed at 12.5%.

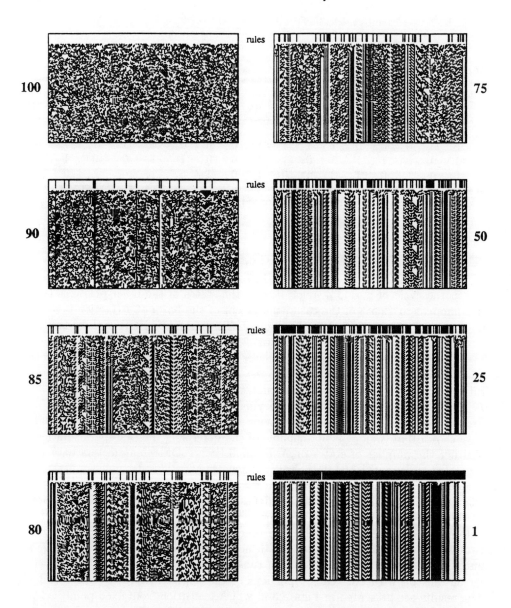

FIGURE 3 The effect of non-forcing rules in a one-dimensional CA ($K = 2$). Time series (100 generations) of a one-dimensional cellular automaton of 200 cells ($K = 2$) with a different percentage of non-forcing rules (rule 6 and 9). On top of every time plot, the rule type of every cell is indicated by a white bar if it obeys a non-forcing rule, or a black bar if it obeys a forcing, half-forcing, or immune rule. Each replicate uses a different initial configuration as well as a different pattern of rules.

Cellular Automata with Non-Uniform Rules

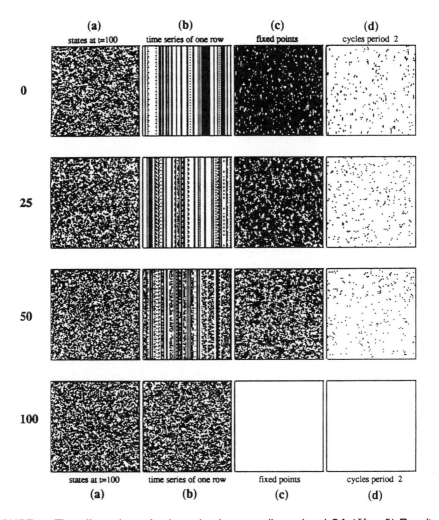

FIGURE 4 The effect of non-forcing rules in a two-dimensional CA ($K = 5$) Results of a two-dimensional cellular automaton of 100×100 cells (on a torus), with $K = 5$ and different percentages of the non-forcing XOR. (a) The states at $t = 100$. (b) Time series (100 generations from $t = 100$ to 200) of row 50. (c) Positions of fixed points, showing the "frozen" regions. (d) Positions of cycles with period 2. The rules which have been used in these simulations are (with N, S, E, and W, representing the neighboring cells and c the center cell): 0% non-forcing— C or E or W or N or S (OR-rule), C and E and W and N and S (AND-rule), C or E or W xor N and S, C or E or W or N XOR S; 25% non-forcing— C or E or W XOR N and S, OR-rule, AND-rule, XOR-rule; 50% non-forcing— OR-rule, AND-rule, XOR-rule; 100% non-forcing— C xor E xor W xor N xor S (XOR-rule). It will take some time before the system will attain its attractor, (continued)

FIGURE 4 (cont'd.) so (c) and (d) are obtained by recording the fixed points and cycles of period 2 from $t = 100$ to $t = 200$ generations. Only in the system with 2% non-forcing rules did we observe a small number of cells (< 10) that didn't settle yet in their local periodic or fixed-point behavior after 200 generations.

TWO-DIMENSIONAL CELLULAR AUTOMATA

The above results extended to two-dimensional cellular automata are shown in Figure 4. For practical reasons, we do not assign every cell a rule from the whole set of possible rules (which is $2^{(2^5)} = 5.9 * 10^8$), but we draw our rules from a small subset, which (of course) includes the non-forcing XOR function.

Figure 4 (a) shows the states at $t = 100$. These static plots do not show any differences. Figures 4 (c) and (d) show the positions of respectively the fixed points (c) and the fixed points or cycles of period 2 (d), in order to show the extent of the "frozen" regions. However, the tie series of the one-dimensional sections (b) provide a much better insight in the qualitative behavior of the two-dimensional system. It is striking that also in these two-dimensional systems, the patterns are extremely localized. Again, increasing the proportion of non-forcing rules yields a transition from fixed-point and periodic behavior to chaotic behavior.

CONCLUSION

The analyses of the rule space of uniform cellular automata showed that one parameter (usually λ) is not sufficient to characterize the rich behavior of these systems. Therefore, we advocate the inclusion of more parameters in the analysis on non-uniform cellular automata. The proportion of non-forcing rules is an important parameter in these systems.

ACKNOWLEDGMENTS

The meetings of the discussion group on cellular automata during the summer school have been an important stimulans for writing this paper (JDvdL). We therefore thank Pedro, Antonio, Erhard, Erik, Paul, Richard, and Sidney. We also thank Julius Wintjes for linguistic advice.

REFERENCES

1. Boerlijst, M. C., and P. Hogeweg. "Self-Structuring and Selection: Spiral Waves as a Substrate for Prebiotic Evolution." In *Artificial Life II*, edited by C. G. Langton, C. Taylor, J. D. Farmer, and S. Rasmussen, 255–276. Santa Fe Institute Studies in the Sciences of Complexity, Proc. Vol. X. Redwood City, CA: Addison-Wesley, 1991.
2. Derrida, B., and D. Stauffer. "Phase-Transitions in Two-Dimensional Kauffman Cellular Automata." *Europhy. Lett.* **2** (1986): 739–745.
3. Derrida, B., and Y. Pomeau. "Random Networks of Automata: A Simple Annealed Approximation." *Europhysics Letters* **1** (1986): 45–49.
4. Hartman, H., and G. Y. Vichniac. "Inhomogeneous Cellular Automata (INCA)." In *Disordered Systems and Biological Organization*, edited by E. Bienenstock, F. Fogelman Soulié, and G. Weisbuch, 53–57. Berlin: Springer-Verlag, 1986.
5. Hartman, H., P. Tamayo, and W. Klein. "Inhomogeneous Cellular Automata and Statistical Mechanics." *Complex Systems* **1** (1987): 245–256.
6. Kauffman, S. A. "Metabolic Stability and Epigenesis in Randomly Constructed Genetic Nets." *J. Theor. Biol.* **22** (1969): 437–467.
7. Kauffman, S. A. "Boolean Systems, Adaptive Automata, Evolution." In *Disordered Systems and Biological Organization*, edited by E. Bienenstock, F. Fogelman Soulié, and G. Weisbuch, 339–360. Berlin: Springer-Verlag, 1986.
8. Kauffman, S. A. "Principles of Adaptation in Complex Systems." In *Lectures in the Sciences of Complexity*, edited by D. Stein, 619–712. Santa Fe Institute Studies in the Sciences of Complexity, Lect. Vol. I. Redwood City, CA: Addison-Wesley, 1989.
9. Kauffman, S. A. "Antichaos and Adaptation." *Sci. Am.* **265(2)** (1991): 64–70.
10. van der Laan, J. D., and P. Hogeweg. "Waves of Crown-of-Thorns Outbreaks—Where do They Come From?" *Coral Reefs* (1992): in preparation.
11. Langton, C. G. "Computation at the Edge of Chaos: Phase Transitions and Emergent Computation." *Physica D* **42** (1990): 12–37.
12. Li, W., and N. H. Packard. "Structure of the Elementary Cellular Automata Rule Space." *Complex Systems* **4** (1990): 281–297.
13. Li, W., N. H. Packard, and C. G. Langton. "Transition Phenomena in Cellular Automata Rule Space." *Physica D.* **45** (1990): 77–94.
14. Stauffer, D. "Random Boolean Networks: Analogy With Percolation." *Philosophical Magazine B.* **56** (1987):901–916.
15. Stauffer, D. "On Forcing Functions in Kauffman's Random Boolean Networks." *Physica D.* **38** (1987): 789–794.
16. Stauffer, D. "Hunting for the Fractal Dimension of the Kauffman Model." *Physica D.* **38** (1989): 341–344.
17. Stauffer, D. "Cellular Automata." In *Fractals and Disordered Systems*, edited by A. Bunde and S. Havlin, 297–321. Berlin: Springer-Verlag, 1991.

18. Weisbuch, G., and D. Stauffer. "Phase Transitions in Cellular Random Boolean Nets." *J. Physique (France)* **48** (1987): 11–18.
19. Wolfram, S. "Statistical Mechanics of Cellular Automata." *Rev. Mod. Phys.* **55** (1983): 601–644.
20. Wolfram, S. "Universality and Complexity in Cellular Automata." *Physica D* **10** (1984): 1–35.
21. Wolfram, S. *Theory and Applications of Cellular Automata*. Singapore: World Scientific, 1986.

Stella Veretnik
Department of Genetics and Cell Biology, University of Minnesota, St Paul, MN 55108

Random Boolean Networks: Comparison Between Randomly Connected and Lattice-Connected Networks

Random Boolean networks exhibit self-organizing properties and can be used as a model of the biological cell. Below I show that network geometry—the way in which nodes are connected—produces networks with different behaviors. Network behavior is affected by the method used to calculate the subsequent state of the network, absence of certain Boolean functions and the size of the network. The behavior of a variety of networks is revealed through computer simulation.

1. INTRODUCTION

Random Boolean networks have been shown to exhibit self-organizing properties; particularly striking results are achieved in the nets with connectivity of two ($K = 2$).[1,2,3,4,5] A network with N nodes (each can take the value 0 or 1) has 2^N potential states, but on the average only a small proportion of those, approximately N states, are stable (i.e., belonging to the cycles). The number of cycles reached by the network and cycle length are extremely short, approximately $\sqrt{N}$.[4] Kauffman showed that the above networks can be used as a model of a biological cell or some aspects of it.[4,5] In this model individual *genes* are represented by the *nodes* of the network, *interactions* between genes are modeled by the *connections* between the nodes. A network *cycle* is then viewed as a *cell type*; the *length* of the cycle is associated with *time between cell divisions*. The majority of the data about the networks is derived from computer simulations,[2,4] although theoretical predictions have been made.[3,7] The results of computer simulation presented in this paper argue that randomly connected and lattice-connected networks exhibit *different* behavior under identical set of parameters. The effect of several basic parameters was studied: the method used for updating the nodes during calculation of the network's next state, exclusion of Contradiction and Tautology Boolean functions, and increasing the size of the network. Lattice-connected networks are very sensitive to changes in these parameters: the number of cycles and the length of the run-in increases dramatically causing loss of some self-organizing properties of the network. Those types of networks cannot, therefore, be a successful model of a biological cell. Furthermore, some combination of parameters—synchronous method of node update and use of only 14 Boolean functions—causes randomly connected networks to go through an extensive number of steps ($< 5N$, where N is number of nodes in the network) before a cycle is found. Only selected types of networks exhibit biologically plausible behavior.

2. DEFINITIONS

2.1 BASIC ELEMENTS AND BEHAVIOR OF THE NETWORK

A network is constructed using basic elements called nodes. Each node is a binary device, taking values of 0 or 1. A node calculates its state based on inputs from other nodes and its internal logical function called a Boolean function. The state of the node serves as an input to other nodes. *The state of the network* can be defined as a joint state of all its nodes at any given time. Starting in any of the possible 2^N initial configurations, the network passes through some sequence of states until it comes to one of the previously encountered states, closing the cycle. From that point on, the network traverses the same subset of states, since the transitions between network states are fully deterministic. The number of states comprising a given cycle will determine its length. Depending on the initial state of the network, different cycles can be reached. The sequence of states that the network traverses before it reaches a cycle is called the run-in. Networks can leave a cycle and move into another cycle upon introduction of noise (noise can be viewed as a temporary switch in the state of one or more nodes).

2.2 CONNECTION BETWEEN NODES

Each node in the network is connected to other nodes. Connectivity of two ($K = 2$) means that each node has two nodes from which it receives inputs and two nodes (possibly different ones) to which it sends its output. Two possible ways of making connections between nodes have been explored: randomly connected[4,5] and lattice-connected networks.[1,2,3] In randomly connected networks the connections between nodes are assigned randomly. In lattice-connected networks the immediate four neighbors of the nodes are used for connection—two serve as input nodes and two as output nodes. The latter method of connection is useful for easy visualization of closely interacting nodes.

2.3 TECHNIQUE OF NODE UPDATE

Each node of the network calculates and updates its state at every iteration of the network. The network develops in discrete time steps: outputs of the nodes at time $(t-1)$ will serve as the inputs into the nodes at time $(t-1)$ and will, therefore, determine the outputs at time t. There are two possible temporal ways in which the node update can be done:

- All nodes in the network update their state simultaneously—at time $(t-1)$ all nodes look at their inputs and calculate the outputs for time t. This can be seen as a synchronous or parallel process (this is a type of update that is claimed to be used in the majority of the simulations.[2,4])
- Nodes update their state asynchronously—that is, there is some order in which the update is performed until *all* nodes calculate their state, at which point the network reaches its subsequent state. This is a sequential process; it can be viewed as multistep (precisely, $N - step$, N is a number of nodes in the network) process. Therefore the time unit t consists of N intermediate steps: $t_1, t_2, \cdots, t_N$, the last step coinciding with time t of the simultaneous update: $t = t_N$. At each step the network's state is affected by a change in the state of only one node (the node that is being updated at that time unit). The $(N-1)$ steps of this process are transient and only a final step is a "real" state of the network.

Both types of update are considered here to be a one-step iteration from the point of view of the network.

2.4 BOOLEAN FUNCTIONS

There are 16 transformations with two inputs, called Boolean functions. Most of the 16 Boolean functions with two inputs are forcing functions. The term "forcing" means that the function's output can be determined by one input value, independent of the value of the second input. The forcing input is the value of the input that determines the output, the forced output (or value) is the outcome of the function under the forcing input. Two out of 16 Boolean functions are not forcing: these are Equal and Exclusive OR. Two other functions (Contradiction and Tautology) output the same value under any input. For these and the rest of the functions, one or both input values can be forcing.

2.5 FORCING STRUCTURES

When a forced output from one node in the network turns out to be a forcing input into another node—the basic unit of a forcing structure appears. Forcing structures vary in length; a forcing input into the forcing structure is guaranteed to propagate along it, forcing all the nodes along the path into their forced values. Closing a forcing structure on itself creates a forcing loop, it is a more powerful structure than a linear forcing structure and its propagating signal eventually reinforces itself.[5] Forcing structures are abundant in the random Boolean networks with two inputs because majority of the Boolean functions (14 out of 16) are forcing functions. Forcing structures (loops, in particular) will "freeze" parts of the network in a particular mode, artificially reducing the number of potential states of the network and, therefore, contributing to the self-organizing behavior.

2.6 PREVIOUSLY PREDICTED BEHAVIOR OF THE NETWORK

Kauffman's simulations[4,5] predicted that for randomly connected networks with two inputs, the average cycle length is $\sqrt{N}$ and average number of cycles $\sqrt{N}$, where N is size of the network. When perturbed, the network is expected to return to its original cycle in approximately 90% of the cases.

3. SIMULATIONS

A set of parameters is chosen for the network simulation and run repeatedly under multiple conditions. The results of the simulations are averaged and are interpreted as a "tendency in the behavior" for a particular type of network.

3.1 INPUT PARAMETERS UNDER INVESTIGATION

All networks have connectivity of two ($k = 2$); a network simulation is selected according to:

- type of connection between nodes (random or lattice)
- technique of node update (synchronous, asynchronous)
- subset of the Boolean functions used (14 or 16)

Every combination of the above parameters specifies a *type* of network; thus, a total of eight different types of networks are simulated.

A network is determined by assigning node connections and Boolean functions. Many different networks of each type are simulated. Each network is simulated under multiple initial node states, and results are averaged to determine behavior of that network.

Results from different networks are later averaged as well in order to come up with *average behavior* of the network type under a specific set of parameters (such as the method of connections, technique of node update, or exclusion of some Boolean functions). This second average represents behavior of the network *independent* of the specific Boolean function or connections between nodes (in the case of randomly connected nets).

3.2 OUTPUT PARAMETERS OF INTEREST

Every type of network can be characterized by several averaged parameters, in particular:

- number of different cycles to which networks arrive
- length of the cycles (weighted (which includes all cycles) and not-weighted (only unique cycles are considered))
- length of the run-in (how long it takes before the network arrives at a cycle)

3.3 SIZE OF THE NETWORKS

Simulations are done on networks of two sizes.

1. **100-node Networks (10×10)**: 100 different networks of each type are searched; each is run under 400 initial conditions. Simulations are done on a Mac II and a Sun4.
2. **900-node Networks (30×30)**: small numbers of networks are used; each run under 600 initial conditions. Simulations are done on a Cray XM-P. Only selected types of networks were simulated on the Cray.

4. RESULTS

The effects of three different parameters on the behavior of the random Boolean networks with two possible geometries of node connection (randomly connected and lattice connected) have been studied:

1. Exclusion of Contradiction and Tautology functions out of the set of 16 possible Boolean functions.

2. Asynchronous vs. synchronous update of nodes in the network.
3. Increasing the number of nodes in the network.

Lattice-connected networks have a different response to the change in these parameters; in particular lattice-connected networks appear to be more "sensitive" to the changes than randomly connected networks. In this section I compare the behavior of lattice-connected and randomly connected networks. Note that the effect of the first two parameters is studied on the networks of 100 nodes.

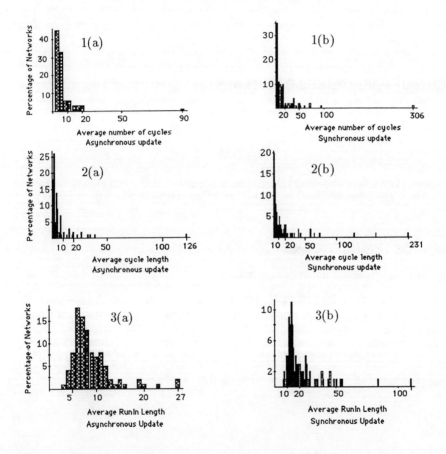

FIGURE 1 Comparison between synchronous and asynchronous update in the *randomly* connected networks.

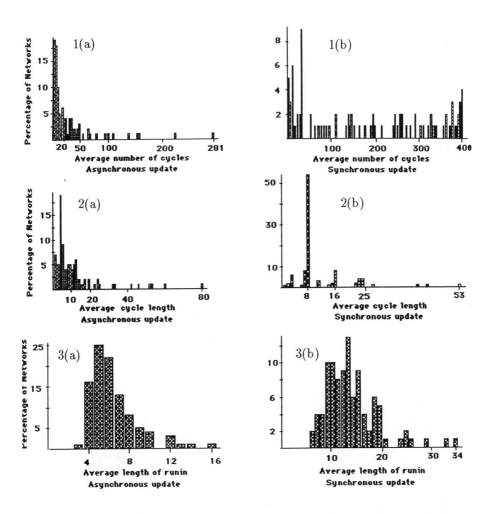

FIGURE 2 Comparison between synchronous and asynchronous update in the *lattice-connected* networks.

TABLE 1 Statistical behavior of randomly connected and lattice-connected networks with 100 nodes.[1]

Number of initial conditions	Technique of update	Geometry of connections	Number of Boolean functions	Number of cycles	Length of cycles	Weighted length of cycle	Length of run-in	No cycle found after 500 steps % of networks
1. 400	Asynchronous	Random	16	5.8	5.8	7.4	8.9	0
2. 400	Asynchronous	Random	14	8.2	20.2	34.3	28.4	10%
3. 400	Synchronous	Random	16	17.3	16.4	16.5	22.0	3%
4. 400	Synchronous	Random	14	25.3	26.8	43.5	56.4	50%
5. 400	Asynchronous	Lattice	16	28.1	8.5	11.1	6.4	0%
6. 400	Asynchronous	Lattice	14	30.9	13.1	18.5	14.1	1%
7. 400	Synchronous	Lattice	16	213.2	11.5	11.4	12.9	1%
8. 400	Synchronous	Lattice	14	246.6	21.1	20.8	23.6	0%
9. 1000	Synchronous	Lattice	16	428.0	11	*	14.0	*
10. 1000	Synchronous	Lattice	14	583.0	18	*	24.0	*

[1] * has not been measured. For each type of network, three parameters can be determined: technique of update, geometry of connections, and number of Boolean functions. Behavior of the network is measured in terms of number of cycles, length of cycles and run-ins. Each type of network is tested under 100 different Boolean assignments; each Boolean assignment is simulated under 400 or 1000 initial conditions.

4.1 EFFECT OF THE NODE UPDATE TECHNIQUE

TYPES OF THE NODE UPDATE There are two different techniques of node update that are used in the simulations:

1. **Synchronous update**: all nodes of the network are updated simultaneously.
2. **Asynchronous update**: nodes are updated in some predetermined sequence. Two different types of sequential update are tried: (1) *lattice-ordered*, that is, nodes are updated in their numerical order (left to right, top to bottom), and (2) *random*, that is, nodes are updated in random order (which is established once for each network).

Results produced under sequential and random update are essentially identical as expected; only the random type of asynchronous update is reported here.

RANDOMLY CONNECTED NETWORKS have, essentially, the same behavior under the synchronous and asynchronous update. All three studied parameters—number of cycles, cycle length, and length of run-in—tend to be 2–3 times longer under the synchronous update. See Figure 1 (compare histograms **A** (asynchronous update) with histograms **B** (synchronous update)); also, compare lines 1 and 3 in Table 1.

LATTICE-CONNECTED NETWORKS show an increase in the *number of cycles* that the network reaches, while the length of the cycle and run-in remain the same; see Figure 2 (compare histograms (a) (asynchronous update) with histograms (b) (synchronous update)). The increase in the number of cycles under synchronous update is rather dramatic—an average of 213 cycles is found in 400 runs (under different initial conditions). When the number of different initial conditions is increased to 1000 the number of found cycles is increased to 428 (see Table 1), indicating that the number of potential cycles of the network is not exhausted yet. Thus, the number of cycles in the lattice-connected networks under synchronous update is $> 4N$ (N is a number of nodes) for 100-node networks.

4.2 EFFECT OF EXCLUSION OF TWO BOOLEAN FUNCTIONS: TAUTOLOGY AND CONTRADICTION

Two sets of Boolean functions are studied:

- The set of 16 functions consists of all possible Boolean functions occurring with equal probability, and
- The set of 14 functions lacks Contradiction and Tautology; the rest of the functions are equally distributed.

Tautology and Contradiction are the most powerful forcing functions: their output is independent of the input. One would expect networks without those functions to possess weaker forcing structures.

BOTH RANDOMLY AND LATTICE-CONNECTED NETWORKS show a tendency toward longer run-ins and longer cycle length in networks with 14 Boolean functions; see Table 1. The increase in the length of the cycle and run-in is approximately twice for lattice-connected nets and 3–4 times for randomly connected networks with 100 nodes under *asynchronous update*. The effect is very different under the synchronous update (see next section).

4.3 JOINT EFFECT OF THE SYNCHRONOUS UPDATE AND EXCLUSION OF CONTRADICTION AND TAUTOLOGY

FOR THE RANDOMLY CONNECTED NETWORKS: The exclusion of Contradiction and Tautology functions can be seen as a destabilizing effect on the network—it takes longer for the network to find a cycle. Synchronous update has a similar effect—it can increase the number of potential cycles or the length of the run-in and of the cycle. Individually those effects are mild—increasing values only by factor of two. However, joining the two effects produces networks with interesting properties—their *run-in* length increases dramatically: from a 9-step average run-in, it increases to more than 450 steps (in 50% of the simulated networks; see Table 1, compare line 1 and 4). It is interesting to note that this effect is specific to the *length of the run-in* while cycle length and number of cycles are affected mildly; see Figure 3. Furthermore, run-in length does not appear to be distributed evenly—networks can be divided into two classes: those with relatively short run-ins (average is 56) and those with very long run-ins (exceed 450). Any network of this type (synchronous update, 14 Boolean functions, randomly connected) has an equal probability to fall into one of the classes.

FOR THE LATTICE CONNECTED NETWORKS: The joint effects of the synchronous update and 14 Boolean functions increase an already very large number of the potential cycles to approximately $6N$ (N is number of nodes) and probably higher, since the ceiling of the number of cycles had not been reached during these simulations. Interestingly, this size of lattice connected networks (100 nodes) does not show a significant change in cycle length or length of the run-in—which is characteristic for randomly connected networks.

Random Boolean Networks

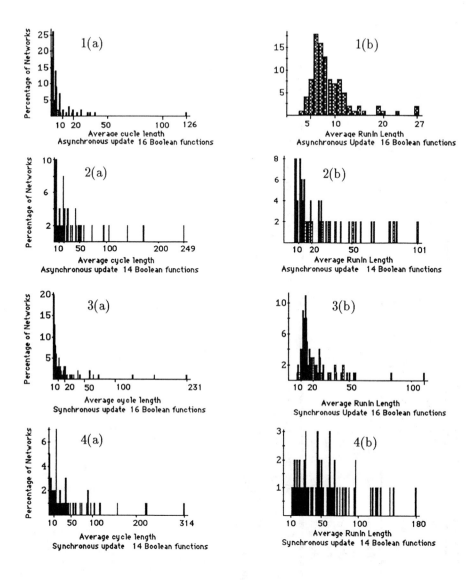

FIGURE 3 Effect of the method of node update and an exclusion of two Boolean functions on the randomly connected networks.

TABLE 2 Behavior of networks with 900 nodes.[1]

number of networks tried	Technique of update	Geometry of connections	Number of Boolean functions	Number of steps allowed for search	Number of cycles	Weighted length of cycle	% of networks cycle is not found
1. 5	Asynchronous	Random	16	350	7.6	24	40%
2. 11	Asynchronous	Random	16	600	141.5	14.5	0
3. 5	Synchronous	Random	16	600	175	20.4	0
4. 11	Asynchronous	Lattice	16	350	526	86	50%
5. 5	Asynchronous	Lattice	16	600	207	69.7	60%

[1] Simulations of 900 nodes networks were done on Cray X-MP. Individual networks had been simulated, each under 600 initial conditions (number of simulated networks is indicated in the left column). Three types of networks are reported here; network was allowed to find a cycle within the first 350 or 600 states.

4.4 INCREASING NUMBER OF NODES IN THE NETWORKS

All of the above results are from the networks with 100 nodes. 900-node networks were simulated on a Cray X-MP supercomputer. Three types of the networks were simulated; only a limited subset of Boolean assignments were tested, but some basic properties of the networks of larger sizes might be discerned; see Table 2.

FOR THE RANDOMLY CONNECTED *SYNCHRONOUSLY* AND *ASYNCHRONOUSLY* UPDATED NETWORKS WITH 16 FUNCTIONS: The number of cycles and the cycle length increase proportionally with the size of the network. All networks found a cycle within first 600 steps of the network. 40% found a cycle after 350 steps of the network.

FOR *LATTICE*-CONNECTED ASYNCHRONOUSLY UPDATED NETWORKS: Fifty percent of the networks could not reach the cycle after 350 or 600 steps. Those networks that do reach a cycle have long cycles and almost every cycle found is identified as unique. Therefore, it is yet undetermined how many potential cycles there are in networks with 900 nodes. This type of the network has a small number of cycles and the *shortest* average run-in from among all types of networks with 100 nodes, but its behavior changes radically with an increase in the size of the network; compare line 5 in Table 1 and lines four and five in Table 2.

4.5 DISTRIBUTION OF CYCLE LENGTH

One of the interesting questions is whether there is a tendency toward a specific cycle length (longer or shorter than average) in the frequently occurring cycles. For that purpose two methods were used for measuring cycle length: weighted and not-weighted cycle length. For the weighted cycle length, each cycle contributes to the average cycle length proportionally to the frequency of its occurrence. In the not-weighted average, every type of cycle is considered only once.

It is interesting to notice that there is a general tendency toward higher values for the weighted cycle length, indicating a correlation between longer cycle length and frequently occurring cycles. The increase is more noticeable in the randomly connected network with 14 Boolean functions—the weighted average cycle length is 40% longer.

5. DISCUSSION

Different types of networks can be ranked according to the degree to which they exhibit characteristics appropriate for a biological model. Both lattice-connected and randomly connected networks perform well under asynchronous update with 16 Boolean functions and the small size of the network (100 nodes here). Their actual number of cycles and cycle length are close to the predicted values of $\sqrt{N}$ (N is the number of nodes). Any deviation from this set of parameters introduces a destabilizing effect into the behavior of the networks.

FOR RANDOMLY CONNECTED NETWORKS The effect of synchronous node update or exclusion of the Contradiction and Tautology Boolean functions is small, but the combination of the two effects causes a dramatic increase in the length of the run-in. There is no clear biological interpretation of the length of run-in; it could be correlated with the length of transition of the network between different cycles in the presence of noise. During the run-in the network is not committed to a specific cycle. In biological terms it means that the transition between different cell types takes a very long time in comparison with the cell's lifetime, which is a poor model of a real cell. It is, therefore, important to test how long it takes for this type of network to move *between* the cycles (if noise is introduced).

Also, only the most "well behaved" type of the network (asynchronous update, 16 Boolean functions) was tested on the larger size networks; the behavior of other types of the randomly connected networks under the increased size of the network is unknown.

FOR LATTICE-CONNECTED NETWORKS Synchronous update has a quite different effect: in the case of randomly connected networks, it increased the length of cycles and run-ins; in the lattice-connected networks, it increases number of potential cycles dramatically. Exclusion of the Contradiction and Tautology functions augments this effect; the network loses some of its self-organizing properties.

When network size was increased from 100 to 900 nodes, even the "best behaved" lattice network (asynchronous update with 16 Boolean functions) loses its self-organizing behavior: its run-ins increased more than 50 times (in 50% of the cases) and the number of cycles is undetermined yet, but is at least on the order of N (N is size of network) and, probably, much higher (see below).

The above results indicate that lattice-connected networks under a very limited subset of possible conditions are able to model biological behavior. Loss of the self-organizing properties with increasing size appear to be the most crucial disadvantage—real biological cells have several thousand genes and therefore should be modeled with at least that many nodes in the network. Lattice-connected networks are clearly incapable of doing so.

THE TEMPORAL WAY IN WHICH NODES OF THE NETWORK CALCULATE their states (local rule) has a strong impact on the global behavior of the lattice-connected network (it affects number of the cycles of network). Asynchronous update results in a smaller number of potential cycles of the network when compared to networks under synchronous update. Why is this so? Asynchronous update allows much faster propagation of the signal through the network; the signal can traverse part of the forcing loop or forcing structure within one update iteration. Forcing structures which are responsible for the self-organizing effect of the network are formed sooner.

Once part of the structure is formed, it becomes insensitive toward all non-forcing inputs. Somehow, rapid formation of the forcing structures limits the number of potential cycles the network can reach. I currently do not have an explanation for the phenomena nor for the fact that it affects lattice-connected networks much stronger than randomly connected networks. My guess is that certain combinations of the oscillating groups (this term is borrowed from Atlan et al.[1,2]) can only be formed in the absence of the nearby forcing structure, which in the case of asynchronous update appears just too fast. Every cell cycle is a combination of the local oscillating groups[2,6]; thus, fewer oscillation groups will mean fewer cycles. This point should be investigated further.

Although asynchronous update does not appear to be a pure one-iteration update, it is a more realistic model from a biological point of view—different components within cell (enzymes, m-RNAs) have different thresholds for synthesis, different stability, etc. A more detailed model is presented by Thomas.[8] It is important to mention that the technique of node update has a rather modest effect on the length of the cycles and the run-in length.

The dramatic change in the behavior of lattice-connected networks with an increase in size could be explained by the unique geometry of connections between the nodes. Lattice connection between nodes forces a formation of the very small local forcing loops, which, in turn, contribute to the formation of oscillating groups. Oscillating groups contribute directly to the number of the potential cycles: the number of potential cycles is proportional to the number of combinations of the oscillating groups. An increase in the network size increases the number of the local loops linearly and, therefore, causes an exponential increase in combinations of the oscillating groups and number of potential cycles. In randomly connected networks random connections between the nodes ensure *longer* forcing loops and a relatively small increase in the number of loops, preventing rapid increase in potential cycles.

Length of a cycle appears to be the most stable property among three measured parameters (cycle length, run-in length, and number of cycles). It showed a significant increase only in the case of large lattice-connected networks.

Even within a particular type of network, the statistical behavior is not uniformal and may depend on the Boolean assignments and node connections within individual network.

SUMMARY

Behavior of lattice-connected and randomly connected networks is different, and it converges under a very small subset of parameters. Randomly connected networks show strong self-organizing behavior under most of the studied conditions. Lattice-connected networks, on the contrary, exhibit very large number of potential cycles under most of the parameters studied and, therefore, are a poor model of the biological cell.

ACKNOWLEDGMENT

I am very grateful to Allen Farquhar whose help and discussions were essential throughout this work. I am particularly thankful for the endless hours of porting the program to Sun and Cray computers. Some of the simulations were done at the Supercomputer Institute at University of Minnesota.

REFERENCES

1. Atlan, H., F. Fogelman-Soulie, J. Salomon, and G. Weisbuch. *Cybernetics and Systems* **12** (1981): 103–121.
2. Atlan, H., E. Ben-Ezra, F. Fogelman-Soulie, D. Pellegrin, and G. Weisbuch. *J. Theor. Biol.* **120** (1986): 371–380.
3. Fogelman-Soulie, F., E. Goleg-Chacc, and G. Weisbuch. *Bull. Math. Biol.* **44(5)** (1982): 715–730.
4. Kauffman, S. A. *J. Theor. Biol.* **22** (1969): 437–467.
5. Kauffman, S. A. *Current Topics in Developmental Biology*, 145–182. New York: Academic Press, 1971.
6. Kauffman, S. A. *Science* **181** (1973): 310–318.
7. Sherlock, R. A. *Bull. Math. Biol.* **41** (1979): 687–724.
8. Thomas, R. "Kinetic Logic: A Boolean Approach to the Analysis of Complex Regulatory Systems." In *Lecture Notes in Biomathematics* **29** (1977): 107–127.

Jin Wang,* Thomas Meyer, and Alfred Hübler
Center for Complex Systems Research, Department of Physics, Beckman Institute, 405 N. Mathews Ave., Urbana, IL 61801; *Mail Box 40, Noyes Laboratory, University of Illinois, Urbana, IL 61801

The Production of Solitons By Optimal Driving Forces

In general, nonlinear waves are not stable in a chain of finite length. Since they have a finite lifetime, it is important to investigate the production of nonlinear waves, e.g., the production of solitons. A general feature of nonlinear waves is the amplitude frequency coupling, which causes the excitation by sinusoidal driving forces to be very inefficient. The response is usually very complex in addition. We present a method[10,15] to calculate special aperiodic driving forces, which generates nonlinear waves very efficiently. The response to these driving forces is very simple.

INTRODUCTION

When a nonlinear oscillator is perturbed by a sinusoidal force, the response is comparatively small in amplitude,[3] and does not fulfill any well-defined resonance condition,[9] even when the frequency of the driving force coincides with a peak (resonance) in the power spectrum of the unperturbed system.[12] Outside the region of entrainment, the response is complicated, in many cases chaotic.[16,17] In order to obtain a large, simple, predictable response, the frequency of the driving force has to be varied in such a way, that it coincides at all amplitudes with the characteristic frequency of the oscillator.[7] Since the characteristic frequencies of nonlinear oscillators usually depends on the amplitude, the optimal driving force has to be aperiodic. Recently a method to calculate those optimal driving forces has been presented.[8] We apply this method in order to calculate optimal driving forces for the creation of solitons.

CREATION OF SOLITONS BY APERIODIC DRIVING FORCES

Nonlinear waves and solitons provide good mathematical models in various fields of science.[18] In most experimental systems solitons have a long but finite lifetime. Therefore we investigate the creation of solitons by external perturbations. We assume that the dynamics of the experimental system can modeled by a sine-Gordon equation

$$u_{xx} - u_{tt} - \sin(u) = F(x,t) \tag{1}$$

where $u(x,t)$ is the field amplitude which depends on space x and time t and where F is an external perturbation which only depends on time and space. In order to calculate resonant driving forces, we integrate according to Hübler and Lüscher[8] the following goal dynamics

$$w_{xx} - w_{tt} - B\sin(u) + w_t\Theta(|x - 50| - 2.5) = 0 \tag{2}$$

where B is a parameter and where Θ is Heavisides step function. We take circular or fixed boundaries at $x = 0$ and $x = 100$. The simulation is finished at time T is when $|w(x,T)| \geq \pi$. The initial conditions are $w(x,0) = .0$ and $w(50,0) = .001$. The driving force results from

$$F(x,t) = -w_t(x,t)\Theta(|x - 50| - 2.5) \tag{3}$$

and $F(x,t) = 0$ for $t \geq T$. The basic idea is that, if the structure of Eqs. (1) and (2) are the same, i.e., $B = 1, u(x,t) = w(x,t)$ is a special solution of Eq. (1). In this case the energy transfer $P(t) = \int_0^{100} F\dot{u}dx$ is positive for all t, i.e., no energy is reflected since F is proportional to w_t. Therefore the coefficient of absorption is 100%, the reaction power is zero, and the perturbation is resonant. The special space, dependence of F was taken in order to create solitons instead of other nonlinear waves. Figure 1(a) shows the result of a numerical simulation of the response of the sine-Gordon system. For the integration we use 100 homogeneously distributed break points. The initial amplitudes of u at these break points are randomly distributed in the interval $[-10^{-5}, 10^{-5}]$ and the initial velocity is set equal zero. Figure 1(a) illustrates that nearly all the transferred energy is used for the creation of a soliton-antisoliton pair since there are no additional waves in the chain. The situation is completely different if we apply a sinusoidal driving force of the same magnitude for the same period of time and in the same region of the chain. In this case no solitons are created (see Figure 1(b)), but a very complicated dynamics results due to the misfit of the driving frequency and the eigen frequency of the system (Figure 2(a)). This example illustrates that the response of a nonlinear system is usually very complicated whereas the response can be well predictable and simple if special aperiodic driving forces are used, since $u(x,t) = w(x,t)$ and $w(x,t)$ can be calculated in advance for an infinite long period of time.

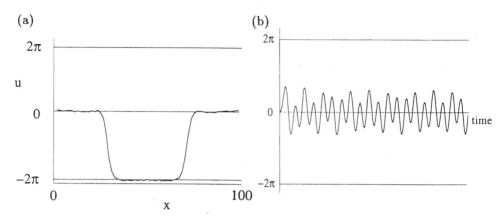

FIGURE 1 The field amplitude u versus x after an aperiodic optimal stimulation (a) and after a sinusoidal stimulation (b).

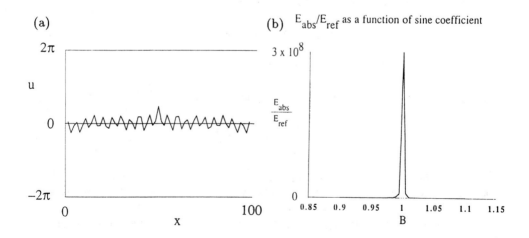

FIGURE 2 The field amplitude $u(50,t)$ versus $time$ for a sinusoidal perturbation (a) and the ratio between the reflected and the absorbed energy versus the parameter B of the model (b).

NONLINEAR RESONANCE SPECTROSCOPY

An essential condition in order to get such a simple response is to have a correct model. Otherwise, u differs from w and usually the dynamics is chaotic and an essential part of the energy is reflected. Figure 2(b) shows the ratio R between the reflected and the absorbed energy versus B. R reaches its maximum value when the parameters of the model and the parameters of the goal dynamics coincide. In this case the response is simple and predictable for an infinite long period of time, while in all other cases including periodic perturbations, a very complicated response was found. By a systematic search for the minimum of the reflected energy as a function of the parameters of the model, the correct magnitude of these parameters can be determined.

MORE GENERAL CONSIDERATIONS

We now calculate how much energy provided by external driving force feeds into the system. The total energy of the system is:

$$E(t) = \sum_o^\pi \left(\frac{u_x^2}{2} + \frac{u_t^2}{2} + \cos(u) \right) dx. \tag{4}$$

The absorption energy and reflection energy of the system is:

$$E_abs = \sum P\theta(P)dt \tag{5}$$

$$E_{ref} = -\sum P\theta(-P)dt \tag{6}$$

whereas $P = \int F\dot{u}dx$ is the power pumped into the system. The efficiency is:

$$\beta = \left(\frac{E_{abs} - E_{ref}}{E(T)} \right). \tag{7}$$

The numerical calculation we did shows that $E_{ref} = 0.0$ and $\beta = 0.975$ which is quite close to one. We see that all the energy pumped in is absorbed by the system. The discrepancy of β from 1 is caused by some errors while doing the calculation using specific numerical methods which are not intrinsic to the problem we are studying.

There are possible difficulties with controlling systems represented by partial differential equations regardless of the method. One major drawback of controlling spatially extended systems is the possible complexity of the driving force. Since this force is spatially dependent, it may not be possible to apply the force to the system once it is calculated.

There are several possible solutions worth exploring.[13] The first is a simplification of the driving term by spatial Fourier decomposition. Control or at least favorable modification of the system might be achieved by applying a simplified $f(x,t)$ on a finite set of points in the domain. In the event that it is not possible to spatially modulate the driving force, a suitable $f(t)$ might be found by comparing the local attractors of the field variable at various points in the domain. Some systems have small localized regions which are extremely sensitive to external perturbations. For situations where these perturbations influence a major portion of the system, our control theory holds promise.

Another possible problem is the magnitude of the driving force. Situations may arise where the energy required to apply $F(x,t)$ forbids its use. The size of the driving force is directly related to how far the goal and model dynamics are separated in function space. Since it is the phase information, not the magnitude of the driving term that is important, we can replace $F(x,t)$ with $\delta F(x,t)$, where $\delta \ll 1$, in the system, provided boundary correction terms are not employed. The result will be that experimental systems will entrain to the goal dynamics more slowly.

If we apply a force locally to a string, the curvature of the string is proportional to the force. Since the driving force we now use is $\dot{w}$, we would expect $w_{xx} = \alpha w_t$ where α is a constant in the middle of the region of applied force. Indeed, our numerical calculation shows that $w_{xx}(50,t) \approx -0.2 w_t(50,t)$. Now if we substitute this relation into model equation, the system becomes:

$$u_{xx} - u_{tt} - \sin(u) = F_1 \tag{8}$$
$$\alpha w_t - w_{tt} - \sin(w) = F \tag{9}$$
$$F_1 = \alpha w_t (2 + \lambda) \tag{10}$$

where λ is a parameter within the range of 1.0 and -1.0. F is at least αw_t in order to overcome the friction; a larger force will drive the system and possibly produce solitons.

Now we can see that the model equation (or goal equation) is an ordinary differential equation, so we can numerically integrate this equation to get the driving force. We apply this homogenous driving force to the experimental system to see if we can drive the system and produce the solitons.

Our numerical simulation shows that if we use initial condition $u(x,0) = 0., w(x,0) = 0.01, w(50,0) = 0.01$, the whole system (except boundary points) is moved together as the time goes on. This is easy to understand since the homogeneous driving force exerted on the system with initial uniform distribution will lead the whole system to move simultaneously. There are apparently no solitons produced (Figure 3).

If we add some noise to the experimental system (that is, to take into account the effect of temperature), the situation is different. The initial distribution of the system is no longer uniform, so there are many different frequency modes in the

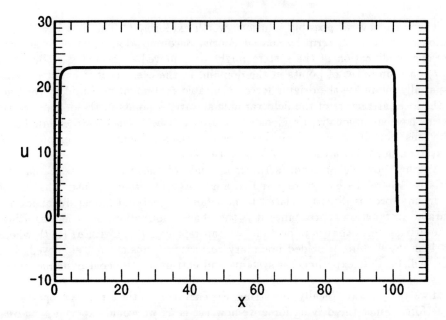

FIGURE 3 The whole system moves up. Here $F(x)$ or $F1(x)$ is $U(x)$.

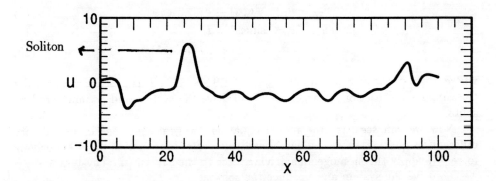

FIGURE 4 The production of a soliton by an optimal driving force.

system compared with just one mode in the uniform system. We would expect that even the homogeneous driving force will excite (or resonate) one or more specific frequency modes of the nonuniform system, amplify them, and therefore produce solitons. So we use a point-wise driving force estimate w, and then we use the resulting driving force as a homogeneous driving force. This might sound inconsistent. In principle we need to apply a very local driving force in order to stay consistent, however if the noise in the system produces local maxima which get resonance with the driving force, the inter-reaction is local even if the driving force is homogeneous.

The numerical simulation results are shown in Figure 4. We find that with the initial distribution $w(x,0) = 0.1, u(x,0) = 0.2(ranf() - .5), F = 0.2\dot{w}, F_1 = 2.4F$, one soliton is produced. The result is quite sensitive to the driving force we choose and noise of the system. Any small deviation of these parameters would not lead a soliton production. We only see one soliton produced since the form of our variational derived force F is a special solution of the differential equation, not the general solution. So it can only excite certain modes of the nonlinear system. The sensitive dependence of the result on the driving force can be understood since a larger driving force will move the whole system and a smaller driving force will not be enough to create a soliton. The sensitive dependence of the result on temperature or noise can be understood since, for lower temperature or noise, the degree of nonuniformity is small, and for higher temperature the noise becomes so large that it overwhelms the system, which is not a realistic situation.

ACKNOWLEDGMENTS

This work was supported in part by ONR grant N00014-88-K-0293 and by NSF grant NSF-PHY 86-58062.

REFERENCES

1. Bullough, R. K. and P. J. Caudrey, eds. *Solitons*. Berlin: Springer, 1980.
2. Davydov, A. S. *Solitons in Molecular Systems*. Dordrecht: D. Reindel, 1985.
3. Eisenhammer, T., T. Hecht, A. Hübler, and E. Luscher. "Skalengesetze für den Maximalen Energieaustausch Nichtlinearer Gekoppelter Systeme." *Naturwissenschaften* **74** (1987): 336.
4. Eisenhammer, T. A., A. Hübler, T. Geisel, and E. Lüscher. "Scaling Behavior of the Maximum Energy Exchange Between Coupled Anharmonic Oscillators." 1992, unpublished.
5. Huberman, B. A., and J. P. Crutchfield. "Chaotic States of Anhermonic Systems in Periodic Fields." Special issue edited by D. D. Humiéres and M. R. Beasley. *Phys. Rev. Lett.* **43** (1979), 1743.
6. Huberman, B. A., and A. Libchaber. "Chaotic States and Routes to Chaos in the Forced Pendulum." *Phys. Rev. A* **26** (1982): 3483.
7. Hübler, A and E. Lüscher. "Resonant Stimulation and Control of Complex Systems." *Helv. Phys. Acta* **61** (1989).
8. Hübler, A and E. Lüscher. "Resonant Stimulation and Control of Nonlinear Oscillators." *Naturwissenschaften* **76** (1989): 67.
9. Hueter, T. F., and R. H. Bolt. *Sonics*, 5th ed., 20. New York: John Wiley & Sons, 1966.
10. Meyer, T., A. Hübler, and N. Packard. "Reduction of Complexity by Optimal Driving Forces." In *Quantitative Measures of Complex Dynamical Systems*, edited by N. B. Abraham. New York: Plenum Press, 1989.
11. Parlitz, U., and W. Lauterborn. "Superstructure in the Bifurcation Set of the Duffing Equation." *Phys. Lett.* **107A** (1985), 351.
12. Ruelle, D. "Resonances of Chaotic Dynamical Systems." *Phys. Rev. Lett.* **56** (1966): 405.
13. Shermer, R., A. Hübler, and N. Packard. "Nonlinear Control of Burgurs Equation." UIUC Preprint, CCSR-90-6, 1990.
14. Takeno, S. *Dynamical Problems in Soliton Systems*. Springer Series in Synergetic. Berlin: Springer, 1985.

Index

A

Abelian, 263
absolute minimum, 73
activation laws, 3
activator, 166
active sliding, 415
actors, 347
adaline, 3, 9, 11
adaline perceptron, 9
adaptation, 8, 223
adaptationism, 397
adaptive, 73
    coding, 25
    filter formalism, 3-4
    resonance theory, 3, 30
    model, 22
aggregation, 160
algorithm, 195
algorithmic process, 93
allosteric enzyme, 163
Amari, S.-I., 27, 372
Amari's continuous model for neural networks, 372
analogy, 121
analysis, 190
analytic continuation, 273
Anderson, J., 17
aperiodic, 458
approximation scheme, 322
arc length, 418
architects of neural structure, 106
ART 1, 31
ART 2, 31
ART module, 32
artificial intelligence, 113, 191, 200
artificial life, 189-190, 200, 389
Ashby nets, 320
associative
    learning, 12, 79, 82, 113
    memory, 3, 32
attracting states, 82, 85
attractors, 83, 85, 121, 320
autoassociative memory, 20
automata networks, 319
avalanche, 3, 24

B

back propagation, 3, 12, 144

back-coupled
    error correction, 3, 9, 12
    perceptron, 8
balance, 105
Barbara McClintock, 125
basal plate, 425
base, 418
basic equation, 419
basin of attraction, 83
bees, 120
behavior, 91, 191, 201
    generators, 191
behavioral choice, 80
behavioral hierarchy, 80-81
belief systems, 347-348
Belousov-Zhabotinski reaction, 115, 160, 179
bend propagation, 416, 427
bending, 416
    angle, 418
    pattern, 415, 420
    resistance, 419
    waves, 416
Bernoulli scaling, 302
biaxial nematics, 264
Bienenstock, E., 27
bifurcation, 95
    analysis, 95
    dynamics and network architecture, 118
    parameters, 85
    induced variation, 87
binocular vision, 48
biological adaptation, 123
biology, 189
biomorphs, 225
birds, 101
boids, 216
Boltzmann machine, 35
bone, 160
Boolean function, 441-443, 445, 449
bottom-up specifications, 203, 216
boundary conditions, 175
brain function, 72
broken
    gauge symmetry, 255
    symmetry, 243-244, 269, 276, 278
    translational symmetry, 269
buccal-cerebral neurons, 76
building blocks, 390
Burger's vector, 256

Index

butterflies, 163, 175, 181

C

calculus of variations, 274
canal surfaces, 271
cats, 58, 95, 355
category, 25
    formation, 25
    learning, 3
cells, 390
    migration, 160
    states, 390
    type, 442
cellism, 399
cellular automata, 179-180, 198, 213, 317, 319, 389
    dynamics, 399
    non-local, 317
cellular basis of behavior, 108
cellular correlates, 53
cellular space, 390
Central Limit Theorem, 297
central pattern generator, 93
cerebro-buccal connectives, 78
Chance, J. E., 355
chaos, 85
    lessons from, 86
chaotic, 83
    attractors, 83
    behavior, 415
    systems, 72
Chay membrane, 118
chimpanzee, 95
Clepsydra, 192
clock face model, 160
cockroaches, 93
coding, 354
coevolution, 347
coevolving strategies, 347-348
cognitive consistency, 348
cognitron, 3, 34
collision, 425
color patterns, 159-161, 181
command neurons, 72
comparative biology, 74
compartmentalization, 183
competency traps, 348
competition, 43-44, 53
competitive interactions, 50
competitive learning, 3, 15, 27

competitive learning (cont'd.)
    module, 28
complex, 347
    behavior, 416
    dynamical behavior, 416
    patterns, 353
computability in practice, 197
computability in principle, 196
computational mapping, 3, 27, 30
computational representative, 227
computer, 196, 200
    simulations, 116
concentric spheres, 272
condensate wave function, 252
condition of continuation, 418
conditioned stimulus (CS), 109
conjugate gradient optimization, 144
connection machine, 227
connections, 441
consensus, 80, 92
    inferences from behavioral choice, 80
    states, 324
constraints, 267, 269, 272, 279
contexts, 79-80, 92
    content-addressable memory, 19
    sensitivity, 212
    window concept, 142
contextual groups, 79
    vs. discrete processes, 119
continuous symmetry breaking, 254
contradiction, 449
control, 72, 461
    input, 323
    mechanisms, 104
    theory, 461
    of visual experience, 355
convection, 162
convergence, 71
Cooper, L., 27
cooperative action, 74
cooperativity, 79
counter-propagation network, 28
coupled selectors, 323
CRAY supercomputer, 143
creation of solitons, 458
critical period, 48
critical bifurcation conditions, 88
critical fluctuations, 88
crossover, 222, 335
    operator, 335

crossover (cont'd.)
    space, 336
crowding, 399
crystals, 251, 267, 281
curl, 279
curvature, 418, 427
    controlled models, 427
cycle, 442
    length, 450
cyclic AMP, 160
cyclides of Dupin, 269, 271

D

Darwin, Charles, 224
Dawkins, Richard, 225
decoding, 354
defect, 258
    entanglement, 262-263
dependency, 435
Descartes, 106
destabilization, 427
developmental, 165
    physiology, 180
    plasticity, 54
dialectical, 106
Dictyostelium, 160
difference, 340
diffusion, 162-163
discrete processes, 119
    vs. continuous processes, 119
dislocation, 258
dislocation lines, 281
dissipative structures, 83
divergence, 71
diversity, 180
DNA, 89
domain, 281
    boundaries, 281
    walls, 282, 284
drive-reinforcement, 113
driving force, 457
    complexity of, 460
duplication, 223
dynamical behavior, 422
    of neural networks, 371
dynamical states, 115
dynamical systems, 317
dynamics, 190, 289, 415

E

economic actors, 348
edge of chaos, 323
electromagnetic radiation, 279
elementary excitations, 244
ellipses, 271, 267-268
embedding, 344
    fields, 3, 19
emergent, 108
    phenomena, 247
    system, 68
energy, 422
    dissipation, 422, 425
    gradients, 120
engrams, 112
environment, 234
epistatic interaction, 345
error, 90
    as an integrative principle, 90
    backpropagation algorithm, 89
Euclid, 106
eukaryotic cells, 416
even powers, 427
evolution, 180, 182-183
evolutionary, 390
    search, 390
    theory, 389
exaptationism, 397, 403
exaptationist standpoint, 389
excitability, 415, 420
excitatory neurons, 371-372
expulsion, 272, 275
external viscous, 418
    drag coefficient, 418, 420
    force, 418
    moment, 418
extrapyramidal system (EPS), 96

F

feeding, 75
    behavior, 75
Feigenbaum, 95
finite component, 308
firm diversification, 347
fitness, 337, 347
    function, 337, 410
    landscape, 410, 412-413

fitness (cont'd.)
    peaks, 228
fixed-action pattern, 72
fixed points, 307
fixed shape, 372
flagella, 415-416
flagellar shape, 418, 428
fluid, 273
flux expulsion, 276
focal conic defect, 271
force, 420
    constant, 420
        balance equation, 418
        distance functions, 420
forced periodic oscillations, 420
forcing functions, 434, 443
forcing structures, 444, 449
formalism, 332
Fortran, 143
fractal space, 299
fractal time, 290
free-end boundary conditions, 420
free energy, 279
Fukushima, Kunihiko, 34
full oscillation, 378
fuzzy control, 112

G

Gacs-Kurdyumov-Levin rule, 325
Gellatt, C.D., 35
gene, 441
    duplication, 231
    network, 427
generalization, 73, 144
generalized reaction-diffusion equation, 428
genetic
    programming paradigm, 220
    algorithms, 218, 220
    search, 390
genotypes, 205, 332
geometrical defects, 269
glasses, 289, 292
Goldstein, A. G., 355
Goldstone boson eats the photon, 280
Goldstone, Jeremy, 254
Goldstone modes, 279-280, 255
Goldstone rotational waves, 284
Goldstone's Theorem, 254
Goodyear, 348
gradient descent, 144

grand unification theories (GUTs), 73
Greene's criterion, 309
Grossberg, Stephen, 19, 22-23, 26-27
group function, 92
group meeting problems, 324
GTYPES, 206, 216, 219
Guillery, R. W., 50

H

halting problem, 196
Hayot, F., 301
Hebbian learning, 3, 14-15
Hebbian type, 374
Hebb's rule, 113, 371, 374
hedgehog, 259
Helisoma, 104
heteroassociative memory, 20
heterochrony, 185
Higgs mechanism, 255, 279
high dimensionality, 72
Hillis, Danny, 226
Hodgkin-Huxley, 22
    equations, 22
    membranes, 117, 85
holistic, 113
Holland, John, 220
holography, 71
homogeneous, 420, 425
homotopy
    group, 259
    theory, 258
human visual system, 353
humans, 355
hyperbolas, 267, 271
hysteresis, 415, 420
Hénon system, 119

I

icosahedral symmetry, 245
illumination, 356
immune
    network, 427
    rule, 435
immunofluorescence, 98
inertial terms, 418
inhibition-excitation, 102
inhibitor, 166
inhibitory neurons, 371-372
initial conditions, 175
insect, 416, 425

instability, 30
instantiated transitions, 400
instar, 3, 19, 25, 27
instar-outstar computational map, 28
instrumental conditioning, 114
integrative mechanisms, 73, 82
interactions, 441
intermittent movements, 416
internal
    elastic moment, 418
    shear force, 418-419, 427
    shear moment, 418
    viscosity, 427
    viscous resistance, 420
intracellular messenger, 104
intrinsic instability, 415, 422, 425, 427-428
inversion, 223, 335
    operator, 336
    space, 336
invertebrates, 98
irregular modes, 428
iterated prisoner's dilemma, 230

J

Jordan curve theorem, 87

K

K_{opt}, 345
KAM curves, 306
Kaplan-Yorke conjecture, 120
Kauffman nets, 320
Kauffman system, 172
Kirkpatrick, S., 35
Klopf, A. H., 113
know-nothing approach, 141
knowledge, 111
Kohonen, T., 17
Koza, John, 220
Kuramoto-Sivashinsky equation, 428

L

lateral inhibition, 161, 166, 173-174, 184-185
lattice-connected networks, 443, 446
lazy synapses, 113, 122
lead, 277
leading edge, 422
learning, 3, 8, 79, 109
    equations, 23

learning (cont'd.)
    laws, 3, 30
    matrix, 3, 15, 27
    rate, 144
    competitive, 3
least mean squared error correction rule, 9
leech locomotion, 93
leopards, 175
Levy, E. I., 355
Levy
    flights, 298
    walks, 300
life, 197
limit cycles, 72, 83, 86
Lindenmayer systems, 209
line dislocations, 284
linear associative memory models, 3, 17-18
    optimal, 17
linear theory, 169
lobsters, 93-94
local lesions, 360
local minima, 73
local neighborhood, 345
locomotion, 416
locust walking, 93
logical structure, 197
logistic, 116
long-range interactions, 422
long-term potentiation (LTP), 121
low-dimensional dynamics, 72
LTP, 56
Lyapunov exponents, 87

M

macroscopic quantities, 321
madaline, 3, 9, 11
    perceptron, 9
magnet, 248-249, 261
magnetic shield, 277
mammals, 95, 101
mantle, 161
    computational, 3
Marr's assertion, 354
mass, 276
massive fields, 267, 273, 280
mathematical modeling of neural networks, 371
McCulloch, W. S., 4, 7
McCulloch-Pitts
    formalism, 7

McCulloch-Pitts (cont'd.)
    linear filter, 12
    model, 7
    neuron, 3-4
mean-field theory, 322
mechanical process, 195
mechanics, 190
Meinhardt system, 167, 177, 185
Meissner-Higgs effects, 267, 277, 284
melanophores, 161
memory
    storage, 71
    associative, 32
    autoassociative, 20
    content addressable, 19
    heteroassociative, 20
    linear associative, 17
Mermin-Wagner Theorem, 256
metaknowledge, 111, 122
metallic glasses, 264
Metropolis algorithm, 35
mexican hat, 276, 278
Michelin, 348
microtubules, 416
minimal coupling, 280
    adaline, 9
    additive, 22
    linear associative memory, 17-18
    madaline, 9
    perceptron, 8
    real-time, 19
    shunting, 22
mollusk shells, 161
moment-balance equation, 418
monkeys, 58, 355
Montroll-Weiss random walks, 293
Moore-Penrose pseudoinverse, 17
    see also linear associative memory, 17
morphogenesis, 160
motor
    learning, 97
    system, 67
movement, 393
multi-level perceptrons, 12
multifunctional, 76
    networks, 91
multilayer perceptrons, 140
multiplexer, 323
Munro, P., 27
muscimol, 56

mutation, 221, 335

N

Nakano, K., 17
Nash equilibria, 347
natural selection, 219
negative diffusion, 427
neighborhood, 390
nematic liquid crystals, 260
nematics, 249
neocognitron, 3, 34
network
    architecture and bifurcation dynamics, 118
    behavior, 441
    cycle, 442
    size, 121
    systems, 427
    state of, 442
    shunting competitive, 3
neural
    networks, 371
    architecture, 74
    crest, 161
    fields, 371
    model, 22, 34
    network, 3-4, 143, 427
    network models, 3
neurocircuits, 91, 104
neuromodulation, 98
neuronal
    architectures, 104
    dynamics, 373
    signalling, 50
neurotransmitters, 98, 104
niobium, 277
NK model, 409-411, 413
NKC models, 331, 413
node, 441-442
node update technique, 443
noise, 357, 442
nondimensionalization, 167
nonlinear
    dynamics, 114
    oscillator, 457
    systems, 179
    waves, 457
normal components, 418
normalization, 142
null clines, 168

O

numerical simulations, 175, 180, 415
nymphalid ground plan, 181-183

occlusion, 356
octopus, 357
ocular dominance columns, 44
ocular dominance stripes, 175, 179-180
odd power, 427
off center, 26
oil and water, 246
on center, 26
one-dimensional crystal, 253
one-mutant neighbor, 410
    operator, 335
    space, 336
ontogenetic transformation, 180
ontogeny, 181
operators, 335
optimal driving forces, 457-458
order parameter, 243-244, 247
    field, 249
organizing centers, 166
orienting subsystem, 31
oscillatory behavior, 373
outliers, 30
outstar, 3, 19, 24, 27
    learning theorems, 23
overtraining, 146

P

pace-maker, 427
parallel processing, 108
parameter space, 168
Parker, David, 12
Parkinson's disease, 96, 102
partial differential equation, 415
passive elastic component, 420
passive region, 427
passive sliding, 415
pattern
    formation, 159, 163, 179, 181
    recognition, 3, 355
    training, 147
patterning, 43
    butterfly wings, 181
    cell movement-mediated, 160
    necessary conditions for reaction-diffusion systems, 166
    of bone formation, 160

patterning (cont'd.)
    of color, 161
    of pigment, 161
    of presynaptic activity, 58
    random and non-random, 175
Pavlovian and avoidance associative learning, 79
Pavlovian conditioning, 114
perceptron, 3
    model, 8
performance, 347
period-doubling bifurcations, 95
periodic behavioral patterns, 433
phase portraits, 83, 307
phase transition, 432
phenotype, 205
pigment patterns, 159, 161
Pitts, W., 4, 7
pituitary, 102
pneumatics, 192
Poincaré-Bendixon, 87
polycrystal, 281-282
polygonalization, 284
Potter, M. C., 355
pre-patterning, 160-162
principle of superposition, 203
problem of time, 142
process control, 194
program, 195
propagating waves, 415-416, 420, 422
proportionality constant, 427
PTYPES, 206, 219
    behavioral, 215
punctuated equilibria, 230, 232
pyramidal system (PS), 96

Q

quasi-crystals, 243, 245
quickprop, 144
quiescent background, 393

R

rabbit, 112
    olfactory bulb, 79
random
    Boolean networks, 431, 441, 444-445
    distribution of fitness, 343
    noise, 89, 428
    perturbation, 175

randomly connected networks, 442-444, 446, 450, 454
rats, 102, 121
Ray, Tom, 235
reaction diffusion, 165-166, 169, 171-173, 175, 177, 180, 185
reaction-diffusion equation, 427
reaction-diffusion systems, 166
  necessary conditions for patterning, 166
real-time models, 19
recognition, 353
  accuracy, 355
  performance, 355
recursive description, 209
recursively generated objects, 208
reduction, 104
reductionist approaches, 67, 106
relatedness, 354
relaxation law, 295
remodeling, 45
reproduction, 221
resonance, 457
response, 457
  optimization, 120
  thresholds, 121
reticular activating system (RAS), 97
retina, 101
retinotectal system, 55
Reynolds number, 418
Reynolds, Craig, 215
Richardson's law, 297
Richardson, Lewis Fry, 297
rigidity of solids, 255
Rosenblatt perceptrons, 8
Rosenblatt, Frank, 8-9, 12
rotational Goldstone mode, 284
rotational symmetry, 246
rubber bands, 267
rugged fitness landscape, 412
ruggedness, 409-410, 412-413
Rumelhart, D., 27
run-in, 442, 445
  length, 450
Rössler system, 116

S

scale, 168, 289
scaling, 290
Schnakenberg system, 167, 177
sea slug, 67, 73, 75, 93, 96
sea urchin, 425
searchable neighborhoods, 335
second Homotopy group, 260
second-order machines, 196
second sound, 255, 279
segmentation problem, 142
selection, 224, 354, 390, 396
selector, 323
self-organization, 3, 190, 402, 441, 454
self-organizing criticality, 84-85
self-organizing feature map, 28
self-oscillatory models, 427
self-reproduction, 198, 392
sensitization, 123
sexual reproduction in cellular automata, 391
shear force, 427
shunting competitive networks, 3, 26
shunting model, 22
  see also additive models, 22
signal propagation, 212
simulated annealing, 3, 35, 89, 123
simulation and mimicry, 180
sliding, 416
  displacement, 418, 420, 427
  filament mechanism, 416
slow human beings, 253
small-amplitude assumption, 418
smectic liquid crystals, 268-269
social sciences, 347
solitons, 422, 457
sound waves, 253
space derivative, 427
  term, 422
space-time diagram, 422
space-time patterns, 24
spatial homogeneity, 372
spatial inversion, 427
spatio-temporal dynamics, 123
spatio-temporal sliding pattern, 415, 422
specificity, 354
speech acoustics, 139
spiky patterns, 422
spin model, 409-410, 413
spontaneity, 278
spontaneous symmetry breaking, 276
St. Petersburg Paradox, 302
stability-plasticity dilemma, 30
stabilization, 427
stable behaviors, 382

starting transients, 416
states of matter, 243
Steinbuch, K., 15, 27
stochastic
    layers, 308
        boundaries, 309
    web, 308
stomatogastric ganglion, 94
stopping, 416
strain matrix, 286
stretched exponential relaxation, 292
structural asymmetry, 427
superconductors, 252, 267, 277, 279
superfluid, 252, 278
    principle of, 203
swarm, 341
switchboard factors, 94
switchboard system, 68
switching functions, 420
symmetry, 246-247, 307, 427
    breaking, 278, 284
    systems, 182-183
synaptic dynamics, 374
synaptic projections, 71
synthesis, 190, 203

T

Takeuchi, A., 27
tautology, 449
terminal piece, 425, 427
    adaptive resonance, 3, 30
theory of embedding fields, 19
theory of superconductivity, 200
theorem
    Goldstone's Theorem, 254
    Jordan curve Theorem, 87
    Mermin-Wagner Theorem, 256
    outstar learning Theorems, 23
Thomas system, 167, 177
threshold, 163-164, 415
Tierra, 235
time evolution, 317
time-independent noise algorithms (TINA), 89
time-invariant noise algorithms, 123
time scale, 296
tongue, 139
topological defects, 244, 256, 258
    importance of, 259
topology, 243, 258, 267

torus, 252
traction-aggregation mechanism, 160
trailing edge, 422
training, 144
    set, 148
transient behavior, 373
transition
    function, 431
    values, 390
translational, 284
    movements, 418
    symmetry, 246
Tritonia, 93
turbulence, 84, 289
turbulent diffusion, 300
Turing, A. M., 166
two-filament system, 416
two-quadrant oscillation, 380
type I superconductors, 277
type II superconductors, 277

U

Ulam, Stan, 198
unbounded acceleration, 306
unconditioned stimulus (UCS), 109
unidirectional propagating waves, 415, 425, 427
unified theory of biological organization, 67
universal constructor, 197

V

variability, 390
variation, 93
variation-dependent optimization, 90
Vaucanson's duck, 192-193
Vecchi, H.P., 35
velocity, 418
vertebrate, 101
    visual system, 43
vibrational modes, 173
vigilance parameter, 31
vision, 353
visual, 357
    deprivation, 48
volume contraction, 84
von der Malsburg, C., 27
von Neumann, John, 7, 197

Index

vortex lines, 277

W

wave, 425
    frequencies, 425
    propagation, 425
    splitting, 422
    splitting mechanism, 425
weak interaction, 280
wavelengths, 420, 425
web map, 306
Weierstrass function, 297, 299
weight transport, 13
Weinberg-Salaam theory, 280
Werbos, Paul, 12
Widrow, Bernard, 9
winding number, 258

wing patterns, 175
wing veins, 185
wiring diagram, 320
    partially local, 320
Wolpert, Lewis, 163
word chunking, 342
work hardening, 259
wrapping number, 260

Y

Young's theory, 173

Z

Zaslavsky's web map, 305
zebras, 175, 180
Zipser, D., 27